RINGS, MODULES AND ALGEBRAS

In homage to my Master

EMIL ARTIN

who being dead yet speaketh

RINGS, MODULES AND ALGEBRAS

IAIN T. ADAMSON
Senior Lecturer in Mathematics
in the University of Dundee

OLIVER & BOYD
EDINBURGH

OLIVER & BOYD
Tweeddale Court
14 High Street
Edinburgh EH1 1YL
A Division of Longman Group Limited

First published 1971

ISBN 0 05 002191 5

Printed in Great Britain by Page Bros. (Norwich) Ltd.

PREFACE

This book is intended to provide an introduction to the basic facts about modules, abelian categories and homological algebra and to apply these in deriving the classical results on Artinian rings and simple and separable algebras. I hope it may be found to give a clear, connected and not over-condensed account of some more or less well-established topics in ring theory from which the reader may proceed without too much difficulty to study the current research work into the structure of rings, category theory and homological algebra. Apart from a few results on fields, for which the reader may refer to my *Introduction to Field Theory*, the exposition is entirely self-contained.

In Chapter 1 we discuss the elementary ideas of rings, ideals, homomorphisms and extensions of rings.

Chapter 2 begins with two sections on modules and their homomorphisms. These are followed by a study of certain important constructions—groups of homomorphisms, direct sums and products, direct and inverse limits and tensor products—and of free, projective and injective modules. The treatment of these topics, with its emphasis on universal and couniversal properties, is intended to provide a foundation for category theory. The chapter concludes with sections on composition series and Artinian and Noetherian modules and rings.

Chapters 1 and 2 are expanded versions of Chapters 1 and 2 of my *Elementary rings and modules.*

Drawing on Chapter 2 for illustrative examples, Chapter 3 introduces the notions of category and functor, with special emphasis on abelian categories. It includes an exposition of Puppe's theory of relations in an abelian category, as first presented in his paper Korrespondenzen in abelsche Kategorien, *Math. Ann.,* **148** (1962).

Chapter 4 presents an account of semisimple and simple modules and the Artin-Wedderburn structure theory for semisimple and simple rings. The last section of this chapter discusses the radical and includes a proof of Hopkins's theorem.

Chapter 5 is concerned mainly with the theory of finite-dimensional simple algebras, including the Brauer group, splitting fields and the Skolem-Noether theorem. The chapter also contains a discussion of separable algebras.

The final chapter develops homological algebra in locally small abelian categories, proceeding as far as the construction of derived functors. It then specialises to the cohomology theory of algebras and the characterization of algebras of dimension zero, concluding with a section on extensions of algebras and the Wedderburn Principal Theorem.

Most of the writing of this book was done while I was on leave from Dundee and I am happy to express my thanks to the University of Western Australia for affording me the hospitality of its beautiful campus and the stimulation of working in its friendly and lively department of Mathematics. Since returning to Dundee I have had the valuable assistance of Dr Arthur Sands, who read the whole manuscript, and of Dr Hamish Anderson, who read the proofs with great care; I am deeply indebted to them both.

Finally I should like to express my gratitude to my wife who has borne with great patience the long period of gestation of this book.

IAIN T. ADAMSON

Dundee.
April 1970.

CONTENTS

CHAPTER 5

ALGEBRAS

CHAPTER 6

HOMOLOGY

Chapter 1

RINGS AND IDEALS

§1. Internal Laws of Composition

Let E be any set; by an *internal law of composition* or *binary operation* on E we mean a mapping from the Cartesian product $E \times E$ to E. If φ is an internal law of composition on E, the image under φ of an ordered pair (a, b) in $E \times E$ is usually not denoted by $\varphi(a, b)$, which is the notation we would expect from our acquaintance with elementary set theory. The various alternative notations used for $\varphi(a, b)$ include $a + b$ (in which case we call $\varphi(a, b)$ the *sum* of a and b, and refer to φ as an *addition operation*) and ab, $a.b$ or $a \times b$ (when we call $\varphi(a, b)$ the *product* of a and b, and describe φ as a *multiplication operation*). Even when we discuss completely arbitrary internal laws of composition we use a notation reminiscent of these long-familiar ones and write $a \circ b$, $a * b$, $a \wedge b$ or $a \vee b$ instead of $\varphi(a, b)$.

The ordinary addition and multiplication operations are internal laws of composition on the sets $\mathbf{N}$, $\mathbf{Z}$, $\mathbf{Q}$, $\mathbf{R}$, $\mathbf{C}$ of natural numbers, integers, rational numbers, real numbers and complex numbers. Subtraction is an internal law of composition on $\mathbf{Z}$, $\mathbf{Q}$, $\mathbf{R}$, $\mathbf{C}$, and so is division on the sets of non-zero elements of $\mathbf{Q}$, $\mathbf{R}$ and $\mathbf{C}$. If X is any set, composition of mappings from X to itself is an internal law of composition in the set Map(X, X).

Let E be a set, φ an internal law of composition on E; for each ordered pair (x, y) in $E \times E$ we shall denote $\varphi(x, y)$ by $x \wedge y$. A subset A of E is said to be *closed* or *stable* under φ (or loosely, under $\wedge$) if for every ordered pair (a, b) in $A \times A$ the element $\varphi(a, b) = a \wedge b$ of E actually belongs to the subset A itself. When A is closed under φ we may say that φ *induces* an internal law of composition φ_A on A; φ_A is given by $\varphi_A(a, b) = \varphi(a, b) = a \wedge b$ for all ordered pairs (a, b) in $A \times A$.

Let n be a positive integer, and let I be the interval $[1, n] = \{i \in \mathbf{Z} : 1 \leqslant i \leqslant n\}$. If $(a_i)_{i \in I}$ is a sequence of elements of E we define the *composition* $\bigwedge_{i \in I} a_i$ of this sequence by induction on n as follows:

(1) If $n = 1$, we set $\bigwedge_{i \in I} a_i = a_1$;

(2) If $n = k + 1$, let $I' = [1, k]$ and set $\bigwedge_{i \in I} a_i = (\bigwedge_{i \in I'} a_i) \wedge a_{k+1}$.

When we use the additive notation for φ we write $\sum_{i \in I} a_i$ for the composition of the sequence $(a_i)_{i \in I}$ and call it the *sum* of the sequence; when we use the multiplicative notation we denote the composition by $\prod_{i \in I} a_i$ and call it the *product* of the sequence.

Now suppose that we have two sets E_1, E_2 and internal laws of composition φ_1, φ_2 on E_1, E_2 respectively; write $\varphi_i(x_i, y_i) = x_i \wedge_i y_i$ for every pair (x_i, y_i) in $E_i \times E_i$ $(i = 1, 2)$. A mapping α from E_1 to E_2 is said to be a *homomorphism* (relative to φ_1 and φ_2) if for every pair (x_1, y_1) in $E_1 \times E_1$ we have

$$\alpha(x_1 \wedge_1 y_1) = \alpha(x_1) \wedge_2 \alpha(y_1).$$

If we introduce the mapping $\alpha \times \alpha$ from $E_1 \times E_1$ to $E_2 \times E_2$ by setting $(\alpha \times \alpha)(x_1, y_1) = (\alpha(x_1), \alpha(y_1))$ for each pair (x_1, y_1) in $E_1 \times E_1$, we see that α is a homomorphism if and only if the diagram

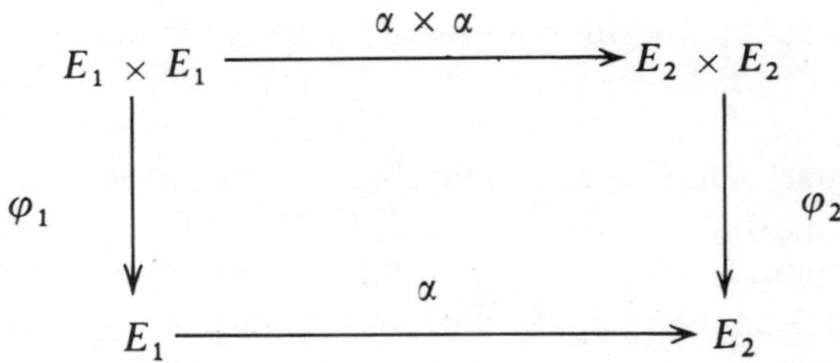

is commutative.

Let α be a homomorphism from E_1 to E_2. If α is surjective we call it an *epimorphism* from E_1 onto E_2; if α is injective we call it a *monomorphism*; and if α is bijective we say that it is an *isomorphism* of E_1 onto E_2. If $E_1 = E_2 = E$ and $\varphi_1 = \varphi_2 = \varphi$, a homomorphism from E to E (relative to φ and φ) is called an *endomorphism* of E (relative to φ); a bijective endomorphism is called an *automorphism* of E. As we have already done in this paragraph we shall usually omit any reference to the laws of composition when they are clear from the context.

We have the following general result concerning homomorphisms.

THEOREM 1.1. *Let α be a homomorphism from E_1 to E_2 relative to internal laws of composition φ_1 and φ_2 on E_1 and E_2 respectively. Then* $\operatorname{Im} \alpha = \alpha(E_1)$ *is closed under φ_2.*

Proof. Let a_2, b_2 be elements of Im α. Then there are elements a_1, b_1 of E_1 such that $a_2 = \alpha(a_1)$, $b_2 = \alpha(b_1)$. Since α is a homomorphism we have $a_2 \wedge_2 b_2 = \alpha(a_1) \wedge_2 \alpha(b_1) = \alpha(a_1 \wedge_1 b_1)$; so $a_2 \wedge_2 b_2 \in \text{Im } \alpha$, as required.

We shall also use from time to time the results of the following two theorems.

THEOREM 1.2. *Let E_1, E_2, E_3 be sets with internal laws of composition φ_1, φ_2, φ_3 respectively. If α and β are homomorphisms from E_1 to E_2 and E_2 to E_3 respectively, then $\beta\alpha$ is a homomorphism from E_1 to E_3.*

Proof. Let a and b be any two elements of E_1. Then we have $\beta\alpha(a \wedge_1 b) = \beta(\alpha(a \wedge_1 b)) = \beta(\alpha(a) \wedge_2 \alpha(b)) = \beta(\alpha(a)) \wedge_3 \beta(\alpha(b)) = \beta\alpha(a) \wedge_3 \beta\alpha(b)$. Thus $\beta\alpha$ is a homomorphism, as asserted.

THEOREM 1.3. *Let E_1 and E_2 be sets with internal laws of composition φ_1 and φ_2 respectively. If α is an isomorphism of E_1 onto E_2 then α^{-1} is an isomorphism of E_2 onto E_1.*

Proof. Since α is a bijection from E_1 onto E_2, its inverse α^{-1} is a bijection from E_2 onto E_1.

Now let c and d be elements of E_2; let $a = \alpha^{-1}(c)$, $b = \alpha^{-1}(d)$. Then, since α is a homomorphism, we have $\alpha(a \wedge_1 b) = \alpha(a) \wedge_2 \alpha(b) = c \wedge_2 d$ and hence $\alpha^{-1}(c \wedge_2 d) = a \wedge_1 b = \alpha^{-1}(c) \wedge_1 \alpha^{-1}(d)$. Thus α^{-1} is a homomorphism, and hence an isomorphism, as required.

Once again let φ be an internal law of composition on a set E; we shall write $\varphi(a, b) = a \wedge b$ for each pair (a, b) in $E \times E$. We say that φ (or loosely, $\wedge$) is *associative* if for all elements a, b, c of E we have

$$a \wedge (b \wedge c) = (a \wedge b) \wedge c. \qquad [1.1]$$

The importance of this property lies in the fact that it allows us to omit parentheses and to write both sides of [1.1] simply as $a \wedge b \wedge c$. From this it follows by an inductive argument that we may omit parentheses in the notation for the composition of arbitrary finite sequences of elements.

We remark that associativity is 'preserved under homomorphisms'; more formally, we have the following theorem, whose proof is a simple exercise.

THEOREM 1.4. *Let α be a homomorphism from E_1 to E_2 relative to* *laws of composition* φ_1 *and* φ_2 *on* E_1 *and* E_2 *respectively. If* φ_1 *is associative then the law of composition induced by* φ_2 *on* Im α *is associative also.*

Keeping the same notation as above, we say that φ (or $\wedge$) is *commutative* if for every ordered pair (a, b) in $E \times E$ we have $a \wedge b = b \wedge a$. Even when φ is not commutative there may nevertheless exist some pairs (a, b) in $E \times E$ for which $a \wedge b = b \wedge a$. If (a, b) is such a pair we say that a *commutes* with b or that a and b *commute*.

We have an obvious analogue of Theorem 1.4.

THEOREM 1.5. *Let* α *be a homomorphism from* E_1 *to* E_2 *relative to laws of composition* φ_1 *and* φ_2 *on* E_1 *and* E_2 *respectively. If* φ_1 *is commutative then the law of composition induced by* φ_2 *on* Im α *is also commutative.*

An element n of E is said to be a *left neutral element* with respect to φ if for every element a of E we have $n \wedge a = a$ and a *right neutral element* if for every element a of E we have $a \wedge n = a$; an element which is both left neutral and right neutral is called simply a *neutral element*. If the law of composition φ is an addition operation a neutral element is usually called a *zero element* and denoted by 0 or z or ζ; if φ is a multiplication operation we refer to a neutral element as an *identity* and denote it by 1 or e or I.

THEOREM 1.6. *Let* φ *be an internal law of composition on a set* E. *If* n_l *and* n_r *are left and right neutral elements respectively with respect to* φ, *then* $n_l = n_r$.

Proof. Since n_l is a left neutral element we have $n_l \wedge n_r = n_r$; but since n_r is a right neutral element we have also $n_l \wedge n_r = n_l$. Hence $n_l = n_r$, as asserted.

COROLLARY. *A set can have at most one neutral element with respect to an internal law of composition.*

We also have a theorem about the 'preservation' of neutral elements under homomorphisms.

THEOREM 1.7. *Let* α *be a homomorphism from* E_1 *to* E_2 *relative to internal laws of composition* φ_1 *and* φ_2 *on* E_1 *and* E_2 *respectively. If* n_1 *is a left (or right) neutral element with respect to* φ_1 *in* E_1 *then*

$\alpha(n_1)$ is a left (or right) neutral element with respect to the law of composition induced by φ_2 on Im α.

Proof. If a_2 is any element of Im α there is an element a_1 of E_1 such that $\alpha(a_1) = a_2$. Then $\alpha(n_1) \wedge_2 a_2 = \alpha(n_1) \wedge_2 \alpha(a_1) = \alpha(n_1 \wedge_1 a_1) = \alpha(a_1) = a_2$ and $\alpha(n_1)$ is a left neutral element, as asserted.

At this point we introduce a useful convention. If the set E has a neutral element n and $(a_i)_{i \in \phi}$ is an empty family of elements of E we define the composition $\wedge_{i \in \phi} a_i$ to be the neutral element n.

Let E be a set which has a neutral element n with respect to an internal law of composition φ (or $\wedge$). An element a of E is said to be *left-invertible* (with respect to φ) if there exists an element a' of E such that $a' \wedge a = n$; in this situation the element a' is called a *left inverse* of a. The terms *right-invertible* and *right inverse* are defined in the obvious way. An element of E which is both left- and right-invertible is said to be *invertible.* If a is invertible and a' is both a left and a right inverse of a, we say simply that a' is an *inverse* of a; in this case, of course, a is an inverse of a'.

THEOREM 1.8. *Let φ be an associative internal law of composition on a set E, with respect to which E has a neutral element n. If an element a of E has a left inverse a'_l and a right inverse a'_r, then $a'_l = a'_r$.*

Proof. We have $a'_l \wedge a = n = a \wedge a'_r$. Since φ is associative, $(a'_l \wedge a) \wedge a'_r = a'_l \wedge (a \wedge a'_r)$; hence $n \wedge a'_r = a'_l \wedge n$, i.e. $a'_r = a'_l$, as asserted.

We can also show that inverses are 'preserved under homomorphisms'. It will be enough to state the result: the proof is clear.

THEOREM 1.9. *Let α be a homomorphism from E_1 to E_2 relative to internal laws of composition φ_1 and φ_2 on E_1 and E_2 respectively. Suppose E_1 has a neutral element n_1. If a'_1 is a left (or right) inverse of an element a_1 of E_1 with respect to φ_1 then $\alpha(a'_1)$ is a left (or right) inverse of $\alpha(a_1)$ with respect to φ_2.*

There are two further useful elementary results concerning invertible elements which we include in the following theorem.

THEOREM 1.10. *Let φ be an associative internal law of composition on a set E, with respect to which E has a neutral element n. If a and b are invertible elements of E then so is their product $a \wedge b$. If, further,*

φ is commutative and $a \wedge b$ is invertible, then a and b are themselves invertible.

Proof. Let us write φ multiplicatively, abbreviating $a \wedge b$ to ab.

(1) Suppose a and b are invertible, with inverses a' and b' respectively. Then $(ab)(b'a') = a(b(b'a')) = a((bb')a') = a(na') = aa' = n$, and similarly $(b'a')(ab) = n$. Thus ab is invertible.

(2) Now suppose that φ is commutative as well as associative and that the product ab is invertible, with inverse c say. Then we have $(ab)c = n$, whence $a(bc) = n$ and $b(ac) = (ba)c = (ab)c = n$. So a and b are invertible, as asserted.

Let now φ and θ be two internal laws of composition on a set E; we shall write $\varphi(a, b) = a \wedge b$ and $\theta(a, b) = a \vee b$ for every pair (a, b) in $E \times E$. Then we say that φ is *left-distributive* over θ (or that $\wedge$ is left-distributive over $\vee$) if for all elements a, b, c of E we have $a \wedge (b \vee c) = (a \wedge b) \vee (a \wedge c)$. Similarly we say that φ is *right-distributive* over θ if for all elements a, b, c of E we have $(b \vee c) \wedge a = (b \wedge a) \vee (c \wedge a)$. If φ is both left- and right-distributive over θ we say simply that it is *distributive over θ*. Clearly if φ is commutative the three properties coincide.

Once again let φ be an internal law of composition on a set E and write as usual $\varphi(a, b) = a \wedge b$. If R is an equivalence relation on E we say that φ (or $\wedge$) is *compatible* with R if for all elements a, a', b, b' of E such that aRa' and bRb' we also have $(a \wedge b)R(a' \wedge b')$. Suppose that φ is compatible with R, and let η be the canonical surjection from E onto E/R; we shall show how to define an internal law of composition φ_* on E/R such that η is a homomorphism from E to E/R relative to φ and φ_*.

To this end, let C_1, C_2 be elements of E/R; choose elements a_1, a_2 from C_1, C_2 respectively, and define $C_1 \wedge_* C_2 = \eta(a_1 \wedge a_2)$. We must check that the equivalence class so obtained depends only on the classes C_1, C_2 and not on the choice of elements a_1, a_2. But if a'_1, a'_2 are also elements of C_1, C_2 respectively, we have $a_1R_1a'_1$ and $a_2Ra'_2$; hence, since $\wedge$ is compatible with R, $(a_1 \wedge a_2)R(a'_1 \wedge a'_2)$ and so $\eta(a_1 \wedge a_2) = \eta(a'_1 \wedge a'_2)$. We say that the law of composition $\wedge_*$ is *induced* by $\wedge$. It is clear that η is a homomorphism from E to E/R relative to $\wedge$ and $\wedge_*$; for if x and y are elements of E they are also elements of the equivalence classes $\eta(x)$ and $\eta(y)$ respectively and so $\eta(x) \wedge_* \eta(y) = \eta(x \wedge y)$ by the definition of $\wedge_*$

§2. Semigroups, Groups and Rings

Let φ be an internal law of composition on a set E and write as usual $\varphi(a, b) = a \wedge b$. If φ is associative we say that E is a *semigroup* under φ (or, under $\wedge$). Usually, once we have described the law of composition φ with which we are to work, we refer simply to 'the semigroup E' without further explicit mention of φ.

Let E be a semigroup under $\wedge$; let a be an element of E. We proceed to define for each non-zero natural number n the nth *iterate* of a, which we denote by $\wedge^n a$; we set

(1) $\wedge^1 a = a$;
(2) $\wedge^{k+1} a = (\wedge^k a) \wedge a$ for all $k \geqslant 1$.

Thus $\wedge^n a$ is the composition under $\wedge$ of a sequence $(a_i)_{i \in I}$ of elements of E such that Card $I = n$ and $a_i = a$ for every index i in I. If E has a neutral element e with respect to $\wedge$, we further define

(3) $\wedge^0 a = e$,

and if a has an inverse a' relative to $\wedge$ then for every natural number n we set

(4) $\wedge^{-n} a = \wedge^n(a')$.

When $\wedge$ is an addition operation we usually write na instead of $\wedge^n a$ and call it the nth *multiple* of a; when $\wedge$ is a multiplication operation we use the notation a^n and call this the nth *power* of a. Fairly straightforward inductive arguments serve to establish the following results which (at least in the additive and multiplicative notations) look very familiar.

Theorem 2.1. *Let E be a semigroup under $\wedge$; let a be an element of E. Then for all integers m and n for which the relevant iterates are defined, we have*

$$\wedge^{m+n} a = (\wedge^m a) \wedge (\wedge^n a), \quad \wedge^m(\wedge^n a) = \wedge^{mn} a.$$

If, further, b is an element of E which commutes with a, then

$$\wedge^m(a \wedge b) = (\wedge^m a) \wedge (\wedge^m b).$$

A semigroup is said to be a *group* if it has a neutral element relative to the law of composition and every element has an inverse relative to the law of composition. A group is said to be *abelian* if its law of composition is commutative.

Now let R be a set equipped with an addition operation and a multiplication operation; that is to say, we are given two internal laws of composition on R, one of which is written additively, the other multiplicatively. We say that R is a *ring* under these laws of composition if

(1) R is an abelian group under the addition operation;
(2) R is a semigroup under the multiplication operation;
(3) The multiplication operation is distributive over the addition operation.

We now spell out in detail what is involved in these conditions: R is a ring under the given addition and multiplication operations if

A1. The addition is associative, i.e. for all elements a, b, c of R we have $a + (b + c) = (a + b) + c$.

A2. The addition is commutative, i.e. for all elements a, b of R we have $a + b = b + a$.

A3. There is a neutral element relative to the addition, i.e. an element of R which we call the zero element and denote by 0, such that for every element a of R we have $a + 0 = a = 0 + a$.

A4. Every element of R has an inverse relative to the addition, i.e. for each element a of R there exists an element of R which we call the negative of a and denote by $-a$ such that $a + (-a) = 0 = (-a) + a$.

M1. The multiplication is associative, i.e. for all elements a, b, c of R we have $a(bc) = (ab)c$.

AM. The multiplication is distributive over the addition, i.e. for all elements a, b, c of R we have $a(b + c) = ab + ac$ and $(b + c)a = ba + ca$.

A ring R is called a *commutative ring* if, in addition to the defining properties, it also satisfies

M2. The multiplication is commutative, i.e. for all elements a, b of R we have $ab = ba$.

A ring R is called a *ring with identity* if it satisfies conditions *A1-4*, *M1*, *AM* and, in addition,

M3. There is a neutral element for the multiplication, i.e. an element e of R which we call the identity of R such that for every element a of R we have $ae = a = ea$.

If a, b, c are elements of a ring R such that $a = bc$ we say that a is a *left multiple* of c and a *right multiple* of b; we say also that c is a *right divisor* of a and that b is a *left divisor* of a. A little later we shall see that for every element a of R we have $a0 = 0$; so, according to the definition we have just given, every element of R is a left divisor of zero. It is conventional, however, to make a further condition in this case, and to say that an element b of a ring R is a *left divisor of zero* if and only if it is itself non-zero and there exists a *non-zero* element c of R such that $bc = 0$. *Right divisors of zero* are defined in a similarly restricted way.

A ring R is said to be an *integral domain* if it is commutative (i.e. it satisfies *M2*) and in addition satisfies

*M4**. There are no divisors of zero in R, i.e. if a and b are elements of R such that $ab = 0$ then either $a = 0$ or $b = 0$.

A ring R is called a *division ring* if it has at least two elements, satisfies the condition *M3* and in addition

M4. Every non-zero element of R has an inverse relative to multiplication, i.e. for each non-zero element a of R there exists an element a^{-1} of R such that $aa^{-1} = e = a^{-1}a$.

Finally, a commutative division ring is called a *field*; thus a field is a ring with at least two elements satisfying conditions *A1-4*, *M1-4*, and *AM*.

Example 1. The set of integers **Z** is an integral domain with identity under the ordinary addition and multiplication operations. The only elements of **Z** which have multiplicative inverses are 1 and -1; so **Z** is not a field under its ordinary addition and multiplication.

Example 2. The set of even integers is an integral domain under the ordinary addition and multiplication operations. There is no identity element and so it does not even make sense to ask whether there are any elements with multiplicative inverses.

Example 3. Let R be any set with an addition operation under which it forms an abelian group. Define a multiplication operation in R by setting $ab = 0$ for every pair of elements a, b of R. Then R is a commutative ring under the given addition and the multiplication we have defined. If R does not consist of the zero element alone there is no identity element and R is not an integral domain.

Example 4. The sets **Q**, **R**, **C** of rational, real and complex numbers are fields under the ordinary addition and multiplication operations in these sets.

Example 5. Let E be any set, A any ring; consider the set $R =$ Map (E, A) of all mappings from E to A. We define operations of addition and multiplication in R as follows: let α and β be mappings from E to A, and define $\alpha + \beta$ and $\alpha . \beta$ to be the mappings from E to A given by

$$(\alpha + \beta)(x) = \alpha(x) + \beta(x)$$
$$(\alpha . \beta)(x) = \alpha(x)\,\beta(x)$$

for every element x of E. Then R is a ring under these operations. The conditions *A1*, *A2*, *M1* and *AM* for R follow from the same conditions for A; the zero element of R is the mapping ζ from E to A defined by setting $\zeta(x) = 0$ for every element x of E; and if α is any element of R its negative is the mapping $-\alpha$ from E to A given by $(-\alpha)(x) = -(\alpha(x))$ for all elements x of E. If the ring A is commutative then so is R. If A has an identity element e, then the mapping ε from E to A, defined by setting $\varepsilon(x) = e$ for every element x of E, is an identity for R. If A is a field then every element α of R such that $\alpha(x) \neq 0$ for every element x of E has an inverse in R: the inverse is the mapping α^{-1} from E to A given by $\alpha^{-1}(x) = (\alpha(x))^{-1}$ for every element x of E.

Example 6. Let A be an additive abelian group, i.e. a set which forms an abelian group under an internal law of composition which is written additively; let End A be the set of endomorphisms of A. If α and β are elements of End A we define mappings $\alpha + \beta$ and $\alpha\beta$ from A to itself by setting

$$(\alpha + \beta)(a) = \alpha(a) + \beta(a) \quad \text{and} \quad (\alpha\beta)(a) = \alpha(\beta(a)) \qquad [2.1]$$

for all elements a of A. We claim that these mappings are in fact endomorphisms of A. So let a and b be any two elements of A; then we have

$$\begin{aligned}(\alpha + \beta)(a + b) &= \alpha(a + b) + \beta(a + b)\\ &= (\alpha(a) + \alpha(b)) + (\beta(a) + \beta(b))\\ &= (\alpha(a) + \beta(a)) + (\alpha(b) + \beta(b))\\ &= (\alpha + \beta)(a) + (\alpha + \beta)(b)\end{aligned}$$

and

$$\begin{aligned}(\alpha\beta)(a + b) = \alpha(\beta(a + b)) &= \alpha(\beta(a) + \beta(b))\\ &= \alpha(\beta(a)) + \alpha(\beta(b)) = (\alpha\beta)(a) + (\alpha\beta)(b).\end{aligned}$$

Thus the prescriptions [2.1] actually define internal laws of composition on End A. We shall now show that under these laws End A

is a ring with identity. The conditions *A1* and *A2* for End A follow from the associativity and commutativity of the group operation in A; the mapping ζ from A to A given by $\zeta(a) = 0$ for all elements a of A is easily seen to be an endomorphism of A and to be a zero element for End A; if α is any endomorphism of A then the mapping $-\alpha$ from A to A given by $(-\alpha)(a) = -(\alpha(a))$ for all elements a of A is quickly shown to be an endomorphism of A and clearly $\alpha + (-\alpha) = \zeta$. Condition *M1* is a basic property of composition of mappings. To show that condition *AM* holds in End A, let α, β, γ be endomorphisms of A and let a be any element of A. Then

$$((\alpha + \beta)\gamma)(a) = (\alpha + \beta)(\gamma(a)) = \alpha(\gamma(a)) + \beta(\gamma(a)) = \alpha\gamma(a) + \beta\gamma(a) = (\alpha\gamma + \beta\gamma)(a),$$

whence $(\alpha + \beta)\gamma = \alpha\gamma + \beta\gamma$; on the other hand,

$$(\gamma(\alpha + \beta))(a) = \gamma((\alpha + \beta)(a)) = \gamma(\alpha(a) + \beta(a)),$$

while

$$(\gamma\alpha + \gamma\beta)(a) = \gamma\alpha(a) + \gamma\beta(a) = \gamma(\alpha(a)) + \gamma(\beta(a)) = \gamma(\alpha(a) + \beta(a)),$$

so that $\gamma(\alpha + \beta) = \gamma\alpha + \gamma\beta$. Thus End A satisfies *AM*. The identity map I_A of A belongs to End A and is an identity element. We call End A the *ring of endomorphisms* of A.

Example 7. Let A be any ring; then for each positive integer n the set $M_n(A)$ of $n \times n$ matrices with coefficients in A is a ring under ordinary matrix addition and multiplication. In general the ring $M_n(A)$ is not commutative; it has an identity if A does.

Example 8. Let $\mathbf{H}$ be the four-dimensional vector space over the real number field $\mathbf{R}$ consisting of all ordered quadruples (a, b, c, d) of real numbers with the vector space addition and scalar multiplication given by

$$(a, b, c, d) + (a', b', c', d') = (a + a', b + b', c + c', d + d'),$$

$$r(a, b, c, d) = (ra, rb, rc, rd)$$

for all real numbers $a, a', b, b', c, c', d, d', r$. We denote the natural basis vectors $(1, 0, 0, 0)$, $(0, 1, 0, 0)$, $(0, 0, 1, 0)$, $(0, 0, 0, 1)$ by $\mathbf{e}, \mathbf{i}, \mathbf{j}, \mathbf{k}$ respectively; so we may write $(a, b, c, d) = a\mathbf{e} + b\mathbf{i} + c\mathbf{j} + d\mathbf{k}$. We now define a multiplication operation in $\mathbf{H}$ by prescribing the products of the basis vectors according to the table

	e	**i**	**j**	**k**
e	**e**	**i**	**j**	**k**
i	**i**	−**e**	**k**	−**j**
j	**j**	−**k**	−**e**	**i**
k	**k**	**j**	−**i**	−**e**

and then defining the product of two arbitrary vectors in the natural way:

$$\begin{aligned}(a\mathbf{e} + b\mathbf{i} + c\mathbf{j} + d\mathbf{k})&(a'\mathbf{e} + b'\mathbf{i} + c'\mathbf{j} + d'\mathbf{k}) \\ &= (aa')\,\mathbf{e}^2 + (ab')\,\mathbf{ei} + (ac')\,\mathbf{ej} + (ad')\,\mathbf{ek} \\ &\quad + (ba')\,\mathbf{ie} + (bb')\,\mathbf{i}^2 + (bc')\,\mathbf{ij} + (bd')\,\mathbf{ik} \\ &\quad + (ca')\,\mathbf{je} + (cb')\,\mathbf{ji} + (cc')\,\mathbf{j}^2 + (cd')\,\mathbf{jk} \\ &\quad + (da')\,\mathbf{ke} + (db')\,\mathbf{ki} + (dc')\,\mathbf{kj} + (dd')\,\mathbf{k}^2. \qquad [2.3]\end{aligned}$$

(We say that we 'extend by distributivity'.) Then it can be verified that under the addition and multiplication defined **H** is a non-commutative ring with identity, the identity being the basis vector **e**. We claim that **H** is in fact a division ring. Let $\mathbf{q} = a\mathbf{e} + b\mathbf{i} + c\mathbf{j} + d\mathbf{k}$ be a non-zero element of **H**; then a, b, c, d are not all zero and hence $a^2 + b^2 + c^2 + d^2 \neq 0$. Let $\mathbf{q}^* = a\mathbf{e} - b\mathbf{i} - c\mathbf{j} - d\mathbf{k}$; then, using [2.3] above we readily compute $\mathbf{qq}^* = (a^2 + b^2 + c^2 + d^2)\,\mathbf{e}$. Thus $(a^2 + b^2 + c^2 + d^2)^{-1}\mathbf{q}^*$ is a multiplicative inverse for **q**. So **H** is a non-commutative division ring; we call it the ring of *real quaternions*.

We now mention some very elementary properties of rings. First of all it follows from Theorem 1.6 that the zero element of a ring and the identity element of a ring with identity are unique, and from Theorem 1.8 that the negatives of all the elements of a ring and the multiplicative inverses of all the elements which are invertible with respect to multiplication are unique.

Next, we show that in a ring subtraction is uniquely possible. So let a and b be elements of a ring R; to subtract b from a we must find an element x of R such that $a = x + b$. Clearly $a + (-b)$ satisfies this requirement; and if x_1 is any element of R such that $a = x_1 + b$ then, on adding $-b$ to both sides and applying *A1* and *A4*, we have $x_1 = a + (-b)$. We shall write $a - b$ instead of $a + (-b)$.

Now let a be any element of R; we shall show that $0a = 0$. First we remark that, by *A3*, $0 + 0 = 0$; whence $(0 + 0)\,a = 0a$. By *AM* this yields $0a + 0a = 0a$; adding $-0a$ to both sides and using *A1* and *A4* we obtain $0a = 0$. Similarly we can show that $a0 = 0$.

The last result shows that if the ring R has more than one element then the zero element of R cannot be an identity; for if e is an identity of R and a is any non-zero element we have $ea = a \neq 0$, while we have just seen that $0a = 0$. We deduce also that if R has more than one element and has an identity e then 0 is not invertible with respect to multiplication; for if a were a multiplicative inverse of 0 we would have $e = a0 = 0$, which is a contradiction.

Next let a and b be any two elements of R; we claim that we have $a(-b) = (-a)\,b = -(ab)$ and that $(-a)(-b) = ab$. Since $b + (-b) = 0$ we deduce that $a(b + (-b)) = a0$, which is zero by what we have just proved. Using *AM* we have $ab + a(-b) = 0$; so $a(-b)$ is an additive inverse for ab. Since additive inverses are unique we deduce that $a(-b) = -(ab)$. The other results follow by similar arguments.

Suppose now that R has an identity e, and let U be the set of elements of R which have multiplicative inverses; these elements are called the *units* of the ring R. We shall show that U forms a group under the multiplication operation in R. First we must show that U is closed under this operation; but this is an immediate consequence of Theorem 1.10. Thus the multiplication in the ring R induces an internal law of composition on U. This law of composition is clearly associative, since the multiplication in R is associative (*M1*). The identity of R is its own multiplicative inverse and hence belongs to U, and it is of course an identity for U. Finally, by definition, every element of U has an inverse in R; but if $a \in U$, the inverse a^{-1} of a also belongs to U, since it is invertible (with inverse a). Thus U forms a group under the internal law of composition induced on U by the multiplication in R. We call U the *group of units* of R. When D is a division ring, the group of units consists of all the non-zero elements; in this case we call it the *multiplicative group* of the division ring D and denote it by D^*.

We remark also that in a division ring there are no divisors of zero. For suppose a and b are elements of a division ring D with identity e such that $ab = 0$; if a is non-zero, it has an inverse a^{-1} in D and hence

$$0 = a^{-1}0 = a^{-1}(ab) = (a^{-1}a)\,b = eb = b.$$

Finally, let R be any ring and as usual let the sum and product of any two elements a and b of R be denoted by $a + b$ and ab respectively. We define a new internal law of composition $*$ in R by setting

$$a * b = ba$$

for all elements a, b of R. Then the set of elements of R, equipped with the original addition operation and this new law of composition as multiplication, is a ring which we call the *opposite ring* of R and denote by R^{op}.

§3. Subrings and Ideals

Let S be a subset of a ring R. If S is closed under the addition and multiplication operations in R, then these operations induce internal laws of composition on S. So it makes sense to ask whether S forms a ring under these induced laws of composition; to answer this question it looks at first sight as if we would have to check whether the defining conditions *A1-4*, *M1* and *AM* of §2 are all satisfied. But a moment's thought should convince the reader that conditions *A1*, *A2*, *M1*, *AM* are automatically satisfied in S.

Suppose now that *A3* is satisfied in S, i.e. suppose that there is a zero element in S, which we denote for a moment by 0_S, such that $s + 0_S = s$ for every element s of S. We claim that 0_S must in fact be the zero element 0 of R; for if a is any element of R we have $a = a + (-(0_S) + 0_S) = (a + (-0_S)) + 0_S$ and hence $a + 0_S = ((a + (-0_S)) + 0_S) + 0_S = (a + (-0_S)) + (0_S + 0_S) = (a + (-0_S)) + 0_S = a$. This shows that 0_S is a neutral element for addition in R, and since neutral elements are unique it follows that $0_S = 0$. Conversely, of course, if 0 belongs to S then it is certainly a zero element of S. Thus *A3* is satisfied in S if and only if 0 belongs to S.

Next suppose that 0 belongs to S and that *A4* is satisfied in S, i.e. suppose that for every element s of S there exists an element s' of S such that $s + s' = 0$. Then s' is an additive inverse of s in S, and hence in R. Since additive inverses are unique it follows that $s' = -s$; so the negative of each element of S in R actually belongs to S. Conversely, if the negative of each element of S belongs to S then *A4* is satisfied in S.

We say that a subset S of a ring R is a *subring* of R if it is closed under the addition and multiplication operations of R and forms a ring under the induced internal laws of composition. We may thus sum up the preceding discussion in the following theorem.

THEOREM 3.1. *A subset S of a ring R is a subring if and only if the following conditions are satisfied: (1) S is closed under the addition and multiplication operations of R, (2) the zero element of R belongs to S, (3) for every element s of S its negative $-s$ belongs to S.*

There is also another criterion for a subset to be a subring, which is often more convenient to apply in practice.

THEOREM 3.2. *A non-empty subset S of a ring R is a subring if and only if for every pair of elements s and t of S the elements $s - t$ and st belong to S.*

Proof. (1) Suppose S is a subring of R and let s and t be any elements of S.

By condition (3) of Theorem 3.1, $-t \in S$. Hence, by condition (1) of 3.1, $s + (-t) = s - t$ belongs to S. Also by condition (1) of 3.1, $st \in S$.

(2) Conversely, suppose that for every pair of elements s, t in S we have $s - t \in S$ and $st \in S$.

The second condition shows that S is closed under the multiplication operation in R.

If s is any element of S it follows from the first condition that $s - s = 0$ belongs to S.

Next, if s is any element of S then since $0 \in S$ and $s \in S$ we deduce that $0 - s = -s$ belongs to S.

Finally, if s and t are any elements of S then $s \in S$ and $-t \in S$; so we have $s - (-t) = s + t \in S$. So S is closed under the addition operation in R.

It now follows from Theorem 3.1 that S is a subring of R.

Suppose that S is a subring of a ring R. It is natural to ask whether S inherits from R any of the additional properties which R may enjoy. It is clear that if R is a commutative ring (i.e. satisfies $M2$) then S is also a commutative ring; similarly if R has no divisors of zero then S has no divisors of zero either. But if R is a division ring we cannot deduce that S is a division ring: for although every non-zero element of S certainly has a multiplicative inverse in R, we cannot assert in general that these inverses belong to S. If R has an identity element e and e belongs to S then it is clearly an identity for S. But R may have an identity while S does not and *vice versa.*

A subring A of a ring R is called a *left ideal* if for every element r of R and every element a of A the product ra belongs to A. Similarly, a subring B of R is called a *right ideal* if for every element r of R and every element b of B the product br belongs to B. A subring which is both a left ideal and a right ideal is called a *two-sided ideal* or simply an *ideal.* A simple adaptation of Theorem 3.2 yields the following criterion for a subset of a ring to be an ideal.

THEOREM 3.3. *A non-empty subset A of a ring R is a left (right) ideal if and only if for all elements a, b of A and r of R the elements $a - b$ and ra (ar) belong to A.*

From this we deduce at once a useful consequence.

COROLLARY. *The intersection of any non-empty collection of left ideals (right ideals, two-sided ideals) of a ring R is a left ideal (right ideal, two-sided ideal) of R.*

The ring R itself is clearly a two-sided ideal of R; the ideals of R (left, right or two-sided) distinct from R itself are called *proper ideals* of R. If R does not consist of the zero element alone then R has at least one proper ideal, for the set $\{0\}$ consisting of the zero element alone is clearly a two-sided ideal of R; we call $\{0\}$ the *zero ideal* of R.

Example 1. Let m be any integer. Then the set of all multiples (positive, negative and zero) of m is a two-sided ideal in the ring of integers $\mathbf{Z}$. We denote this ideal by (m).

We now show that, conversely, every ideal of $\mathbf{Z}$ is of this form. Let I be any ideal of $\mathbf{Z}$. If I is the zero ideal then $I = (0)$. If I is not the zero ideal, then I contains non-zero integers, and hence positive integers (for if z is a non-zero integer in I, $0 - z = -z$ also belongs to I and certainly one of the integers z, $-z$ is positive). Let m be the least positive integer in I; we claim that $I = (m)$. Certainly since I is an ideal all the multiples of m belong to I, i.e. $(m) \subseteq I$. On the other hand, if n is any integer in I we may write $n = qm + r$ where q and r are integers and $0 \leqslant r < m$; since $n \in I$ and $qm \in I$ it follows that $n - qm = r \in I$. But m is the least positive integer in I and $0 \leqslant r < m$; hence $r = 0$ and $n = qm$. Thus $I \subseteq (m)$ and so, finally, $I = (m)$ as asserted.

Example 2. Let E be any set, A any ring; and let R be the ring of mappings from E to A (see §2, Example 5). If x is an element of E, then the set $I(x)$ consisting of all mappings α from E to A such that $\alpha(x) = 0$ is a two-sided ideal of R.

Example 3. Let A be any ring, $R = M_n(A)$ the ring of $n \times n$ matrices with elements in A. Let L_i be the subset of R consisting of matrices which have zero elements in all columns except possibly the ith; then L_i is a left ideal of R. Similarly the subset R_j of R consisting of matrices with zero elements in all rows except possibly the jth is a right ideal of R.

We remark that if R is a ring with identity and I is a proper (left, right or two-sided) ideal of R, then I cannot contain any unit of R. Suppose, for example, that I is a proper left ideal of R and u is a unit of R which belongs to I; then, since $u^{-1} \in R$ and $u \in I$ it follows that $u^{-1}u = e$ belongs to I, and hence if a is any element of R we deduce that $ae = a$ belongs to I; thus $I = R$, contradicting the hypothesis that I is a proper ideal. This discussion shows in particular that if R is a division ring the only ideals (left, right or two-sided) of R are R itself and the zero ideal.

Let X be any subset of a ring R. If E is the set of left ideals of R which include the subset X, then E is certainly non-empty, since R itself belongs to E. According to the Corollary of Theorem 3.3, the intersection $l(X)$ of all the left ideals in the set E is also a left ideal of R; $l(X)$ certainly includes X, and every left ideal which includes X also includes $l(X)$. So $l(X)$ is the smallest left ideal of R which includes X; we call $l(X)$ the left ideal *generated* by X. The right ideal $r(X)$ and the two-sided ideal $t(X)$ of R generated by X are defined in the obvious way; they are respectively the smallest right ideal and the smallest two-sided ideal of R which include X. Since a two-sided ideal is of course both a left and a right ideal, we have $l(X) \subseteq t(X)$ and $r(X) \subseteq t(X)$. If the subset X consists of a single element x of R, we write $l(x)$, $r(x)$, $t(x)$ instead of $l(\{x\})$, $r(\{x\})$, $t(\{x\})$, and we call these ideals the *principal* left, right and two-sided ideals generated by the element x. The next theorem gives a description of the elements of $t(X)$.

THEOREM 3.4. *Let X be a subset of a ring R. Then the two-sided ideal $t(X)$ of R generated by X consists precisely of those elements of R which can be expressed in the form*

$$\sum_{i \in I} n_i x_i + \sum_{j \in J} r_j x_j + \sum_{k \in K} x_k r'_k + \sum_{l \in L} s_l x_l s'_l \qquad [3.1]$$

where I, J, K, L are finite sets (possibly empty), $(x_i)_{i\in I}$, $(x_j)_{j\in J}$, $(x_k)_{k\in K}$ and $(x_l)_{l\in L}$ are families of elements of X, $(n_i)_{i\in I}$ is a family of integers, $(r_j)_{j\in J}$, $(r'_k)_{k\in K}$, $(s_l)_{l\in L}$ and $(s'_l)_{l\in L}$ are families of elements of R, and the elements n_ix_i are the multiples defined in §2.

Proof. Let T be the set of all elements of R which are expressible in the form [3.1].

Then clearly T is included in every two-sided ideal of R which includes X; hence $T \subseteq t(X)$.

But it is easy to check, using the analogue of Theorem 3.3 for two-sided ideals, that T is itself a two-sided ideal of R. Since T includes X, it follows that $T \supseteq t(X)$.

Hence $T = t(X)$, as asserted.

COROLLARY 1. *Let x be an element of a ring R. Then the principal two-sided ideal $t(x)$ generated by x consists of all the elements of R which can be expressed in the form*

$$nx + rx + xr' + \sum_{i\in L} r_ixr'_i \qquad [3.2]$$

where L is a finite set, n is an integer, r and r' are elements of R and $(r_i)_{i\in L}$ and $(r'_i)_{i\in L}$ are families of elements of R.

Similar descriptions can be given for the left and right ideals of R generated by a subset X or an element x of R. For example it is easy to establish that the left ideal $l(X)$ generated by X consists of all elements of R of the form

$$\sum_{i\in I} n_ix_i + \sum_{j\in J} r_jx_j \qquad [3.3]$$

where I and J are finite sets, $(x_i)_{i\in I}$ and $(x_j)_{j\in J}$ are families of elements of X, $(n_i)_{i\in I}$ is a family of integers and $(r_j)_{j\in J}$ is a family of elements of R. In the same way we deduce that the principal right ideal generated by the element x of R is made up of all elements of the form

$$nx + xr \qquad [3.4]$$

where n is an integer and r is an element of R.

When the ring R has an identity element e, it is easily established by an inductive argument that for every element x of R and every integer n we have

$$nx = (ne)\,x = x(ne).$$

Thus when R has an identity the sums $\sum_{i \in I} n_i x_i$ may be dropped from [3.1] and [3.3] and the term nx may be dropped from [3.2] and [3.4], since they may be absorbed in the other terms of the various expressions.

Combining this last remark with Corollary 1 and our descriptions of the left and right ideals generated by an element x of R we have the following useful consequence.

COROLLARY 2. *Let R be a ring with identity, u a unit of R. If x is any element of R then $l(x) = l(ux)$, $r(x) = r(xu)$, $t(x) = t(ux) = t(xu)$.*

If A is a left ideal of a ring R and X is a subset of A such that $l(X) = A$ we say that X is a *generating system* for the left ideal A; if A has a finite generating system we say that A is a *finitely generated* left ideal. In particular, of course, every principal left ideal is finitely generated. The notions of generating system and of finitely generated ideals are defined in the obvious way for right and two-sided ideals.

Let E_* be the set of non-zero left ideals in R. Then the inclusion relation is a relation of order in E_*; if L is a minimal element of E_* under this relation it is called a *minimal left ideal* of R. We similarly define minimal right ideals and minimal two-sided ideals. We remark that a ring R need not have any minimal left, right or two-sided ideals.

Similarly, let E^* be the set of proper left ideals in R. Again the inclusion relation is a relation of order in E^*; a maximal element of E^* under this relation is called a *maximal left ideal* of R. Maximal right ideals and maximal two-sided ideals are defined in the obvious way. If the ring R has no identity element then R need not have any maximal left, right or two-sided ideals; but if R has an identity we deduce as a special case of the next theorem that it must have maximal ideals.

THEOREM 3.5. *Let R be a ring with identity. If A is any proper left ideal of R there exists a maximal left ideal of R which includes A.*

Proof. Let E be the set of proper left ideals which include A; E is certainly non-empty, since A itself belongs to E. The inclusion relation is a relation of order on E; we shall show that E is inductively ordered by this relation.

So let E' be any totally ordered subset of E; we have to show that E' has a least upper bound in E. Let L be the union of all the left ideals in the set E'; we claim that L is a proper left ideal of R which includes A. To this end, let x_1, x_2 be any two elements of L, r any element of R. Then there are left ideals L_1, L_2 say, in the set E' such that $x_1 \in L_1$, $x_2 \in L_2$. Since E' is totally ordered, either $L_1 \subseteq L_2$ or $L_2 \subseteq L_1$, say $L_2 \subseteq L_1$; then x_1 and x_2 both belong to L_1. It follows (by Theorem 3.3) that $x_1 - x_2 \in L_1$ and $rx_1 \in L_1$; hence $x_1 - x_2 \in L$ and $rx_1 \in L$. Thus L is a left ideal. To show that L is a proper left ideal, we have only to remark that, since the ideals in the set E' are all proper, none of them contains the identity of R; hence L does not contain the identity of R; since L clearly includes A it follows that L belongs to E. It is obvious that L is the least upper bound of E'.

The desired result now follows at once by an application of Zorn's Lemma.

COROLLARY. *Every ring with identity has at least one maximal left ideal.*

Now let A and B be left ideals of a ring R. We define the *sum* of A and B to be the subset $A + B$ of R consisting of all elements x of R which can be expressed in the form $x = a + b$ where $a \in A$ and $b \in B$. It is easy to show that $A + B$ is a left ideal of R. First suppose that x_1 and x_2 are elements of $A + B$; then there exist elements a_1, a_2 of A and b_1, b_2 of B such that $x_1 = a_1 + b_1$ and $x_2 = a_2 + b_2$, and hence

$$x_1 - x_2 = (a_1 + b_1) - (a_2 + b_2) = (a_1 - a_2) + (b_1 - b_2).$$

Since $a_1 - a_2 \in A$ and $b_1 - b_2 \in B$ (because A and B are ideals), it follows that $x_1 - x_2 \in A + B$. Next let $x = a + b$ (with a in A and b in B) be any element of $A + B$, r any element of R; then

$$rx = r(a + b) = ra + rb,$$

and so $rx \in A + B$, since $ra \in A$ and $rb \in B$. Thus, by Theorem 3.3, $A + B$ is a left ideal of R. The reader may like to verify that the sum $A + B$ which we have just defined coincides with the left ideal generated by the union $A \cup B$. In the same way we define the sum of two right ideals and two two-sided ideals and show that these sums are respectively right and two-sided ideals.

Again let A and B be left ideals of R. We define the *product* AB of A and B to be the subset of R consisting of all elements which can be expressed in the form $\sum_{i \in I} a_i b_i$ where I is a finite set and $(a_i)_{i \in I}$,

$(b_i)_{i \in I}$ are families of elements of A and B respectively. Using Theorem 3.3 again we can easily show that AB is a left ideal of R. In the same way we can define the product of two right ideals or of two two-sided ideals and show that these products are ideals of the same kind. Having defined the product of two ideals we may apply the procedure described in § 2 to define the powers A^n of an ideal A for all natural numbers n: we check easily that A^0 is the ring R itself, which is a neutral element for multiplication of ideals, and that for each non-zero natural number n the nth power A^n consists of all elements of R which can be expressed in the form $\sum_{i \in I} a_{i1}a_{i2} \ldots a_{in}$, where I is a finite set and $(a_{i1})_{i \in I}, \ldots, (a_{in})_{i \in I}$ are families of elements of A.

The following properties of addition and multiplication of ideals in a ring are all easy consequences of the definitions.

THEOREM 3.6. *Let A, B, C be ideals (left, right or two-sided) of a ring R. Then*

A1. $A + (B + C) = (A + B) + C$, *M1.* $A(BC) = (AB)C$,
A2. $A + B = B + A$,
A3. $A + \{0\} = A$,
AM. $A(B + C) = AB + AC$ *and* $(A + B)C = AC + BC$.

It follows at once from the definition of ideal multiplication that if A and B are left ideals then $AB \subseteq B$; similarly if A and B are right ideals we have $AB \subseteq A$. Hence, of course, if A and B are two-sided ideals we have $AB \subseteq A$ and $AB \subseteq B$; so $AB \subseteq A \cap B$.

Let R be a ring, I a two-sided ideal of R. Consider the relation defined on R by setting $x \equiv y$ (mod. I) if and only if $x - y \in I$. We claim that this relation, which we read 'x is *congruent* to y *modulo* I', is an equivalence relation on R. First let x be any element of R; then $x - x = 0$ belongs to I since I is a subring, i.e. $x \equiv x$ (mod. I). Next let x and y be elements of R such that $x \equiv y$ (mod. I); then $x - y \in I$ and so, since I is a subring, $-(x - y) = y - x$ belongs to I, i.e. $y \equiv x$ (mod. I). Finally let x, y, z be elements of R such that $x \equiv y$ (mod. I) and $y \equiv z$ (mod. I). Then $x - y \in I$ and $y - z \in I$; so $(x - y) + (y - z) = x - z$ belongs to I, i.e. $x \equiv z$ (mod. I). Hence the relation is indeed an equivalence relation.

We now show that the addition and multiplication operations in R are compatible with this relation. So suppose $x \equiv x'$ (mod. I) and $y \equiv y'$ (mod. I); then $x - x' \in I$, $y - y' \in I$ and hence we have

$(x - x') + (y - y') = (x + y) - (x' + y') \in I$; so $x + y \equiv x' + y'$ (mod. I) and the addition in R is compatible with the relation. To prove that the multiplication is compatible we remark that

$$xy - x'y' = xy - xy' + xy' - x'y' = x(y - y') + (x - x')y'.$$

Now $y - y' \in I$ and $x - x' \in I$; and hence, since I is a two-sided ideal, $x(y - y') \in I$ and $(x - x')y' \in I$; it follows that $xy - x'y' \in I$, i.e. $xy \equiv x'y'$ (mod. I).

The quotient set of R with respect to the equivalence relation we have been discussing is denoted by R/I and the equivalence classes are called the *residue classes* modulo the two-sided ideal I; as usual we denote the canonical surjection of R onto R/I by η. According to §1 the addition and multiplication operations in R induce internal laws of composition in R/I as follows: if C_1, C_2 are residue classes modulo I and a_1, a_2 are elements of R in C_1, C_2 respectively, then

$$C_1 + C_2 = \eta(a_1 + a_2) \text{ and } C_1C_2 = \eta(a_1a_2). \qquad [3.5]$$

(The compatibility of the operations in R with the equivalence relation assures us that $C_1 + C_2$ and C_1C_2 depend only on C_1 and C_2, not on the choices of a_1 and a_2.) It is now an easy matter to verify that R/I is a ring under the laws of composition induced on it by those in R. The conditions *A1, A2, M1, AM* for R/I all follow directly from the definitions [3.5] and the corresponding conditions for R; the residue class of 0 (which is, of course, the ideal I itself) is a zero element for R/I. Finally, if C is a residue class modulo I, let a be any element of C; then the residue class $\eta(-a)$ containing $-a$ is an additive inverse for C. We call R/I the *residue class ring* of R modulo I. It is clear that if R is a commutative ring then so is R/I, that if R has an identity element e then the residue class $\eta(e)$ is an identity in R/I, and that if the element a of R is invertible in R with inverse a^{-1} then the residue class $\eta(a)$ is invertible in R/I with inverse $\eta(a^{-1})$. But there may be divisors of zero in R/I even when there are none in R itself.

Example 4. Let m be any non-zero integer; let (m) be the two-sided ideal of $\mathbf{Z}$ consisting of all the multiples of m. If a and b are integers we see at once that a is congruent to b modulo the ideal (m) if and only if a is congruent to b modulo the integer m. Since every integer is congruent modulo m to exactly one of the integers $0, 1, \ldots, m - 1$, it follows that the residue class ring $\mathbf{Z}/(m)$ consists of the m

residue classes $C_0, C_1, \ldots, C_{m-1}$ where, for $i = 0, 1, \ldots, m-1$, C_i consists of all the integers of the form $i + km$ (with k in $\mathbf{Z}$). The residue class C_0 is the zero element of $\mathbf{Z}/(m)$ and the residue class C_1 is the identity. It follows easily from the definition of addition and multiplication in $\mathbf{Z}/(m)$ that if C_a and C_b are any two of the m residue classes then

$$C_a + C_b = C_s \text{ and } C_aC_b = C_p,$$

where s and p are the remainders $(0 \leqslant s, p < m)$ when $a + b$ and ab respectively are divided by m. We notice in particular that if $m = ab$, where $1 < a, b < m$ then $C_aC_b = C_0$; thus if m is not a prime number, $\mathbf{Z}/(m)$ has divisors of zero, even though $\mathbf{Z}$ does not.

Suppose now that p is a prime number; we claim that the residue class ring $\mathbf{Z}/(p)$ is a field. All that remains to be shown is that every non-zero residue class has a multiplicative inverse in $\mathbf{Z}/(p)$. So let C_a be a non-zero element of $\mathbf{Z}/(p)$; then $0 < a < p$ and so, since p is a prime number, the highest common factor of a and p is 1. Hence there exist integers b and q such that $0 < b < p$ and $ab + pq = 1$. Then we have $C_aC_b = C_1$, i.e. C_b is a multiplicative inverse for C_a.

§4. Homomorphisms of Rings

Let R and S be rings. A mapping α from R to S is called a *ring homomorphism* (or simply a homomorphism) if for all elements a, b of R we have

$$\alpha(a + b) = \alpha(a) + \alpha(b) \text{ and } \alpha(ab) = \alpha(a)\,\alpha(b).$$

The other terms, monomorphism epimorphism, ..., defined in §1 are all used in an obvious way.

Example 1. Let R and S be rings, ζ the mapping from R to S defined by setting $\zeta(a) = 0$ for every element a of R. Then ζ is a homomorphism from R to S, which we call the *zero homomorphism.*

Example 2. Let R be a subring of a ring S, ι the canonical injection of R into S. Then ι is a monomorphism from R to S which we call the *inclusion monomorphism.* If $R = S$ and I_R is the identity mapping from R onto itself, I_R is an automorphism, which we call the *identity automorphism* of R.

Example 3. Let R be any ring, I a two-sided ideal of R and R/I the residue class ring. The addition and multiplication operations in R/I are defined in such a way that the canonical surjection η from R onto R/I is a homomorphism. From now on we shall call η the *canonical epimorphism* from R onto R/I.

Gathering together the general results of Theorems 1.1, 1.5, 1.7, 1.9 and 3.2 we can at once make the following statement about ring homomorphisms.

THEOREM 4.1. *Let R and S be rings, α a homomorphism from R to S. Then* $\operatorname{Im} \alpha = \alpha(R)$ *is a subring of S. If 0 is the zero element of R then $\alpha(0)$ is the zero element of S; if a is any element of R then $\alpha(-a)$ is the additive inverse of $\alpha(a)$; if R is commutative, so is* $\operatorname{Im} \alpha$; *if R has an identity element e then $\alpha(e)$ is an identity element of* $\operatorname{Im} \alpha$; *if an element a of R has a multiplicative inverse, then $\alpha(a)$ has a multiplicative inverse in* $\operatorname{Im} \alpha$, *namely $\alpha(a^{-1})$.*

Theorems 1.2 and 1.3 show that the composition of two ring homomorphisms is a homomorphism and that the inverse mapping of a ring isomorphism is also an isomorphism.

The *kernel* of a homomorphism α from a ring R to a ring S is the set of elements of R which are mapped by α onto the zero element of S; we denote the kernel of α by $\operatorname{Ker} \alpha$. Thus

$$\operatorname{Ker} \alpha = \{x \in R \mid \alpha(x) = 0\} = \alpha^{-1}(0).$$

The kernel of α is clearly non-empty since, as we mentioned in Theorem 4.1, $\alpha(0) = 0$; so the zero element of R belongs to the kernel of every homomorphism of R. Our next theorem shows that if the kernel of a homomorphism contains no other element then the homomorphism is injective.

THEOREM 4.2. *Let α be a homomorphism from a ring R to a ring S. Then α is a monomorphism if and only if* $\operatorname{Ker} \alpha = \{0\}$.

Proof. (1) Suppose α is a monomorphism. Then for every element s of $\operatorname{Im} \alpha$ the set $\alpha^{-1}(s)$ consists of a single element. In particular $\operatorname{Ker} \alpha = \alpha^{-1}(0)$ consists of a single element; and since $\alpha(0) = 0$, this single element is the zero element of R.

(2) Conversely, suppose $\operatorname{Ker} \alpha = \{0\}$. We shall show that α is a monomorphism. So let a and b be elements of R such that $\alpha(a) = \alpha(b)$. Since α is a homomorphism, we deduce that $\alpha(a - b) = \alpha(a + (-b))$

$= \alpha(a) + \alpha(-b) = \alpha(a) - \alpha(b) = 0$; so $a - b \in \operatorname{Ker} \alpha$ and hence $a - b = 0$, i.e. $a = b$.

Now we come to the fundamental theorem on homomorphisms of rings.

THEOREM 4.3. (*First homomorphism theorem*). *Let α be a homomorphism from a ring R to a ring S with kernel K. Then K is a two-sided ideal of R and if η is the canonical epimorphism from R to R/K, there exists a unique monomorphism α_* from R/K to S such that $\alpha_*\eta = \alpha$; further, α_* is an isomorphism if and only if α is an epimorphism.*

Proof. Let k and l be any two elements of K, a any element of R. Then, since α is a homomorphism we have, as above

$$\alpha(k - l) = \alpha(k) - \alpha(l) = 0$$

and also

$$\alpha(ak) = \alpha(a)\,\alpha(k) = 0 = \alpha(k)\,\alpha(a) = \alpha(ka).$$

So $k - l$, ak and ka belong to K, which is therefore a two-sided ideal by Theorem 3.3.

We now define a mapping α_* from the residue class ring R/K to S as follows. Let C be any element of R/K and choose any element a of R lying in the residue class C; then set $\alpha_*(C) = \alpha(a)$. To show that $\alpha_*(C)$ so defined depends only on C and not on the choice of a, let a_1 be another element of C; then $a - a_1 \in K$ and hence $0 = \alpha(a - a_1) = \alpha(a) - \alpha(a_1)$, i.e. $\alpha(a) = \alpha(a_1)$, as required. It is clear from the definition of α_* that $\alpha_*\eta = \alpha$; since η is surjective it follows that α_* is the only mapping φ such that $\varphi\eta = \alpha$.

We prove next that α_* is a homomorphism. So let C and C' be any two elements of R/K and choose elements a, a' from C, C' respectively. Then

$$\begin{aligned}\alpha_*(C) + \alpha_*(C') = \alpha_*(\eta(a)) + \alpha_*(\eta(a')) = \alpha(a) + \alpha(a') = \alpha(a + a')\\ = \alpha_*(\eta(a + a')) = \alpha_*(C + C').\end{aligned}$$

So α_* is a homomorphism, as asserted.

To show that α_* is a monomorphism, let C be any element of $\operatorname{Ker} \alpha_*$. If a is any element of C we have $0 = \alpha_*(C) = \alpha(a)$; so a belongs to $\operatorname{Ker} \alpha = K$. It follows that C is the zero residue class, and hence, by Theorem 4.2, that α_* is a monomorphism.

It is clear from the definition of α_* that Im α_* = Im α. Thus α_* is an epimorphism, and hence an isomorphism, if and only if α is an epimorphism.

We call α_* the monomorphism *induced* by α.

Example 4. Let E be any set, A any ring and R the ring of mappings from E to A. Let x be an element of E and consider the mapping σ_x from R to A defined by setting $\sigma_x(\varphi) = \varphi(x)$ for every mapping φ from E to A. Then σ_x is a homomorphism from R to A; it is actually an epimorphism, since for every element a of A there exists at least one mapping φ from E to A such that $\varphi(x) = a$, namely the constant mapping φ_a. The kernel of σ_x is the ideal $I(x)$ described in Example 2 of §3. Thus we deduce from Theorem 4.3 that $R/I(x)$ is isomorphic to A.

The next homomorphism theorem is a simple application of Theorem 4.3.

THEOREM 4.4. (*Second homomorphism theorem*). *Let A and B be two-sided ideals of a ring R. Then there exists an isomorphism from $(A + B)/B$ onto $A/(A \cap B)$.*

Proof. We recall that $A + B$ is the subset of R consisting of all elements of the form $a + b$ with a in A and b in B. Clearly B is a two-sided ideal of $A + B$, and $A \cap B$ is a two-sided ideal of A.

Let η be the canonical epimorphism from A onto $A/(A \cap B)$, and consider the mapping α from $A + B$ to $A/(A \cap B)$ defined as follows: for each element x of $A + B$ write x in the form $x = a + b$, where $a \in A$ and $b \in B$, and set $\alpha(x) = \eta(a)$. We must check that the right hand member depends only on x and not on the particular representation of x. So suppose $x = a + b = a' + b'$ where $a, a' \in A$ and $b, b' \in B$. Then $a - a' = b' - b$; hence $a - a' \in A \cap B$ and so $\eta(a) = \eta(a')$, as we hoped.

To show that α is a homomorphism from $A + B$ to $A/(A \cap B)$ let $x_1 = a_1 + b_1, x_2 = a_2 + b_2$ be elements of $A + B$ (where $a_1, a_2 \in A$, $b_1, b_2 \in B$). Then we have

$$x_1 + x_2 = (a_1 + a_2) + (b_1 + b_2),$$

$$x_1x_2 = a_1a_2 + (a_1b_2 + a_2b_1 + b_1b_2);$$

and, since A and B are two-sided ideals we have $a_1 + a_2 \in A$,

$a_1a_2 \in A$, $b_1 + b_2 \in B$ and $a_1b_2 + a_2b_1 + b_1b_2 \in B$. Hence we have

$$\alpha(x_1 + x_2) = \eta(a_1 + a_2) = \eta(a_1) + \eta(a_2) = \alpha(x_1) + \alpha(x_2)$$

and

$$\alpha(x_1x_2) = \eta(a_1a_2) = \eta(a_1)\,\eta(a_2) = \alpha(x_1)\,\alpha(x_2).$$

Since η is an epimorphism and $\alpha(a) = \eta(a)$ for every element a of A it follows that α is an epimorphism.

Now the element $x = a + b$ (where $a \in A$, $b \in B$) belongs to Ker α if and only if $\alpha(x) = \eta(a)$ is the zero element of $A/(A \cap B)$, i.e. if and only if $a \in A \cap B$. But since a is an element of A, we have $a \in A \cap B$ if and only if $a \in B$, and this happens if and only if $x = a + b$ belongs to B. Thus Ker $\alpha = B$.

The induced map α_* from $(A + B)/B$ to $A/(A \cap B)$ is then the required isomorphism.

The next main theorem on homomorphisms gives us a description of the ideals of any homomorphic image of a ring in terms of the ideals in the ring itself.

THEOREM 4.5. (*Third homomorphism theorem*). *Let α be an epimorphism from a ring R to a ring S with kernel K. Then there is a one-to-one correspondence between the set of ideals of R (left, right and two-sided) which include K and the set of all ideals of S. Further, if I is any two-sided ideal of R which includes K then R/I is isomorphic to $S/\alpha(I)$.*

Proof. Let $\mathscr{R}_K$ be the set of ideals of R which include K, $\mathscr{S}$ the set of all ideals of S. The mapping α from R to S gives rise to a mapping from $\mathscr{R}_K$ to the set of subsets of S; this is the mapping which assigns to each ideal I in $\mathscr{R}_K$ its image under the mapping α; we shall denote this mapping by α also. We have also a mapping β from $\mathscr{S}$ to the set of subsets of R defined by setting $\beta(J) = \alpha^{-1}(J)$ for each ideal J in $\mathscr{S}$. We shall show that α and β set up the required one-to-one correspondence between $\mathscr{R}_K$ and $\mathscr{S}$.

First we show that α actually maps $\mathscr{R}_K$ into $\mathscr{S}$. So let I be any ideal of R which includes K; let b_1, b_2 be elements of $\alpha(I)$, s any element of $S = \alpha(R)$. Then there are elements a_1, a_2 of I and r of R such that $b_1 = \alpha(a_1)$, $b_2 = \alpha(a_2)$, $s = \alpha(r)$. Hence we have

$$b_1 - b_2 = \alpha(a_1) - \alpha(a_2) = \alpha(a_1 - a_2),$$
$$sb_1 = \alpha(r)\,\alpha(a_1) = \alpha(ra_1),\ b_1s = \alpha(a_1)\,\alpha(r) = \alpha(a_1r),$$

from which it follows that $\alpha(I)$ is a left, right or two-sided ideal of S according as I is a left, right or two-sided ideal of R.

Next we show that β maps $\mathscr{S}$ into $\mathscr{R}_K$. First let J be any left ideal of S. Then certainly $\beta(J) = \alpha^{-1}(J)$ includes $\alpha^{-1}(0) = K$. Now let a_1, a_2 be elements of $\beta(J)$, r any element of R; then, by definition of β, $\alpha(a_1) \in J$, $\alpha(a_2) \in J$ and of course $\alpha(r) \in S$. We have $\alpha(a_1 - a_2) = \alpha(a_1) - \alpha(a_2) \in J$, so $a_1 - a_2 \in \beta(J)$; similarly $\alpha(ra_1) = \alpha(r)\,\alpha(a_1) \in J$ and so $ra_1 \in \beta(J)$. Thus $\beta(J)$ is a left ideal of R. Similar arguments show that if J is a right or two-sided ideal of S then $\beta(J)$ is a right or two-sided ideal of I which includes K.

Now we show that $\alpha\beta$ and $\beta\alpha$ are the identity mappings of $\mathscr{S}$ and $\mathscr{R}_K$ respectively.

If J is any ideal in $\mathscr{S}$ we certainly have $\alpha\beta(J) = \alpha(\alpha^{-1}(J)) \subseteq J$. Conversely, if b is any element of J then, since α is surjective, there is an element a of R such that $b = \alpha(a)$; but then $a \in \alpha^{-1}(b) \subseteq \alpha^{-1}(J)$ and so $b \in \alpha(\alpha^{-1}(J)) = \alpha\beta(J)$. Thus $J \subseteq \alpha\beta(J)$ and hence, as asserted, $\alpha\beta(J) = J$.

On the other hand, if I is any ideal in $\mathscr{R}_K$, we have $\beta\alpha(I) = \alpha^{-1}(\alpha(I)) \supseteq I$. If, conversely, $r \in \beta\alpha(I)$ it follows that $\alpha(r) \in \alpha(I)$; so there is an element a of I such that $\alpha(r) = \alpha(a)$, whence $\alpha(r - a) = 0$ and so $r - a \in K$. Since K is included in I it follows that $r - a \in I$ and hence that $r \in I$. Thus $\beta\alpha(I) \subseteq I$ and consequently $\beta\alpha(I) = I$.

It now follows that $\mathscr{R}_K$ and $\mathscr{S}$ are in one-to-one correspondence.

Finally let I be any two-sided ideal of R which includes K. Let η be the canonical epimorphism from S onto $S/\alpha(I)$, and consider the mapping $\varphi = \eta\alpha$ from R to $S/\alpha(I)$. It is easily verified that in fact φ is an epimorphism from R onto $S/\alpha(I)$. Clearly I is included in Ker φ; but conversely, if $a \in$ Ker φ we have $\alpha(a) \in \alpha(I)$ and as above we deduce that $a \in I$. So Ker $\varphi = I$ and it follows from Theorem 4.3 that R/I is isomorphic to $S/\alpha(I)$, as asserted.

This completes the proof of the theorem.

Let R be a ring with identity element e. Consider the mapping φ from the ring of integers $\mathbf{Z}$ to R defined by setting $\varphi(z) = ze$ for every integer z. (For the definition of the integral multiples ze see §2.) Then it is easy to verify that φ is a homomorphism. The kernel of φ is an ideal of $\mathbf{Z}$ and hence, by Example 1 of §3, a principal ideal (n) where n is a positive integer or zero. This integer n is called the *characteristic* of the ring R.

Example 5. The rings $\mathbf{Z}$, $\mathbf{Q}$, $\mathbf{R}$, $\mathbf{C}$, $\mathbf{H}$ all have characteristic zero. For each positive integer m the residue class ring $\mathbf{Z}/(m)$ has characteristic m.

It follows that there are rings of all possible characteristics. For fields, division rings and integral domains, however, the possible characteristics are restricted, as the next theorem shows.

THEOREM 4.6. *Let R be a ring with identity. If R has no divisors of zero then the characteristic of R is either zero or a prime number.*

Proof. Suppose, to the contrary, that the characteristic n of R is neither zero nor a prime number. Then $n = n_1 n_2$ where $1 < n_1$, $n_2 < n$, and we have $0 = ne = (n_1 n_2)e = (n_1 e)(n_2 e)$. Hence, since R has no divisors of zero, either $n_1 e = 0$ or $n_2 e = 0$, i.e. either $n_1 \in \text{Ker } \varphi$ or $n_2 \in \text{Ker } \varphi$, where φ is the homomorphism described above. But this is a contradiction, since $\text{Ker } \varphi = (n)$ and n_1 and n_2 are not divisible by n. It follows that n is either zero or a prime number.

§5. Extensions of Rings

Let R be a ring. An *extension* of R is an ordered pair (S, ι) consisting of a ring S and a monomorphism ι from R to S. If (S, ι) is an extension of R, the image $\iota(R)$ is a subring of S, and (since ι is a monomorphism) this subring is actually isomorphic to R. Frequently, when once we have given a formal description of an extension (S, ι) we shall refer to the ring S itself as the extension of R, and we shall regard R as being a subring of S, making no further mention of the monomorphism ι. When we do this we shall say that we are *identifying* R and $\iota(R)$.

In this section we shall give a number of important ring-theoretical constructions all of which may be viewed as examples of extensions.

Example 1. Let R be a subring of the ring S, and let ι be the inclusion monomorphism of R in S. Then (S, ι) is an extension of R. Whenever we talk of a ring as an extension of one of its subrings the monomorphism of the extension will always be the appropriate inclusion monomorphism unless explicit mention is made to the contrary.

Example 2. Let R be any ring; let $R_e = R \times \mathbf{Z}$, the Cartesian product of R with the ring of integers $\mathbf{Z}$. We define addition and multiplication operations in R_e by setting

$$\begin{aligned}(a_1, n_1) + (a_2, n_2) &= (a_1 + a_2, n_1 + n_2),\\ (a_1, n_1) \times (a_2, n_2) &= (a_1 a_2 + n_1 a_2 + n_2 a_1, n_1 n_2)\end{aligned} \qquad [5.1]$$

for all (a_1, n_1), (a_2, n_2) in R_e. It is easily verified that R_e forms a ring under these operations; the element $e = (0, 1)$ is an identity for R_e. The mapping ι from R to R_e defined by setting $\iota(a) = (a, 0)$ for each element a of R is a monomorphism from R to R_e; so (R_e, ι) is an extension of R. When we identify R with the subring $\iota(R)$ of R_e we see that we have 'embedded R in a ring with identity'. It was, of course, with this aim in view that we made the definitions [5.1] in the way we did; for if R were a subring of a ring S with identity e, then S would contain all elements of the form $a + ne$ with a in R and n in $\mathbf{Z}$, and we would have

$$(a_1 + n_1e) + (a_2 + n_2e) = (a_1 + a_2) + (n_1 + n_2)e,$$
$$(a_1 + n_1e)(a_2 + n_2e) = (a_1a_2 + n_1a_2 + n_2a_1) + (n_1n_2)e.$$

Example 3. Let R be any ring, m and n any two positive integers. Let $M_{m,n}(R)$ be the set of mappings from the Cartesian product set $[1, m] \times [1, n]$ to R; the elements of $M_{m,n}(R)$ are called $m \times n$ *matrices with coefficients* (or *elements*) in R. If $\alpha \in M_{m,n}(R)$ let us write $\alpha(i, j) = a_{ij}$ for $1 \leqslant i \leqslant m$, $1 \leqslant j \leqslant n$; we then write

$$\alpha = \begin{bmatrix} a_{11}\ a_{12} \ldots a_{1n} \\ a_{21}\ a_{22} \ldots a_{2n} \\ \ldots\ldots\ldots \\ a_{m1}\ a_{m2} \ldots a_{mn} \end{bmatrix}.$$

We define an addition operation in $M_{m,n}(R)$ by setting, for all α, β in $M_{m,n}(R)$,

$$(\alpha + \beta)(i, j) = \alpha(i, j) + \beta(i, j) \quad (i = 1, \ldots, m;\ j = 1, \ldots, n).$$

Under this operation $M_{m,n}(R)$ forms an abelian group: the associativity and commutativity follow at once from the corresponding properties of the addition operation in R; the matrix ζ given by $\zeta(i, j) = 0$ for $i = 1, \ldots, m$; $j = 1, \ldots, n$ is a zero element; for each matrix α in $M_{m,n}(R)$ the matrix $-\alpha$ given by $(-\alpha)(i, j) = -(\alpha(i, j))$ for $i = 1, \ldots, m$; $j = 1, \ldots, n$ is an additive inverse for α.

Now suppose $m = n$. We shall write $M_n(R)$ as an abbreviation for $M_{n,n}(R)$ and define a multiplication operation in $M_n(R)$ as follows. Let α and β be $n \times n$ matrices with coefficients in R; define $\alpha\beta$ to be the $n \times n$ matrix given by

$$(\alpha\beta)(i, j) = \sum_{k=1}^{n} \alpha(i, k)\,\beta(k, j) \quad (i, j = 1, \ldots, n).$$

It is now easy to verify that $M_n(R)$ is a ring under the addition operation defined above and this multiplication operation. We call this the *ring of* $n \times n$ *matrices* with coefficients in R.

Consider the *diagonal mapping* δ from R to $M_n(R)$ defined as follows. Let a be any element of R and let $\delta(a)$ be the $n \times n$ matrix given by

$$(\delta(a))(i,j) = \begin{cases} a & \text{if } i = j \\ 0 & \text{if } i \neq j \end{cases} \qquad (i,j = 1, \ldots, n).$$

The matrices of the form $\delta(a)$ are called *diagonal matrices.* It is easy to show that δ is a monomorphism from R to $M_n(R)$. Thus $(M_n(R), \delta)$ is an extension of R; in this case, however, it is not customary to identify the elements a of R with the corresponding diagonal matrices $\delta(a)$. If R has an identity element e then $\delta(e)$ is an identity for $M_n(R)$; the ring of matrices is in general not commutative even when R is commutative.

Most readers will of course be familiar with the results and techniques of matrix algebra, and we shall use any of these that we may require without giving any detailed explanation. We include this example mainly to show how matrices, which are usually introduced as 'rectangular arrays of numbers', fit into the abstract set-theoretic framework of modern algebra.

Example 4. Let R be a ring with identity element e. Let $P(R)$ be the subset of Map($\mathbf{N}$, R) consisting of those mappings from $\mathbf{N}$ to R which take the value zero for all but a finite set of natural numbers. Thus a mapping α from $\mathbf{N}$ to R belongs to $P(R)$ if and only if there is a natural number N_α such that $\alpha(n) = 0$ for all natural numbers $n > N_\alpha$. We define an addition operation in $P(R)$ as follows: let α and β be elements of $P(R)$ and consider the mapping $\alpha + \beta$ from $\mathbf{N}$ to R defined by setting $(\alpha + \beta)(n) = \alpha(n) + \beta(n)$ for each natural number n. Now $(\alpha + \beta)(n) = 0$ for all natural numbers $n > \sup(N_\alpha, N_\beta)$; so $\alpha + \beta$ actually belongs to $P(R)$. Completely standard arguments show that $P(R)$ forms an abelian group under the addition so defined.

Next we define a multiplication operation; for each pair of elements α, β of $P(R)$ let $\alpha\beta$ be the mapping from $\mathbf{N}$ to R defined by setting $\alpha\beta(n) = \sum_{i=0}^{n} \alpha(i)\beta(n-i)$ for all natural numbers n. Since $\alpha\beta(n) = 0$ for all natural numbers $n > N_\alpha + N_\beta$, it follows that $\alpha\beta$

belongs to $P(R)$. It is now a routine matter to check that $P(R)$ forms a ring under the addition and multiplication operations we have defined.

Let κ be the mapping from R to $P(R)$ given as follows: for each element a of R let $\kappa(a)$ be the mapping from $\mathbf{N}$ to R defined by setting

$$(\kappa(a))(n) = \begin{cases} a & \text{if } n = 0 \\ 0 & \text{if } n \neq 0. \end{cases}$$

Clearly $\kappa(a)$ belongs to $P(R)$, and it is easy to check that κ is a monomorphism from R to $P(R)$; thus $(P(R), \kappa)$ is an extension of R. Further, the element $\kappa(e)$ is an identity for $P(R)$.

Consider now the mapping X from $\mathbf{N}$ to R defined by setting

$$X(n) = \begin{cases} e & \text{if } n = 1 \\ 0 & \text{if } n \neq 1. \end{cases}$$

Certainly X belongs to $P(R)$ and it is easy to establish by induction that for every natural number k we have

$$X^k(n) = \begin{cases} e & \text{if } n = k \\ 0 & \text{if } n \neq k. \end{cases}$$

These powers X^k clearly commute with every element α of $P(R)$. An easy computation shows that for each element a of R and each natural number k we have

$$(\kappa(a)\, X^k)(n) = \begin{cases} a & \text{if } n = k \\ 0 & \text{if } n \neq k. \end{cases}$$

Let α be any element of $P(R)$; for each natural number k we write $\alpha(k) = a_k$. Then, as we have just remarked,

$$(\kappa(a_k)\, X^k)(n) = \begin{cases} \alpha(n) & \text{if } n = k \\ 0 & \text{if } n \neq k. \end{cases}$$

Hence for each natural number n the sum $\sum_{k \in \mathbf{N}} (\kappa(a_k)\, X^k)(n)$ has only one non-zero term, namely $\alpha(n)$. We are thus led to write formally

$$\alpha = \sum_{k \in \mathbf{N}} \kappa(a_k)\, X^k$$

and to call $P(R)$ the *ring of polynomials with coefficients in R*. If we identify R with its image under κ then of course the element α of $P(R)$ is written simply as $\sum_{k \in \mathbf{N}} a_k X^k$ or, of course, $\sum_{k=0}^{N_\alpha} a_k X^k$ since $a_k = \alpha(k) = 0$ for all natural numbers $k > N_\alpha$.

Let α be a non-zero element of $P(R)$; if $\alpha(N) \neq 0$ while $\alpha(n) = 0$ for all integers $n > N$ we may write

$$\alpha = a_0 + a_1 X + \ldots + a_N X^N$$

(with the usual understanding about identifying elements of R with their images under κ). We call N the *degree* of α and write $\partial\alpha = N$; the element a_N is called the *leading coefficient* of α; if $a_N = e$ we say that α is a *monic* polynomial. We define the degree of the zero polynomial to be $-\infty$, and make the usual convention that $-\infty < n$ and $(-\infty) + n = -\infty$ for all integers n. Then for all polynomials α and β in $P(R)$ we have

$$\partial(\alpha + \beta) \leqslant \sup(\partial\alpha, \partial\beta)$$

and

$$\partial(\alpha\beta) \leqslant \partial\alpha + \partial\beta.$$

If $\partial\alpha \neq \partial\beta$ then we have $\partial(\alpha + \beta) = \sup(\partial\alpha, \partial\beta)$, and if R has no divisors of zero we have $\partial(\alpha\beta) = \partial\alpha + \partial\beta$. (It follows at once from this remark that if R is an integral domain then so also is $P(R)$.) Polynomials of degree zero are called *constant* polynomials—these are of course the non-zero elements of the subring $\kappa(R)$.

Example 5. Let R be a ring with identity; suppose R is a subring of a ring S, with inclusion monomorphism ι. Let b be an element of S. Then, for every natural number n and every family $(a_i)_{i \in [0, n]}$ of elements of R, we may form the element $a_0 + a_1 b + a_2 b^2 + \ldots + a_n b^n$ of S. It is not hard to verify that the set of all elements of S of this form is a subring of S, which we call the ring of *polynomials in b with coefficients in R*; it is clearly the smallest subring of S containing R and b, and we denote it by $R[b]$. Since the inclusion monomorphism ι obviously maps R into $R[b]$ it follows that $(R[b], \iota)$ is an extension of R.

Consider now the mapping σ_b from $P(R)$ to $R[b]$ defined by setting $\sigma_b(\alpha) = \sigma_b(a_0 + a_1 X + \ldots + a_n X^n) = a_0 + a_1 b + \ldots + a_n b^n$ for every polynomial $\alpha = a_0 + a_1 X + \ldots + a_n X^n$ of $P(R)$. We call

σ_b the operation of *substituting* b for X; it is clearly a surjection from $P(R)$ to $R[b]$. When b commutes with every element of R we can show that σ_b is also a homomorphism.

Example 6. Let R be any ring with identity; we define for each natural number $n \geqslant 1$ the *nth order polynomial ring* with coefficients in R, which we shall denote by $P_n(R)$. The ring $P_1(R)$ is just the polynomial ring $P(R)$ as we described it in Example 4; then for every natural number $n > 1$ we set $P_n(R) = P(P_{n-1}(R))$. We define monomorphisms κ_n from R to the rings $P_n(R)$ in the obvious way: κ_1 is the monomorphism κ of Example 4, and for every natural number $n > 1$ we set $\kappa_n = \lambda_{n-1}\kappa_{n-1}$, where λ_{n-1} is the obvious monomorphism from $P_{n-1}(R)$ to $P(P_{n-1}(R))$.

It can be shown that by suitable choice of elements $X_1, \ldots, X_n$ and the usual identification procedure every element of $P_n(R)$ can be expressed in the form

$$\sum_{i_1 \in \mathrm{N}} \sum_{i_2 \in \mathrm{N}} \cdots \sum_{i_n \in \mathrm{N}} a_{i_1 i_2 \cdots i_n} X_1^{i_1} X_2^{i_2} \ldots X_n^{i_n},$$

where the coefficients $a_{i_1 i_2 \cdots i_n}$ all belong to R and only finitely many of them are non-zero.

Example 7. Let R, S and ι have the same meaning as in Example 5. If $\{b_1, \ldots, b_n\}$ is a finite set of elements of S, then we readily check that the smallest subring of S which includes $R \cup \{b_1, \ldots, b_n\}$ consists precisely of those elements of S which are expressible in the form

$$\sum_{i_1 \in \mathrm{N}} \cdots \sum_{i_n \in \mathrm{N}} a_{i_1 i_2 \cdots i_n} b_1^{i_1} \ldots b_n^{i_n}$$

where the coefficients $a_{i_1 \cdots i_n}$ belong to R and only finitely many are non-zero. We call this subring the ring of *polynomials in* $\{b_1, \ldots, b_n\}$ *with coefficients in* R, and denote it by $R[b_1, \ldots, b_n]$. It is clear that $(R[b_1, \ldots, b_n], \iota)$ is an extension of R.

Example 8. Let R be any ring with identity. For every pair of natural numbers m and n such that $m \leqslant n$ it is obvious how we may identify $P_m(R)$ with a subring of $P_n(R)$. If we form the union $P_\infty(R) = \cup_{k \in \mathrm{N}} P_k(R)$ this observation allows us to define operations of addition and multiplication in $P_\infty(R)$. Namely, let α and β be elements of $P_\infty(R)$; then there are natural numbers m and n such that $\alpha \in P_m(R)$ and $\beta \in P_n(R)$. Either $m \leqslant n$ or $n \leqslant m$; say $m \leqslant n$.

Then, provided we make the identification mentioned above, α and β are both elements of $P_n(R)$ and we define $\alpha + \beta$ and $\alpha\beta$ to be the sum and product of α and β in the ring $P_n(R)$. Under the laws of composition so defined, $P_\infty(R)$ forms a ring.

Chapter 2

MODULES

§6. Modules, Submodules and Factor Modules

Let A and E be sets; by an *external law of composition on E with A as set of operators* we mean a mapping from the Cartesian product $A \times E$ to E. If μ is such an external law of composition and (a, x) is any element of $A \times E$, we shall usually denote the image $\mu(a, x)$ in E by ax or $a.x$, in which case μ is called *scalar multiplication on the left* of E by elements of A, or by xa or $x.a$ when μ is called *scalar multiplication on the right.*

Example 1. Let $A = \mathbf{N}$, the set of natural numbers, E a semigroup with identity under the internal law of composition $\wedge$. Then the mapping μ from $\mathbf{N} \times E$ to E given by $\mu(n, x) = \wedge^n x$ (as defined in §2) for every natural number n and every element x of E is an external law of composition on E with $\mathbf{N}$ as set of operators.

Example 2. Let $A = \mathbf{R}$, the set of real numbers, E the set of vectors in three-dimensional Euclidean space. For every non-zero real number a and every non-zero vector x in E let $\mu(a, x)$ be the vector whose magnitude is $|a| \, \|x\|$ (where $\|x\|$ is the magnitude of x), whose direction is the same as that of x and whose sense is the same as that of x or opposite to it according as a is positive or negative; if either a or x is zero, let $\mu(a, x)$ be the zero vector. Then the mapping μ from $\mathbf{R} \times E$ to E defined in this way is an external law of composition on E with $\mathbf{R}$ as set of operators.

Let R be a ring with identity element e; let V be a set with an internal law of composition φ which we write additively and an external law of composition μ with R as set of operators which we write as scalar multiplication on the left. Then we say that V is a *left R-module* under φ and μ or that (V, φ, μ) is a left R-module if the following conditions are satisfied:

A. V is an abelian group under the addition operation φ;
LM. For all elements a, b of R and x, y of V,

$$(1)\ (a + b)\, x = ax + bx,$$

(2) $a(x + y) = ax + ay$,
(3) $(ab)x = a(bx)$,
(4) $ex = x$.

If V has an addition operation φ and a right scalar multiplication μ by elements of R then we say that V is a *right R-module* under φ and μ if:

A. V is an abelian group under φ;
RM. For all elements a, b of R and x, y of V,

(1) $x(a + b) = xa + xb$,
(2) $(x + y)a = xa + ya$,
(3) $x(ab) = (xa)b$,
(4) $xe = x$.

If R were an arbitrary ring (i.e. not necessarily with an identity element) we might have defined left R-modules by postulating all the above requirements except for $LM(4)$; in the case where R had an identity left R-modules satisfying $LM(4)$ would be called unitary modules. Similar remarks apply of course to right R-modules with reference to $RM(4)$. We have chosen to assume from now on that *all rings have identity elements and all modules are unitary*; so we have built $LM(4)$ and $RM(4)$ into our basic definition of modules.

If R is a division ring (or, in particular, a field) left R-modules are usually referred to as *left vector spaces* over R and right R-modules are called *right vector spaces* over R.

Example 3. Let A be any additive abelian group (i.e. an abelian group with its law of composition written additively). Let μ be the mapping from $\mathbf{Z} \times A$ to A given by setting $\mu(z, a) = za$ (as defined in §2) for every integer z and every element a of A. Then A is a left $\mathbf{Z}$-module under the addition in A and the mapping μ.

Example 4. Let E be the set of vectors in three-dimensional Euclidean space. Then E is a left vector space over the real number field $\mathbf{R}$ under ordinary vector addition and the scalar multiplication μ defined in Example 2.

Example 5. Let A be any additive abelian group, and let $R = \text{End } A$ be the ring of endomorphisms of A as defined in §2, Example 6. Define the mapping μ from $R \times A$ to A by setting $\mu(\alpha, a) = \alpha(a)$ for every endomorphism α and every element a of A. Then it is an easy matter to verify that A is a left R-module under its addition operation and the external law μ.

Example 6. Let R be any ring with identity. Consider the mappings λ and ρ from $R \times R$ to R defined by setting $\lambda(a, x) = ax$ and $\rho(a, x) = xa$ for all elements a and x of R. Then R is a left R-module under its addition operation and the 'external law' λ, and a right R-module under its addition operation and ρ. When we consider R as a left or right R-module in this way we shall denote it by R_l or R_r respectively.

During the elementary development of the theory of modules we shall give our definitions and results for left modules, leaving it to the reader to make the obvious modifications necessary to deal with right modules.

When we are working with a left R-module V two zero elements will occur in our discussions—the zero element of the ring R and the zero element of the additive group V. We shall usually denote both these elements simply by 0; but, if it is necessary to emphasise which we are dealing with we may denote them by 0_R and 0_V respectively.

If x is any element of V we claim that $0_R x = 0_V$. To see this we remark that $0_R + 0_R = 0_R$ and hence $(0_R + 0_R)\,x = 0_R x$; by $LM(1)$ we deduce that $0_R x + 0_R x = 0_R x$. Adding the negative $-(0_R x)$ to both sides and using the associativity of the addition operation in V, we obtain $0_R x = 0_V$. A similar argument shows that if a is any element of R then $a0_V = 0_V$. Now let a be any element of R, x any element of V; we claim that $(-a)\,x = -(ax)$. To show this we have only to remark that $ax + (-a)\,x = (a + (-a))\,x = 0_R x = 0_V$; hence $(-a)\,x$ is an inverse for ax with respect to the addition in V. But so is $-(ax)$; and since inverses in a group are unique, we have $(-a)\,x = -(ax)$. Similarly $a(-x) = -(ax)$.

In §1 we described how to form the sum $\sum_{i \in I} x_i$ of a *finite* family $(x_i)_{i \in I}$ of elements of a left R-module V. Now, suppose that the index set I is quite arbitrary, i.e. not necessarily finite, and let $(x_i)_{i \in I}$ be a family of elements of V. Let I^* be the subset of I consisting of indices i for which x_i is non-zero. Then if I^* is finite but non-empty, $\sum_{i \in I} x_i$ is defined to mean $\sum_{i \in I^*} x_i$ (as described in §1); if I^* is empty, $\sum_{i \in I} x_i$ is defined to be the zero element of V; and if I^* is infinite $\sum_{i \in I} x_i$ is left undefined.

Let V be a left R-module, W a subset of V. We say that W is *closed* or *stable* under the scalar multiplication μ of V by R if for every element a of R and every element w of W we have $\mu(a, w) = aw \in W$. In this situation we say that μ *induces* an external law of composition on W. If, further, W is closed under the addition operation in V (which therefore induces an addition on W) we may sensibly ask whether W forms a left R-module under these induced laws of composition. If it does, we say that W is a *submodule of* V.

Example 7. Let A be any additive abelian group considered as a left **Z**-module according to the description in Example 3. Then every subgroup of A is a submodule of A.

Example 8. As in Example 6 let R_l be the ring R itself considered as a left R-module. Then it is easily verified that the submodules of R_l are just the left ideals of the ring R considered as left R-modules.

The situation here is reminiscent of that in §3 where we considered subsets of a ring which were closed under the laws of composition of the ring. We saw that certain additional conditions, over and above the closure, were necessary and sufficient for such a subset to be a ring. In the case of modules, however, we shall show that the closure conditions are already themselves necessary and sufficient for a subset to be a submodule.

THEOREM 6.1. *A non-empty subset W of a left R-module V is a submodule if and only if it is closed under the addition in V and the scalar multiplication of V by R.*

Proof. (1) If W is a submodule of V it is certainly closed under the addition and scalar multiplication.

(2) Conversely, suppose W is closed under these laws of composition. The conditions $LM(1)$–(4) are certainly satisfied by the induced scalar multiplication on W since they are satisfied by the original scalar multiplication in V.

All that now remains to be shown is that W is an abelian group under the induced addition. This induced addition is certainly associative and commutative, since the original addition in V is associative and commutative. Let w be any element of W; since W is closed under the scalar multiplication of V by R, $0_R w = 0_V$ belongs to W, which therefore has a neutral element for addition.

Again let w be any element of W; using once more the fact that W is closed under the scalar multiplication of V by R we deduce that $(-e)w = -(ew) = -w$ belongs to W. Thus W is indeed an abelian group under the induced addition.

This completes the proof.

COROLLARY. *The intersection of any non-empty family of submodules of a left R-module V is also a submodule of V.*

The module V itself is clearly a submodule of V; the submodules of V distinct from V are called *proper submodules* of V. If V does not consist of the zero element 0_V alone then $\{0_V\}$ is a proper submodule of V, which we call the *zero submodule*. A left R-module V is said to be *simple* or *irreducible* if (1) it does not consist of the zero element alone and (2) its only submodules are V itself and the zero submodule.

Let X be any subset of a left R-module V. Then the set of all submodules of V which include X is certainly non-empty, since the module V itself belongs to this set. According to the Corollary of Theorem 6.1, the intersection of this set is a submodule of V, and it certainly includes X; this is clearly the smallest submodule of V which includes X. We denote it by RX (or by Rx if X consists of a single element x of V) and call it the submodule of V *generated by* X (or by x). If X is a subset of V such that $RX = V$ we call X a *generating system* for V; if V has a finite generating system we say that V is a *finitely generated* module.

In order to describe the elements of RX it is convenient to introduce the following terminology: an element y of V is said to be expressed as a *linear combination* of elements of X with coefficients in R if there exists a family $(a_x)_{x \in X}$ of elements of R such that only finitely many of the elements a_x are non-zero and $y = \sum_{x \in X} a_x x$.

We shall have to mention very frequently families of elements of R only finitely many of which are non-zero: we call such families *quasi-finite*. Thus a family $(a_i)_{i \in I}$ of elements of R is quasi-finite if and only if the subset I^* of I consisting of those indices i for which a_i is non-zero is a finite set.

THEOREM 6.2. *Let X be a subset of a left R-module V. Then the submodule RX of V generated by X consists of all the elements of V which can be expressed as linear combinations of elements of X with coefficients in R.*

Proof. Let W be the subset of V consisting of all these linear combinations. Then certainly W is included in every submodule of V which includes X and hence $W \subseteq RX$.

But it follows at once from Theorem 6.1 that W is itself a submodule of V, and it clearly includes X. So $W \supseteq RX$.

Hence $W = RX$ as asserted.

By specialising to the case where X consists of a single element x of V we obtain the following consequence.

COROLLARY. *Let x be an element of a left R-module V. Then the submodule Rx of V generated by x consists of all the elements of V of the form ax where a is an element of R.*

Let $(W_i)_{i \in I}$ be a family of submodules of a left R-module V. We may form the union $\cup_{i \in I} W_i$ of this family; this union is of course a subset of V, but it is not in general a submodule. Nevertheless $\cup_{i \in I} W_i$ generates a submodule of V, which we denote by $\sum_{i \in I} W_i$ and call the *sum* of the family $(W_i)_{i \in I}$. In the same way if W_1 and W_2 are submodules of V the submodule generated by $W_1 \cup W_2$ is denoted by $W_1 + W_2$. By using an argument similar to that in the proof of Theorem 6.2 we obtain the following description of the elements of $\sum_{i \in I} W_i$.

THEOREM 6.3. *Let $(W_i)_{i \in I}$ be a family of submodules of a left R-module V. Then the sum $\sum_{i \in I} W_i$ of this family consists of all elements of V of the form $\sum_{i \in I} w_i$ where $w_i \in W_i$ for each index i in I and only finitely many of the elements w_i are non-zero.*

A finite subset X of a left R-module V is said to be *linearly dependent* if there exists a family $(a_x)_{x \in X}$ of elements of R, *not all zero*, such that $\sum_{x \in X} a_x x = 0$. An arbitrary subset is said to be linearly dependent if it has a finite subset which is linearly dependent; that is to say, X is linearly dependent if there is a linear combination of elements of X with non-zero coefficients in R which is equal to the zero element of W. The subset X is said to be *linearly independent* or *free* if it is not linearly dependent; thus X is free if and only if the only linear combination of elements of X with coefficients in R which can be equal to zero is that in which all the coefficients are zero. We make the convention that the empty set is free. It is clear that a subset of V is free if and only if all its finite subsets are free.

If $RX = V$ and X is linearly independent, i.e. if X is a free generating system for V, we say that X is a *basis* for V. The next theorem gives a useful criterion for a subset of V to be a basis.

THEOREM 6.4. *A subset X of a left R-module V is a basis for V if and only if every element of V can be expressed uniquely as a linear combination of elements of X with coefficients in R.*

Proof. (1) Suppose X is a basis for V.

Since $RX = V$ it follows from Theorem 6.2 that every element of V is expressible as a linear combination of elements of X.

To show that the expression is unique, let y be any element of V and suppose there are quasi-finite families $(a_x)_{x \in X}$ and $(b_x)_{x \in X}$ of elements of R such that $y = \sum_{x \in X} a_x x = \sum_{x \in X} b_x x$. Then we have $\sum_{x \in X} (a_x - b_x)\, x = 0$ and hence, since X is linearly independent, $a_x - b_x = 0$, i.e. $a_x = b_x$ for all elements x of X as required.

(2) Conversely, suppose every element of V can be expressed uniquely as a linear combination of elements of X with coefficients in R.

It follows at once from Theorem 6.2 that $RX = V$.

To show that X is linearly independent, let $(a_x)_{x \in X}$ be a quasi-finite family of elements of R such that $\sum_{x \in X} a_x x = 0$. It is clear, however, that if $b_x = 0$ for every element x of X we have $\sum_{x \in X} b_x x = 0$ also. It now follows from the uniqueness assumption that $a_x = b_x = 0$ for every element x of X; so X is linearly independent.

Example 9. Let $V = \mathbf{Z}_l$, i.e., the ring of integers considered as a left $\mathbf{Z}$-module. Then the set $X = \{1\}$ is a basis for V. First of all X generates V; for if $z \in \mathbf{Z}$ we have $z = z1 \in \mathbf{Z}X$. Secondly, X is free; for if we have $z1 = 0$ it follows that $z = 0$. We may show in the same way that $\{-1\}$ is a basis for V.

Example 10. Let V be a finite abelian group, considered as a $\mathbf{Z}$-module in the way we described in Example 3. Then V has no basis. For if n is the order of V, then, according to a well-known result of elementary group theory, we have $nx = 0$ for every element x of V. Thus V cannot have any non-empty free subsets.

This last example shows that we cannot assert in general that every left R-module has a basis. We shall prove, however, that if R is a division ring (or a field) then every R-module (i.e. every vector space over R) has a basis.

THEOREM 6.5. *Let V be a left R-module over a division ring R. If X is a free subset of V there exists a basis for V including X.*

Proof. Let F be the set of linearly independent subsets of V which include X. Then the inclusion relation between subsets of V is an order relation on F. We claim that F is inductively ordered by this relation.

So let F_0 be a totally ordered subset of F; we must show that F_0 has a least upper bound in F. If X_0 is the union of the sets in F_0, then X_0 is clearly the least upper bound of F_0 in the set of all subsets of V which include X; we claim that in fact X_0 belongs to F.

Suppose, to the contrary, that X_0 is not free. Then there is a finite subset of X_0, say $\{x_1, \ldots, x_k\}$, which is linearly dependent. Since X_0 is the union of the sets in F_0, it follows that there are sets $X_{i_1}, \ldots, X_{i_k}$ in F_0 such that $x_j \in X_{i_j}$ $(j = 1, \ldots, k)$. Now F_0 is totally ordered; so one of the sets $X_{i_1}, \ldots, X_{i_k}$ in F_0, say X_{i_1}, includes all the others. But then the free subset X_{i_1} includes the linearly dependent set $\{x_1, \ldots, x_k\}$; and this is a contradiction. Hence X_0 is free.

Applying Zorn's Lemma we deduce that F has a maximal element, i.e. that there is at least one maximal linearly independent subset of V which includes X; choose one such subset and call it B. The argument so far has made no use of the fact that R is a division ring; but this fact is crucial for our next step in which we show that B is a basis for V.

Since B is certainly free, all that remains is to show that B is a generating system for V, in other words that $RB = V$. Suppose then that $RB \neq V$ and let y be any element of V which does not belong to RB. We claim that $B \cup \{y\}$ is linearly independent; so let $ay + \sum_{x \in B} a_x x = 0$, where $a \in R$ and $(a_x)_{x \in B}$ is a quasi-finite family of elements of R. If $a \neq 0$ then, since R is a division ring, a has a multiplicative inverse and we deduce that $y = \sum_{x \in B} (-a^{-1}a_x)x \in RB$, contradicting our assumption that $y \notin RB$. Hence $a = 0$ and so $\sum_{x \in B} a_x x = 0$, from which it follows that all the coefficients a_x are zero (since B is free). Thus $B \cup \{y\}$ is linearly independent. But this contradicts the maximal property of B. Hence $RB = V$ and so B is a basis for V, as required.

COROLLARY. *If R is a division ring then every left R-module has a basis.*

Proof. We have only to apply the theorem in the case where $X = \phi$, remembering that ϕ is defined to be a free subset.

Let V be a left R-module, W a submodule of V. We define a relation on V by setting $x \equiv y$ (mod. W) if and only if $x - y \in W$, and verify that this is an equivalence relation. The addition operation in V is compatible with this relation, for if $x \equiv x'$ (mod. W) and $y \equiv y'$ (mod. W) we have $x - x' \in W$ and $y - y' \in W$, whence $(x + y) - (x' + y') = (x - x') + (y - y') \in W$ (since the submodule W is closed under the addition in V); thus $x + y \equiv x' + y'$ (mod. W). The quotient set of V with respect to this equivalence relation is denoted by V/W and the equivalence classes are called the *cosets* of V modulo the submodule W; the canonical surjection from V onto V/W is denoted by η.

According to §1 the addition operation in V induces an addition operation in V/W as follows: if C_1, C_2 are cosets modulo W and x_1, x_2 are elements of V in C_1, C_2 respectively, then $C_1 + C_2 = \eta(x_1 + x_2)$; we recall that since the addition in V is compatible with our equivalence relation the coset $C_1 + C_2$ depends only on C_1 and C_2, not on the choices of x_1 and x_2. It is easily verified that V/W forms an abelian group under the induced addition: the associativity and commutativity follow from the corresponding conditions in V; $\eta(0_V) = W$ is a zero element for V/W, and if $C = \eta(x)$ is any coset of V modulo W, $\eta(-x)$ is an additive inverse for C.

We now define a left scalar multiplication of V/W by elements of R. Let C be any coset of V modulo W; choose any element x of C. Then for each element a of R define aC to be the coset $\eta(ax)$. We claim that this coset depends only on a and C, not on the choice of x. For if x' is another element of V in C, we have $x \equiv x'$ (mod. W), i.e. $x - x' \in W$; hence, since W is closed under the scalar multiplication in V, $ax - ax' = a(x - x') \in W$. So $ax \equiv ax'$ (mod. W) and $\eta(ax) = \eta(ax')$ as required. A routine verification shows that V/W is a left R-module under the induced addition and the scalar multiplication we have just defined. We call V/W the *factor module* or *quotient module* of V modulo W.

Example 11. Let A be an additive abelian group considered as a $\mathbf{Z}$-module according to Example 3. If B is any subgroup of A (submodule of A as $\mathbf{Z}$-module) the factor module is simply the factor group of A modulo B regarded as a $\mathbf{Z}$-module in the usual way.

Example 12. If R is any ring, A any left ideal of R, then A is a submodule of the left R-module R_l. Thus we may form the factor module R_l/A, which is a left R-module. We must distinguish this construction carefully from that of the residue class ring R/I described in §3—the latter can be carried out only if I is a two-sided ideal of R.

§7. Homomorphisms of Modules

Let R be a ring (with identity, as usual), and let V and V' be (unitary) left R-modules. Then a mapping α from V to V' is called an *R-module homomorphism* (or an R-homomorphism or simply a homomorphism), if for all elements x, y of V and all elements a of R we have

$$\alpha(x + y) = \alpha(x) + \alpha(y) \qquad \text{and} \qquad \alpha(ax) = a\alpha(x).$$

The other terms monomorphism, epimorphism, ... defined in §1 are all applied in the obvious way to special types of module homomorphisms. As usual the composition of two homomorphisms is a homomorphism and the inverse of an isomorphism is an isomorphism.

Example 1. Let V and V' be any two left R-modules. The mapping ζ from V to V' defined by setting $\zeta(x) = 0$ for all elements x of V is a module homomorphism from V to V', which we call the *zero homomorphism.*

Example 2. Let A and A' be abelian groups, α a group homomorphism from A to A'. Then α is also a $\mathbf{Z}$-module homomorphism from A to A' when these groups are considered as $\mathbf{Z}$-modules in the usual way.

Example 3. Let V be any left R-module, W a submodule of V, ι the canonical injection from W to V. Then ι is a monomorphism from W to V, which we call the *inclusion monomorphism.* In particular, if $W = V$ the identity mapping I_V is an automorphism of V.

Example 4. Let V be any left R-module, W a submodule of V. Then the addition operation and scalar multiplication in the factor module V/W are defined in such a way that the canonical surjection η from V onto V/W is a module homomorphism. From now on we shall call η the *canonical epimorphism* from V to V/W.

The basic results about module homomorphisms are closely analogous to those for ring homomorphisms and we shall omit most of the proofs; the interested reader will find it easy to construct them for himself if he follows, with minor modifications, the models given in §4.

THEOREM 7.1. *Let V and V' be left R-modules, α a homomorphism from V to V'. Then* Im $\alpha = \alpha(V)$ *is a submodule of V'. If 0_V is the zero element of V then $\alpha(0_V) = 0_{V'}$, the zero element of V'; if x is any element of V then $\alpha(-x)$ is the additive inverse of $\alpha(x)$ in V'.*

The *kernel* of a module homomorphism α from V to V' is defined as for ring homomorphisms: it consists of the elements of V which are mapped by α onto the zero element of V'; we denote the kernel of α by Ker α. Since $0_V \in$ Ker α for every homomorphism α, the kernel of a homomorphism is always non-empty. We have the familiar criterion for a homomorphism to be injective.

THEOREM 7.2. *Let V and V' be left R-modules, α a homomorphism from V to V'. Then α is a monomorphism if and only if* Ker $\alpha = \{0_V\}$.

Next we have the fundamental theorem on module homomorphisms.

THEOREM 7.3. (*First homomorphism theorem*). *Let V and V' be left R-modules, α a homomorphism from V to V' with kernel V_0. Then V_0 is a submodule of V and if η is the canonical epimorphism from V onto V/V_0 there exists a unique monomorphism $\alpha*$ from V/V_0 to V' such that $\alpha_*\eta = \alpha$; α_* is an isomorphism if and only if α is an epimorphism.*

Proof. We define a mapping α_* from V/V_0 to V' in the familiar way: for each coset C of V modulo V_0 choose an element x from C and set $\alpha_*(C) = \alpha(x)$. The right hand member $\alpha(x)$ depends only on C, not on the choice of x; for if $x' \in C$, then $x \equiv x'$ (mod. V_0), whence $x - x' \in V_0$, and so eventually $\alpha(x) = \alpha(x')$. If x is any element of V then since $x \in \eta(x)$ we certainly have $\alpha_*(\eta(x)) = \alpha(x)$; thus $\alpha_*\eta = \alpha$ as required.

We have now to show that α_* is a module homomorphism from

V/V_0 to V'. So let C_1, C_2 be any two cosets of V modulo V_0, x_1, x_2 elements of C_1, C_2 respectively, and a any element of R. Then

$$\alpha_*(C_1 + C_2) = \alpha_*(\eta(x_1) + \eta(x_2)) = \alpha_*(\eta(x_1 + x_2)) = \alpha(x_1 + x_2)$$

while

$$\alpha_*(C_1) + \alpha_*(C_2) = \alpha_*(\eta(x_1)) + \alpha_*(\eta(x_2)) = \alpha(x_1) + \alpha(x_2) = \alpha(x_1 + x_2),$$

and

$$\alpha_*(aC_1) = \alpha_*(a\eta(x_1)) = \alpha_*(\eta(ax_1)) = \alpha(ax_1)$$
$$= a\alpha(x_1) = a\alpha_*(\eta(x_1)) = a\alpha_*(C_1).$$

Now the coset $C = \eta(x)$ belongs to the kernel of α_* if and only if $\alpha_*(C) = \alpha_*(\eta(x)) = \alpha(x) = 0$, i.e. if and only if $x \in \text{Ker } \alpha$ and so $\eta(x)$ is the zero coset modulo V_0. Hence, by Theorem 7.2, α_* is a monomorphism.

The uniqueness of α_* and the fact that α_* is surjective if and only if α is surjective are established by the arguments used in the proof of Theorem 4.3.

As in §4, we deduce the other two basic homomorphism theorems.

THEOREM 7.4. (*Second homomorphism theorem*). *Let W_1 and W_2 be submodules of a left R-module V. Then there exists an isomorphism from $(W_1 + W_2)/W_2$ onto $W_1/(W_1 \cap W_2)$.*

THEOREM 7.5. (*Third homomorphism theorem*). *Let V and V' be left R-modules, α an epimorphism from V onto V' with kernel V_0. Then there is a one-to-one correspondence between the set of submodules of V which include V_0 and the set of all submodules of V'. Further, if V_1 is any submodule of V which includes V_0, then V/V_1 is isomorphic to $V'/\alpha(V_1)$.*

If V and V' are left R-modules and α is a homomorphism from V to V' the factor modules $V/\text{Ker } \alpha$ and $V'/\text{Im } \alpha$ are called the *coimage* and *cokernel* of the homomorphism α and are denoted by Coim α and Coker α respectively. Clearly α is an epimorphism if and only if Coker $\alpha = \{0_{V'}\}$ (cf. Theorem 7.2). Further, according to the First Homomorphism Theorem there exists an isomorphism α_* from Coim α to Im α.

Let A, B, C be three left R-modules and let α, β be homomorphisms from A to B and B to C respectively. Then the diagram

$$A \xrightarrow{\alpha} B \xrightarrow{\beta} C$$

is said to be *exact* if Im $\alpha =$ Ker β. If $n \geqslant 2$ the diagram

$$A_0 \xrightarrow{\alpha_0} A_1 \xrightarrow{\alpha_1} A_2 \xrightarrow{\alpha_2} A_3 \xrightarrow{\alpha_3} \dots \xrightarrow{\alpha_{n-1}} A_n \qquad [7.1]$$

of left R-modules and homomorphisms is said to be *exact at* A_i $(i = 1, 2, \dots, n-1)$ if the diagram

$$A_{i-1} \xrightarrow{\alpha_{i-1}} A_i \xrightarrow{\alpha_i} A_{i+1} \qquad [7.2]$$

is exact. The diagram [7.1] is said to be an *exact sequence* if it is exact at each module A_i $(i = 1, 2, \dots, n-1)$. In the same way an infinite diagram

$$A_0 \xrightarrow{\alpha_0} A_1 \xrightarrow{\alpha_1} A_2 \xrightarrow{\alpha_2} A_3 \xrightarrow{\alpha_3} \dots \xrightarrow{\alpha_{n-1}} A_n \xrightarrow{\alpha_n} \dots$$

of left R-modules and homomorphisms is said to be exact at A_i $(i \geqslant 1)$ if the diagram [7.2] is exact, and to be an exact sequence if it is exact at every module A_i $(i \geqslant 1)$. We make similar definitions for infinite diagrams such as

$$\dots \xrightarrow{\alpha_{-n-1}} A_{-n} \xrightarrow{\alpha_{-n}} \dots \xrightarrow{\alpha_{-4}} A_{-3} \xrightarrow{\alpha_{-3}} A_{-2} \xrightarrow{\alpha_{-2}} A_{-1} \xrightarrow{\alpha_{-1}} A_0$$

and

$$\dots \xrightarrow{\alpha_{-n-1}} A_{-n} \xrightarrow{\alpha_{-n}} \dots \xrightarrow{\alpha_{-2}} A_{-1} \xrightarrow{\alpha_{-1}} A_0 \xrightarrow{\alpha_0} A_1 \xrightarrow{\alpha_1} \dots \xrightarrow{\alpha_{n-1}} A_n \xrightarrow{\alpha_n} \dots$$

Before proceeding to give examples of exact sequences we introduce some conventions of notation which we shall use constantly. First of all, we agree that when a left R-module V_0 consists of a zero element 0 alone then we shall write $V_0 = 0$ instead of the more strictly correct $\{0\}$. Next, if α is a homomorphism from such a zero module to any left R-module V we must have $\alpha(0) = 0_V$; that is to say, there is only one possible homomorphism from 0 to V. So we agree that whenever a homomorphism from 0 to V occurs in any of our diagrams of modules and homomorphisms we shall represent it simply by $0 \to V$ without giving a name to the homomorphism. Finally, if β is a homomorphism from any module V to 0, it must be the zero homomorphism, given by $\beta(x) = 0$ for every element x in V. We shall represent this by $V \to 0$, again without naming the homomorphism.

Example 5. Let V' and V be left R-modules, α a homomorphism from V' to V. Consider the diagram

$$0 \to V' \xrightarrow{\alpha} V. \tag{7.3}$$

If this is an exact sequence the kernel of α coincides with the image of the first homomorphism, which consists of the zero element of V' alone. Thus, by Theorem 7.2, α is a monomorphism. Conversely, if α is a monomorphism, [7.3] is an exact sequence.

Example 6. Let V and V'' be left R-modules, β a homomorphism from V to V''. Consider the diagram

$$V \xrightarrow{\beta} V'' \to 0. \tag{7.4}$$

If this sequence is exact the image of β coincides with the kernel of the second homomorphism, which is of course V'' itself. Thus $\operatorname{Im} \beta = V''$, i.e. β is an epimorphism. Conversely, if β is an epimorphism, [7.4] is an exact sequence.

A diagram of the form

$$0 \to V' \xrightarrow{\alpha} V \xrightarrow{\beta} V'' \to 0 \tag{7.5}$$

which is exact at V', V and V'' is called a *short exact sequence*. According to Examples 5 and 6 the homomorphisms α and β of [7.5] are respectively a monomorphism and an epimorphism.

Example 7. Let W be a submodule of a left R-module V, ι the inclusion monomorphism from W to V and η the canonical epimorphism from V onto V/W. Then the diagram

$$0 \to W \xrightarrow{\iota} V \xrightarrow{\eta} V/W \to 0$$

is a short exact sequence. It is easy to show that all short exact sequences are 'essentially' of this form; more formally, if we are given a short exact sequence [7.5], we can show that V' is isomorphic to a submodule W of V and V'' is isomorphic to the corresponding factor module V/W.

Example 8. Let V_1 and V_2 be left R-modules, α a homomorphism from V_1 to V_2. Let ι be the inclusion monomorphism from $\operatorname{Ker} \alpha$ to V_1 and η the canonical epimorphism from V_2 onto $\operatorname{Coker} \alpha$. Then the sequence

$$0 \to \operatorname{Ker} \alpha \xrightarrow{\iota} V_1 \xrightarrow{\alpha} V_2 \xrightarrow{\eta} \operatorname{Coker} \alpha \to 0$$

is exact.

Example 9. As an illustration of the technique known as 'diagram-chasing' we prove half of the so-called *Five Lemma.* Consider the diagram

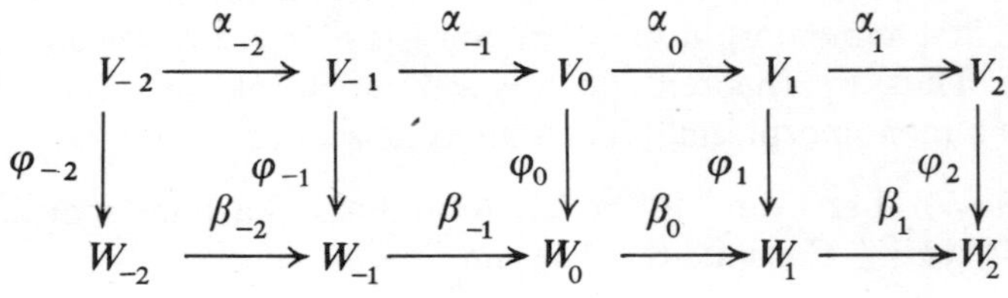

$$\begin{array}{ccccccccc}
V_{-2} & \xrightarrow{\alpha_{-2}} & V_{-1} & \xrightarrow{\alpha_{-1}} & V_0 & \xrightarrow{\alpha_0} & V_1 & \xrightarrow{\alpha_1} & V_2 \\
\downarrow{\scriptstyle \varphi_{-2}} & & \downarrow{\scriptstyle \varphi_{-1}} & & \downarrow{\scriptstyle \varphi_0} & & \downarrow{\scriptstyle \varphi_1} & & \downarrow{\scriptstyle \varphi_2} \\
W_{-2} & \xrightarrow{\beta_{-2}} & W_{-1} & \xrightarrow{\beta_{-1}} & W_0 & \xrightarrow{\beta_0} & W_1 & \xrightarrow{\beta_1} & W_2
\end{array}$$

in which both rows are exact sequences of left R-modules and homomorphisms, and all four squares are commutative (i.e. $\varphi_{i+1}\alpha_i = \beta_i\varphi_i$ for $i = -2, -1, 0, 1$). Suppose we are given that φ_2 is a monomorphism and φ_{-1} and φ_1 are epimorphisms; we shall show that φ_0 is an epimorphism.

So let w_0 be any element of W_0. Since φ_1 is an epimorphism there is an element v_1 of V_1 such that $\varphi_1(v_1) = \beta_0(w_0)$. Then we have $\beta_1\varphi_1(v_1) = \beta_1\beta_0(w_0) = 0$, since the lower row is exact; hence $\varphi_2\alpha_1(v_1) = 0$, by the commutativity in the right-hand square. Now φ_2 is a monomorphism, so $\alpha_1(v_1) = 0$; thus $v_1 \in \operatorname{Ker} \alpha_1 = \operatorname{Im} \alpha_0$, and there is an element v_0 of V_0 such that $v_1 = \alpha_0(v_0)$. Then $\beta_0(w_0) = \varphi_1(v_1) = \varphi_1\alpha_0(v_0) = \beta_0\varphi_0(v_0)$, from which we deduce that $w_0 - \varphi_0(v_0) \in \operatorname{Ker} \beta_0 = \operatorname{Im} \beta_{-1}$. Hence there is an element w_{-1} of W_{-1} such that $w_0 - \varphi_0(v_0) = \beta_{-1}(w_{-1})$. Since φ_{-1} is an epimorphism, there is an element v_{-1} of V_{-1} such that $w_{-1} = \varphi_{-1}(v_{-1})$. Consequently we have $w_0 = \varphi_0(v_0) + \beta_{-1}(w_{-1}) = \varphi_0(v_0) + \beta_{-1}\varphi_{-1}(v_{-1}) = \varphi_0(v_0) + \varphi_0\alpha_{-1}(v_{-1}) \in \operatorname{Im} \varphi_0$. That is to say φ_0 is an epimorphism.

The reader is invited to try his hand at proving the other half of the Five Lemma: if φ_{-2} is an epimorphism and φ_{-1} and φ_1 are monomorphisms then φ_0 is a monomorphism.

§8. Groups of Homomorphisms

Let R be a ring with identity, V and W two left R-modules. We denote the set of R-homomorphisms from V to W by $\operatorname{Hom}_R(V, W)$ or simply by $\operatorname{Hom}(V, W)$ when the ring R is clear from the context. We define an operation of addition in $\operatorname{Hom}(V, W)$ as follows: for every pair α, β of homomorphisms from V to W consider the mapping

$\alpha + \beta$ from V to W defined by setting

$$(\alpha + \beta)(x) = \alpha(x) + \beta(x)$$

for each element x of V. It is easy to verify that this mapping $\alpha + \beta$ is actually a homomorphism from V to W; so we have in fact defined an internal law of composition in Hom (V, W). Routine calculation shows that this law of composition is associative and commutative; the zero homomorphism ζ from V to W is clearly a neutral element; and for each homomorphism α from V to W the mapping $-\alpha$ from V to W given by

$$(-\alpha)(x) = -\alpha(x)$$

for each element x of V is a homomorphism and $\alpha + (-\alpha) = \zeta$. Hence, under the addition operation we have defined, Hom (V, W) is an abelian group.

Suppose that R is a commutative ring and that V and W are left R-modules. In this case we define a scalar multiplication on the left of Hom (V, W) by elements of R as follows: for each element a of R and each homomorphism α from V to W, let $a\alpha$ be the mapping from V to W defined by setting

$$(a\alpha)(x) = a\alpha(x)$$

for each element x of V. We contend that this mapping $a\alpha$ is a homomorphism from V to W. To see this, let x and y be any two elements of V, b any element of R; then

$$(a\alpha)(x + y) = a\alpha(x + y) = a(\alpha(x) + \alpha(y)) = \\ a\alpha(x) + a\alpha(y) = (a\alpha)(x) + (a\alpha)(y)$$

and

$$(a\alpha)(bx) = a\alpha(bx) = a(b\alpha(x)) = b(a\alpha(x)) = b((a\alpha)(x)).$$

It is easily verified that with the addition defined in the preceding paragraph and this scalar multiplication Hom (V, W) becomes a left R-module. We observe that in proving that $a\alpha$ is an R-homomorphism we made essential use of the commutativity of R. Thus in general (i.e. when R is not necessarily commutative) Hom (V, W) is only an additive abelian group, not a left R-module.

There are two special cases, however, even when R is not commutative, in which additional structure can be imposed on the abelian group Hom (V, W): these are the cases in which $V = R_l$

(i.e. the ring R itself, considered as left R-module) and $W = R_l$. We consider these two cases in turn.

If a is any element of R, α any homomorphism from R_l to W, we consider the mapping $a\alpha$ from R to W defined by setting

$$(a\alpha)(r) = \alpha(ra) \qquad [8.1]$$

for every element r of R. Then $a\alpha$ is also a homomorphism from R_l to W; for if r_1, r_2, r, a_1 are any elements of R we have

$$(a\alpha)(r_1 + r_2) = \alpha((r_1 + r_2)a) = \alpha(r_1 a) + \alpha(r_2 a) = (a\alpha)(r_1) + (a\alpha)(r_2)$$

and

$$(a\alpha)(a_1 r) = \alpha((a_1 r)a) = \alpha(a_1(ra)) = a_1\alpha(ra) = a_1((a\alpha)(r)).$$

Thus [8.1] serves to define a scalar multiplication on the left of Hom (R_l, W) by elements of R, and it is an easy matter to show that this scalar multiplication gives the abelian group Hom (R_l, W) the structure of a left R-module.

THEOREM 8.1. *Let R be a ring with identity element e, W a left R-module. Then the left R-module* Hom (R_l, W) *is isomorphic to W.*

Proof. Consider the mapping f from Hom (R_l, W) to W defined by setting $f(\alpha) = \alpha(e)$ for every homomorphism α from R_l to W. We contend that f is an R-homomorphism; so let α and β be elements of Hom (R_l, W), a any element of R. Then we have

$$f(\alpha + \beta) = (\alpha + \beta)(e) = \alpha(e) + \beta(e) = f(\alpha) + f(\beta)$$

and

$$f(a\alpha) = (a\alpha)(e) = \alpha(ea) = \alpha(ae) = a\alpha(e) = af(\alpha).$$

To show that the homomorphism f is actually the required isomorphism, we notice first that $\alpha \in \text{Ker } f$ if and only if $\alpha(e) = 0$. But if $\alpha(e) = 0$ then $\alpha(r) = \alpha(re) = r\alpha(e) = 0$ for all elements r of R, i.e. α is the zero homomorphism. Thus Ker $f = \{\zeta\}$, and so f is a monomorphism. Let x be any element of W; if α_x is the mapping from R to W defined by setting $\alpha_x(r) = rx$ for all elements r of R, then $\alpha_x \in$ Hom (R_l, W) and clearly $f(\alpha_x) = x$. Thus f is an epimorphism.

This completes the proof.

Now let V be any left R-module and consider the group

Hom (V, R_l); the elements of this group are called *linear forms* or *linear functionals* on V. We propose now to give Hom (V, R_l) the structure of a *right* R-module. So let α be any element of Hom (V, R_l), a any element of R and consider the mapping αa from V to R defined by setting

$$(\alpha a)(x) = \alpha(x)\, a \qquad [8.2]$$

for all elements x of V. Routine verification shows that $\alpha a \in$ Hom (V, R_l) and that under the right multiplication so defined Hom (V, R_l) is a right R-module. (To grasp the reason for the left-right interchange, the reader is invited to check that if we tried defining $a\alpha$ by setting $(a\alpha)(x) = a\alpha(x)$, then $a\alpha$ would not be an R-homomorphism unless R were commutative; if we wrote $(a\alpha)(x) = \alpha(x)a$, we should have $(ab)\alpha = b(a\alpha)$, not $a(b\alpha)$ as we require for a left R-module; but [8.2] gives $\alpha(ab) = (\alpha a)\, b$, which is the appropriate condition for a right R-module.) When we consider Hom (V, R_l) as a right R-module in this way we call it the *dual module* of V and denote it by V^*. Clearly we may start with a right R-module and form its dual, which will be a left R-module. In particular, if V is a left R-module we may form the dual module of the right R-module V^*; this is again a left R-module, which we denote by V^{**} and call the *bidual* of V.

We digress for a moment to consider the special case in which R is a division ring D and V is a left vector space over D with a finite basis $\{x_1, \ldots, x_n\}$. Then, according to Theorem 6.4, every element x of V can be expressed uniquely in the form $x = a_1x_1 + \ldots + a_nx_n$ where $a_1, \ldots, a_n$ are elements of D. For $i = 1, \ldots, n$ let ξ_i be the mapping from V to D defined by setting

$$\xi_i(x) = \xi_i(a_1x_1 + \ldots + a_nx_n) = a_i$$

for each element x of V; it is easily verified that each of these mappings ξ_i is a linear functional on V. We claim that $\{\xi_1, \ldots, \xi_n\}$ is a basis for the right vector space V^* over D. To see this, first let $b_1, \ldots, b_n$ be elements of D such that $\xi_1b_1 + \ldots + \xi_nb_n = \zeta$, the zero linear functional; then for $j = 1, \ldots, n$ we have

$$\begin{aligned} 0 = (\xi_1b_1 + \ldots + \xi_nb_n)(x_j) &= (\xi_1b_1)(x_j) + \ldots + (\xi_nb_n)(x_j) \\ &= \xi_1(x_j)\, b_1 + \ldots + \xi_n(x_j)\, b_n = b_j. \end{aligned}$$

Thus $\{\xi_1, \ldots, \xi_n\}$ is linearly independent. Next let α be any linear functional on V; for $i = 1, \ldots, n$ let $\alpha(x_i) = b_i$. We claim that $\alpha = \xi_1 b_1 + \ldots + \xi_n b_n$, and verify this by computing

$$\begin{aligned}\alpha(x) = \alpha(a_1 x_1 + \ldots + a_n x_n) &= a_1 \alpha(x_1) + \ldots + a_n \alpha(x_n) \\ &= a_1 b_1 + \ldots + a_n b_n \\ &= a_1(\xi_1 b_1 + \ldots + \xi_n b_n)(x_1) \\ &\quad + \ldots + a_n(\xi_1 b_1 + \ldots + \xi_n b_n)(x_n) \\ &= (\xi_1 b_1 + \ldots + \xi_n b_n)(x).\end{aligned}$$

So $\{\xi_1, \ldots, \xi_n\}$ is a generating system for V^*, and hence a basis as asserted. We call $\{\xi_1, \ldots, \xi_n\}$ the basis of V^* *dual* to the basis $\{x_1, \ldots, x_n\}$ of V.

We now return to the general situation, in which R is an arbitrary ring with identity.

Let V_1 and V_2 be two left R-modules, α a module homomorphism from V_1 to V_2. If M is any left R-module we shall show how to define abelian group homomorphisms α^* from Hom (V_2, M) to Hom (V_1, M) and α_* from Hom (M, V_1) to Hom (M, V_2). Namely, if φ is any element of Hom (V_2, M), the mapping $\varphi\alpha$ from V_1 to M is easily seen to be an R-homomorphism and if we set

$$\alpha^*(\varphi) = \varphi\alpha$$

for every element φ of Hom (V_2, M) it turns out that the mapping α^* from Hom (V_2, M) to Hom (V_1, M) so defined is a homomorphism; we call it the homomorphism *induced* by α. Again, if θ is any element of Hom (M, V_1), the mapping $\alpha\theta$ from M to V_2 belongs to Hom (M, V_2) and the mapping α_* from Hom (M, V_1) to Hom (M, V_2) defined by setting

$$\alpha_*(\theta) = \alpha\theta$$

for every element θ of Hom (M, V_1) is a homomorphism, which is also said to be *induced* by α.

The constructions of α^* and α_* which we have just described are special cases of a more general procedure. Suppose V, V', W, W' are left R-modules and that α and β are homomorphisms from V' to V, W to W' respectively; then for each element φ of Hom (V, W) the mapping $\beta\varphi\alpha$ is a homomorphism from V' to W'. This comment allows us to define a mapping from Hom (V, W) to Hom (V', W')

which we denote by Hom (α, β): for each element φ of Hom (V, W) we set

$$(\text{Hom}\,(\alpha, \beta))\,(\varphi) = \beta\varphi\alpha.$$

It is a routine matter to verify that Hom (α, β) is an abelian group homomorphism. The mappings α^* and α_* of the previous paragraph become in the present notation simply Hom (α, I_M) and Hom (I_M, α) where I_M is the identity mapping of M. The next two theorems establish fundamental properties of the homomorphisms Hom (α, β).

THEOREM 8.2. *Let V, V', V'', W, W', W'' be left R-modules and let α', α, β, β' be homomorphisms from V'' to V', V' to V, W to W', W' to W'' respectively. Then* Hom $(\alpha\alpha', \beta'\beta) =$ Hom (α', β') Hom (α, β).

Proof. Let φ be any element of Hom (V, W). Then we have

$$\begin{aligned}(\text{Hom}\,(\alpha\alpha', \beta'\beta))\,(\varphi) &= (\beta'\beta)\,\varphi(\alpha\alpha')\\ &= \beta'(\beta\varphi\alpha)\,\alpha'\\ &= (\text{Hom}\,(\alpha', \beta'))\,(\beta\varphi\alpha)\\ &= (\text{Hom}\,(\alpha', \beta'))\,((\text{Hom}\,(\alpha, \beta))\,(\varphi)).\end{aligned}$$

Thus Hom $(\alpha\alpha', \beta'\beta) =$ Hom (α', β') Hom (a, β) as required.

COROLLARY. *Let V_1, V_2, V_3 be left R-modules and let α and β be homomorphisms from V_1 to V_2 and V_2 to V_3 respectively. Let M be any left R-module. Then $(\beta\alpha)^* = \alpha^*\beta^*$ and $(\beta\alpha)_* = \alpha_*\beta_*$.*

THEOREM 8.3. *Let V, V', W, W' be left R-modules; let α, α_1, α_2 be homomorphisms from V' to V and β, β_1, β_2 homomorphisms from W to W'. Then*

$$\text{Hom}\,(\alpha, \beta_1 + \beta_2) = \text{Hom}\,(\alpha, \beta_1) + \text{Hom}\,(\alpha, \beta_2)$$

and

$$\text{Hom}\,(\alpha_1 + \alpha_2, \beta) = \text{Hom}\,(\alpha_1, \beta) + \text{Hom}\,(\alpha_2, \beta).$$

Proof. Let φ be any element of Hom (V, W). Then we have

$$(\text{Hom}\,(\alpha, \beta_1 + \beta_2))\,(\varphi) = (\beta_1 + \beta_2)\,\varphi\alpha$$

while

$$\begin{aligned}(\text{Hom}\,(\alpha, \beta_1) + \text{Hom}\,(\alpha, \beta_2))\,(\varphi) &= (\text{Hom}\,(\alpha, \beta_1))\,(\varphi) + (\text{Hom}\,(\alpha, \beta_2))\,(\varphi)\\ &= \beta_1\varphi\alpha + \beta_2\varphi\alpha.\end{aligned}$$

Now if x' is any element of V' we have

$$\begin{aligned}[(\beta_1 + \beta_2)\varphi\alpha](x') &= (\beta_1 + \beta_2)(\varphi\alpha(x')) \\ &= \beta_1(\varphi\alpha(x')) + \beta_2(\varphi\alpha(x')) \\ &= \beta_1\varphi\alpha(x') + \beta_2\varphi\alpha(x') = [\beta_1\varphi\alpha + \beta_2\varphi\alpha](x').\end{aligned}$$

Hence Hom $(\alpha, \beta_1 + \beta_2) =$ Hom $(\alpha, \beta_1) +$ Hom (α, β_2).

The other result follows similarly.

If R is a commutative ring and the various groups of homomorphisms are given the R-module structure defined earlier, the various mappings Hom (α, β), α^*, α_* are all R-module homomorphisms.

We now examine what happens to short exact sequences of left R-modules when we form the homomorphism groups of the terms from and to some fixed left R-module. In the course of our discussions various zero homomorphisms will occur; if it is necessary to emphasise that we are dealing (for example) with the zero homomorphism from V to W we shall denote it by ζ_{VW}, but usually we shall denote all zero homomorphisms simply by ζ since the context in most cases will make it clear which particular zero homomorphism is meant.

THEOREM 8.4. *Let*

$$0 \to V' \xrightarrow{\alpha} V \xrightarrow{\beta} V'' \to 0 \qquad [8.3]$$

be a short exact sequence of left R-modules, M a left R-module. Then the sequences

$$0 \to \text{Hom}(V'', M) \xrightarrow{\beta^*} \text{Hom}(V, M) \xrightarrow{\alpha^*} \text{Hom}(V', M) \qquad [8.4]$$

and

$$0 \to \text{Hom}(M, V') \xrightarrow{\alpha_*} \text{Hom}(M, V) \xrightarrow{\beta_*} \text{Hom}(M, V'') \qquad [8.5]$$

are exact.

Proof. (1) To show that [8.4] is exact we must show that (a) β^* is a monomorphism, (b) Im $\beta^* \subseteq$ Ker α^* and (c) Ker $\alpha^* \subseteq$ Im β^*.

(a) Let θ'' be an element of Ker β^*; then $\theta''\beta = \beta^*(\theta'') = \zeta_{VM}$. If x'' is any element of V'' then there is an element x of V such that $x'' = \beta(x)$. It follows that $\theta''(x'') = \theta''(\beta(x)) = 0$. Thus $\theta'' = \zeta_{V''M}$ and hence β^* is a monomorphism.

(b) Now let θ be any element of Im β^*; thus there is an element θ''

of Hom (V'', M) such that $\theta = \beta^*(\theta'') = \theta''\beta$. Then $\alpha^*(\theta) = \theta\alpha = (\theta''\beta)\,\alpha = \theta''(\beta\alpha) = \theta''\zeta = \zeta$. Hence Im $\beta^* \subseteq$ Ker α^*. (We have $\beta\alpha = \zeta$ since [8.3] is exact.)

(c) Next let θ be any element of Ker α^*. Then for every element x' of V' we have $\theta\alpha(x') = [\alpha^*(\theta)]\,(x') = 0$. Consider now the mapping θ'' from V'' to M defined as follows: for each element x'' of V'' choose an element x from $\beta^{-1}(x'')$—which is non-empty since β is an epimorphism—and form $\theta(x)$. This element depends only on x'', not on the choice of x; for if x and x_1 are elements of $\beta^{-1}(x'')$ we have $x - x_1 \in \text{Ker}\,\beta = \text{Im}\,\alpha$ and hence $\theta(x - x_1) = 0$. So we may properly define $\theta''(x'')$ to be $\theta(x)$. It is easy to verify that in fact $\theta'' \in$ Hom (V'', M), and clearly $\beta^*(\theta'') = \theta''\beta = \theta$. Hence Ker $\alpha^* \subseteq$ Im β^*.

Thus [8.4] is exact.

(2) To show that [8.5] is exact we must establish that (a) α_* is a monomorphism, (b) Im $\alpha_* \subseteq$ Ker β_* and (c) Ker $\beta_* \subseteq$ Im α_*.

(a) Let θ' be an element of Ker α_* so that $\alpha\theta' = \alpha_*(\theta') = \zeta_{MV}$. Thus if m is any element of M, we have $\alpha(\theta'(m)) = 0$. Since [8.3] is exact, α is a monomorphism and hence $\theta'(m) = 0$. Thus $\theta' = \zeta_{MV'}$ and hence α_* is a monomorphism.

(b) Let θ be an element of Im α_*; thus there is an element θ' of Hom (M, V') such that $\theta = \alpha_*(\theta')$. Then we have $\beta_*(\theta) = \beta_*(\alpha_*(\theta')) = \beta(\alpha\theta') = (\beta\alpha)\,\theta' = \zeta$. Hence Im $\alpha_* \subseteq$ Ker β_*. (We use here again the fact that $\beta\alpha = \zeta$, which follows from the exactness of [8.3].)

(c) Finally let θ be any element of Ker β_*; then for every element m of M we have $\beta(\theta(m)) = \beta\theta(m) = [\beta_*(\theta)](m) = 0$, i.e. $\theta(m) \in \text{Ker}\beta$. Since [8.3] is exact, it follows that $\theta(m) \in \text{Im}\,\alpha$. Thus there is an element x' of V' such that $\theta(m) = \alpha(x')$; and x' is unique since α is a monomorphism. Set $\theta'(m) = x'$; in this way we define a mapping θ' from M to V', which is clearly a homomorphism. Since $\theta = \alpha\theta' = \alpha_*(\theta')$ it follows that $\theta \in \text{Im}\,\alpha_*$. Hence Ker $\beta_* \subseteq$ Im α_*.

Thus [8.5] is exact.

Given a short exact sequence [8.3] of left R-modules, we are naturally interested in the sequence of homomorphism groups

$$0 \to \text{Hom}\,(V'', M) \xrightarrow{\beta^*} \text{Hom}\,(V, M) \xrightarrow{\alpha^*} \text{Hom}\,(V', M) \to 0. \qquad [8.6]$$

In Theorem 11.8 we shall obtain necessary and sufficient conditions on the module M in order that [8.6] be exact for every short exact sequence [8.3]; in Theorem 11.6 we carry out a similar programme in connexion with the sequence

$$0 \to \operatorname{Hom}(M, V') \xrightarrow{\alpha *} \operatorname{Hom}(M, V) \xrightarrow{\beta *} \operatorname{Hom}(M, V'') \to 0. \qquad [8.7]$$

In Theorem 9.5 we obtain an important class of sequences [8.3] such that [8.6] and [8.7] are exact for every module M.

If V is any left R-module, the set $\operatorname{Hom}_R(V, V)$ of R-endomorphisms of V admits not only the addition operation described at the beginning of this section but also the composition operation. As in Example 6 of §2 we may verify that $\operatorname{Hom}_R(V, V)$ forms a ring with these two operations as addition and multiplication respectively. There is an important special case in which we can give a 'concrete' representation of this ring.

THEOREM 8.5. *Let V be a left vector space over a division ring D. If V has a basis consisting of n elements then* $\operatorname{Hom}_D(V, V)$ *is isomorphic to the ring $M_n(D^{\mathrm{op}})$ of $n \times n$ matrices with coefficients in the ring opposite to D.*

Proof. Let $\{x_1, \ldots, x_n\}$ be a basis for V.

If α is any element of $\operatorname{Hom}_D(V, V)$ then for $j = 1, \ldots, n$ we may express $\alpha(x_j)$ uniquely in the form $\alpha(x_j) = \sum_{i=1}^{n} a_{ij}x_i$. Thus we may define a mapping φ from $\operatorname{Hom}_D(V, V)$ to $M_n(D^{\mathrm{op}})$ by setting $\varphi(\alpha) = [a_{ij}]$ for each D-endomorphism α of V. It is almost immediate that φ is bijective. To see that it is an isomorphism, let β be another element of $\operatorname{Hom}_D(V, V)$ and set $\varphi(\beta) = [b_{ij}]$. Then for $j = 1, \ldots, n$ we have

$$\begin{aligned}(\alpha + \beta)(x_j) &= \alpha(x_j) + \beta(x_j) \\ &= \sum_{i=1}^{n} a_{ij}x_i + \sum_{i=1}^{n} b_{ij}x_i \\ &= \sum_{i=1}^{n} (a_{ij} + b_{ij})\, x_i;\end{aligned}$$

so $\varphi(\alpha + \beta) = [a_{ij} + b_{ij}] = [a_{ij}] + [b_{ij}] = \varphi(\alpha) + \varphi(\beta)$. Further, for $j = 1, \ldots, n$ we have

$$\begin{aligned}(\beta\alpha)(x_j) &= \beta(\alpha(x_j)) \\ &= \beta\Big(\sum_{k=1}^{n} a_{kj}x_k\Big)\end{aligned}$$

$$= \sum_{k=1}^{n} a_{kj}\beta(x_k)$$

$$= \sum_{k=1}^{n} a_{kj}\,(\sum_{i=1}^{n} b_{ik}x_i)$$

$$= \sum_{i=1}^{n} (\sum_{k=1}^{n} a_{kj}b_{ik})\, x_i.$$

Hence $\varphi(\beta\alpha)$ is the matrix with (i, j)-th element $\sum_{k=1}^{n} a_{kj}b_{ik} = \sum_{k=1}^{n} b_{ik} * a_{kj}$ (where $*$ denotes the multiplication operation in D^{op}); this shows that $\varphi(\beta\alpha) = \varphi(\beta) * \varphi(\alpha)$. So φ is an isomorphism, as asserted.

§9. Direct Products and Sums

Let R be a ring (with identity), $(V_k)_{k \in K}$ a family of (unitary) left R-modules. (There is no sinister reason for using K to denote the index set instead of the more familiar I; we shall have to consider homomorphisms which are conventionally denoted by ι, and we wish to avoid confusion between mappings ι and indices i—the notation ι_i would be a particularly unhappy one.)

We form the product $P = \prod_{k \in K} V_k$ of the family $(V_k)_{k \in K}$; if x is any element of P we write $x_k = x(k)$ for each index k in K and denote x by $(x_k)_{k \in K}$ or simply by (x_k). For each index k we have the kth projection π_k which maps P onto V_k according to the prescription $\pi_k(x) = x_k$ for every element x of P.

We define an addition operation in P as follows. If $x = (x_k)$ and $y = (y_k)$ are any elements of P we set $x + y = (z_k)$ where $z_k = x_k + y_k$ $(\in V_k)$ for every index k in K. Next we define a scalar multiplication on the left of P by elements of R. If $x = (x_k)$ is any element of P and a is any element of R, set $ax = (t_k)$ where $t_k = ax_k$ $(\in V_k)$ for each k in K. It is an easy matter to verify that P is a left R-module under the addition and scalar multiplication we have just defined, and that the mappings π_k are all homomorphisms. We call P the *direct product module* of the family $(V_k)_{k \in K}$ and the homomorphisms π_k are called the *canonical projection epimorphisms* from P onto the modules V_k. For each index j in K we define a mapping ι_j from V_j to P by setting for each element x_j of V_j

$$\iota_j(x_j) = y = (y_k)_{k \in K}$$

where $y_j = x_j$ and $y_k = 0$ for $k \neq j$. Clearly each of these mappings is injective, and it is easy to verify that they are all homomorphisms; we call ι_k the *canonical injection monomorphism* from V_k to P. It follows at once from the definitions that for each index k in K we have $\pi_k \iota_k = I_k$ (the identity mapping of V_k) while for each pair of distinct indices k, l in K we have $\pi_k \iota_l = \zeta_{lk}$ (the zero homomorphism from V_l to V_k).

Let $(V_k)_{k \in K}$ be a family of left R-modules. Let $(P', (\pi'_k))$ be a pair consisting of a left R-module P' and a family $(\pi'_k)_{k \in K}$ of homomorphisms such that for each index k in K the homomorphism π'_k maps P' into V_k. We say that $(P', (\pi'_k))$ is *couniversal* for homomorphisms to the family (V_k) if for every left R-module V and every family (φ_k) of homomorphisms from V to the modules V_k, there exists a unique homomorphism α from V to P' such that the

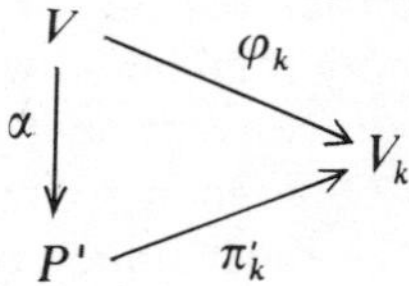

are commutative for every index k in K. (We say in this case that the homomorphisms φ_k can be uniquely factored through P'.)

THEOREM 9.1. *If* $(V_k)_{k \in K}$ *is any family of left R-modules, then the pair* $(P, (\pi_k))$ *consisting of the direct product module P of the family* (V_k) *and the family of canonical projection epimorphisms is couniversal for homomorphisms to the family* (V_k).

Proof. Let V be any left R-module, (φ_k) a family of homomorphisms from V to the modules V_k.

We define a mapping α from V to P by setting for each element x of V

$$\alpha(x) = (y_k)_{k \in K}$$

where $y_k = \varphi_k(x) \in V_k$ for each index k in K. Then α is easily seen to be a homomorphism, and for each index k we have

$$\pi_k \alpha(x) = y_k = \varphi_k(x)$$

for all elements x of V. So $\pi_k \alpha = \varphi_k$ for all indices k in K.

We claim that α is the only homomorphism from V to P with this property. So let α' be another homomorphism from V to P such that $\pi_k\alpha' = \varphi_k$ for every index k. Let x be any element of V. Then for each index k in K we have

$$(\alpha(x))(k) = \pi_k(\alpha(x)) = \varphi_k(x) = \pi_k(\alpha'(x)) = (\alpha'(x))(k).$$

So $\alpha(x) = \alpha'(x)$ for every element x of V and hence $\alpha = \alpha'$.

Thus $(P,(\pi_k))$ is couniversal for homomorphisms to the family (V_k).

It will follow from a result to be established later (Theorem 15.1) that $(P,(\pi_k))$ is (in a sense which we shall make precise) essentially the only pair which is couniversal for homomorphisms to (V_k).

Let V be any left R-module, (φ_k) a family of homomorphisms from V to (V_k). Then, as we have just seen, there exists a unique homomorphism α from V to P such that $\pi_k\alpha = \varphi_k$ for all indices k in K. If this homomorphism is actually an isomorphism then the homomorphisms φ_k are all surjective and we say that the family (φ_k) yields a *projective representation* of V as a direct product of the family (V_k). In this situation the inverse α^{-1} of the isomorphism α is an isomorphism from P to V and so for each index k in K we can define a homomorphism η_k from V_k to V by setting $\eta_k = \alpha^{-1}\iota_k$, where ι_k is the canonical injection monomorphism from V_k to P. These homomorphisms η_k are all injective, and for each index k in K we have $\varphi_k\eta_k = (\pi_k\alpha)(\alpha^{-1}\iota_k) = \pi_k\iota_k = I_k$, while for each pair of distinct indices k, l in K we have $\varphi_k\eta_l = (\pi_k\alpha)(\alpha^{-1}\iota_l) = \pi_k\iota_l = \zeta_{lk}$. Thus the families (φ_k) and (η_k) stand in the same relation to V as the families (π_k) and (ι_k) to P.

Still let $(V_k)_{k\in K}$ be a family of left R-modules, P their direct product module. Let S be the subset of P consisting of those elements $x = (x_k)$ such that the coordinates x_k are non-zero for at most finitely many indices k. It is easily verified that S is a submodule of P, that the canonical injection monomorphism ι_k actually maps V_k into S and that the restriction to S of each canonical projection epimorphism π_k is an epimorphism from S onto V_k. We call S the *external direct sum* of the family (V_k) and denote it by $\bigoplus_{k\in K} V_k$.

If x is any element of S the elements $\iota_k\pi_k(x)$ of S are non-zero for at most finitely many indices k in K. Thus we may form the sum $\sum_{k\in K} \iota_k\pi_k(x)$, and it is clear that in fact $\sum_{k\in K} \iota_k\pi_k(x) = x$.

Let S' be a left R-module, $(\iota'_k)_{k\in K}$ a family of homomorphisms such that for each index k in K the homomorphism ι'_k maps V_k into S'. We say that the pair $(S',(\iota'_k))$ is *universal* for homomorphisms from the family (V_k) if for every left R-module V and every family $(\eta_k)_{k \in K}$ of homomorphisms from the modules V_k to V there exists a unique homomorphism β from S' to V such that the diagrams

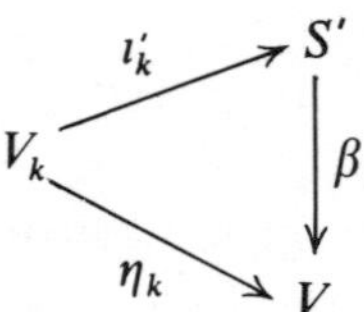

are commutative for every index k in K. (We say that the homomorphisms can be uniquely factored through S'.)

THEOREM 9.2. *If $(V_k)_{k \in K}$ is any family of left R-modules, then the pair $(S,(\iota_k))$ consisting of the external direct sum of the family (V_k) and the family of canonical injection monomorphisms is universal for homomorphisms from the family (V_k).*

Proof. Let V be any left R-module, (η_k) a family of homomorphisms from the modules V_k to V.

We define a mapping β from S to V by setting

$$\beta(x) = \sum_{k \in K} \eta_k(x_k) = \sum_{k \in K} \eta_k \pi_k(x)$$

for each element $x = (x_k)$ of S. (These sums are well-defined since only finitely many of the elements x_k are non-zero). Then β is easily seen to be a homomorphism. For each index l in K and each element x_l of V_l we have

$$\beta \iota_l(x_l) = \sum_{k \in K} \eta_k \pi_k \iota_l(x_l) = \eta_l(x_l)$$

since $\pi_k \iota_l = \zeta_{lk}$ when $k \neq l$ and $\pi_l \iota_l = I_l$; thus $\beta \iota_l = \eta_l$.

We claim that β is the only homomorphism from S to V with this property. So let β' be another homomorphism from S to V such that $\beta' \iota_k = \eta_k$ for all indices k in K. If x is any element of S we have

$$\beta(x) = \beta(\sum \iota_k \pi_k(x)) = \sum \beta \iota_k \pi_k(x) = \sum \eta_k \pi_k(x)$$

and similarly

$$\beta'(x) = \sum \eta_k \pi_k(x).$$

Hence $\beta = \beta'$ as asserted.

Thus $(S, (\iota_k))$ is universal for homomorphisms from (V_k).

We shall show later (Theorem 15.1*) that in a quite precise sense $(S, (\iota_k))$ is essentially the only pair which is universal for homomorphisms from (V_k).

Let V be any left R-module, (η_k) a family of homomorphisms from (V_k) to V. As we have just proved in Theorem 9.2, there exists a unique homomorphism β from S to V such that $\beta\iota_k = \eta_k$ for every index k in K. If this homomorphism is actually an isomorphism, then the homomorphisms η_k are all injective and we say that the family (η_k) yields an *injective representation* of V as a direct sum of the family (V_k). The inverse β^{-1} of β is an isomorphism from V to S and so for every index k in K we have a homomorphism $\varphi_k = \pi_k\beta^{-1}$ from V to V_k; these homomorphisms are all surjective. Now for each index k in K we have $\varphi_k\eta_k = (\pi_k\beta^{-1})(\beta\iota_k) = \pi_k\iota_k = I_k$, and for every pair of distinct indices k, l in K we have $\varphi_k\eta_l = (\pi_k\beta^{-1})(\beta\iota_l) = \pi_k\iota_l = \zeta_{lk}$. Further, for every element x of V we have

$$\sum \eta_k\varphi_k(x) = \sum \beta\iota_k\pi_k(\beta^{-1}(x)) = \beta(\sum \iota_k\pi_k(\beta^{-1}(x))) = \beta(\beta^{-1}(x)) = x.$$

Hence in this situation the families (φ_k) and (η_k) play the same roles for V as the families (π_k) and (ι_k) do for S.

Let V be a left R-module, $(W_k)_{k\in K}$ a family of submodules of V. In §6 we defined the sum $W = \sum_{k\in K} W_k$ of this family and showed that it is the submodule of V consisting of all elements of V of the form $\sum_{k\in K} w_k$ where $w_k \in W_k$ for each index k in K and only finitely many of the elements w_k are non-zero. For each index k in K the inclusion monomorphism from W_k to V, which we shall denote here by η_k, actually maps W_k into W. Using now the universal property of the external direct sum $S = \bigoplus_{k\in K} W_k$ it follows that there exists a unique homomorphism β from S to W such that $\beta\iota_k = \eta_k$ for every index k in K. If β is an isomorphism from S to W, so that the family (η_k) yields an injective representation of W as a direct sum of the family (W_k), we say that W is the *internal direct sum* of the family (W_k) and that the sum $\sum_{k\in K} W_k$ is *direct*. We now derive necessary and sufficient conditions for the sum of a family of submodules to be direct.

THEOREM 9.3. *Let* $(W_k)_{k\in K}$ *be a family of submodules of a left* R-*module* V. *Then the following conditions are equivalent*:

(a) *the sum* $W = \sum_{k\in K} W_k$ *is direct*;

(b) *every element of* W *can be expressed uniquely in the form* $\sum_{k\in K} w_k$ where $w_k \in W_k$ *for each index* k *in* K *and only finitely many of the elements* w_k *are non-zero*;

(c) *for each index* k_0 *in* K *the intersection* $W_{k_0} \cap \sum_{k\in K, k\neq k_0} W_k$ *consists of the zero element alone.*

Proof. (1) First we prove that (a) implies (b). So we suppose that the sum $W = \sum W_k$ is direct.

There is thus a unique isomorphism β from $S = \oplus W_k$ onto W; according to Theorem 9.2 β is defined by setting, for each element x of S,

$$\beta(x) = \sum_{k\in K} \eta_k \pi_k(x). \qquad [9.1]$$

Suppose now $w = \sum w_k = \sum w_k'$ where $w_k', w_k' \in W_k$ for each index k and only finitely many of the elements w_k, w_k' are non-zero. Consider the elements $x = (w_k)$ and $x' = (w_k')$ of S. Then

$$\beta(x) = \sum \eta_k \pi_k(x) = \sum \eta_k(w_k) = \sum w_k = w$$

and similarly $\beta(x') = w$. Since β is an isomorphism it follows that $x = x'$ and hence $w_k = w_k'$ for each index k, as required.

(2) Next suppose that every element of W can be expressed uniquely in the form described.

Let k_0 be any index in K, w any element of $W_{k_0} \cap \sum_{k\neq k_0} W_k$. Since $w \in W_{k_0}$, it can be expressed in the form $w = \sum_{k\in K} v_k$ where $v_{k_0} = w$ and $v_k = 0$ for $k \neq k_0$; since $w \in \sum_{k\neq k_0} W_k$ it can be expressed in the form $w = \sum_{k\in K} v_k'$ where $v_{k_0}' = 0$, and for $k \neq k_0$ we have $v'_k \in W_k$ (and only finitely many of the elements v_k' are non-zero). It now follows from the uniqueness hypothesis that $w = v_{k_0} = v_{k_0}' = 0$.

Thus (b) implies (c).

(3) Finally suppose that for each index k_0 in K we have

$W_{k_0} \cap \sum_{k \neq k_0} W_k = \{0\}$ and consider the homomorphism β defined by the equation [9.1].

Clearly β is an epimorphism; as we saw in (1) above, if $w = \sum w_k$ is any element of W then $w = \beta(x)$ where $x = (w_k)$.

To show that β is a monomorphism, let x be any element of S such that $\beta(x) = \sum \eta_k \pi_k(x) = 0$. If x is non-zero there is at least one index k_0 in K such that $\pi_{k_0}(x) \neq 0$ and hence $\eta_{k_0}\pi_{k_0}(x)$ is a non-zero element of W_{k_0}. But we also have $\eta_{k_0}\pi_{k_0}(x) = -\sum_{k \neq k_0} \eta_k \pi_k(x) \in \sum_{k \neq k_0} W_k$. This contradicts our hypothesis that $W_{k_0} \cap \sum_{k \neq k_0} W_k = \{0\}$. Hence $x = 0$ and β is a monomorphism.

Thus the sum $W = \sum W_k$ is direct; so (c) implies (a).

This completes the proof.

We now specialise to the case where $(V_k)_{k \in K}$ is a finite family of left R-modules. In this case the direct product $P = \prod_{k \in K} V_k$ and the external direct sum $S = \bigoplus_{k \in K} V_k$ coincide. If $K = [1, n]$ we denote the module $P = S$ by

$$V_1 \oplus V_2 \oplus \ldots \oplus V_n$$

and we may write the elements of this module in the form $x = (x_1, \ldots, x_n)$ where $x_k \in V_k$ $(k = 1, \ldots, n)$. Suppose that the family (η_k) of monomorphisms from (V_k) to a module V yields an injective representation of V as a direct sum of the family (V_k). Then, as we have seen (just after Theorem 9.2.), there is a family (φ_k) of epimorphisms from V onto the modules of the family (V_k). It is easy to verify that this family (φ_k) provides a projective representation of V as a direct product of the family (V_k). We can show similarly that if a family (φ_k) of epimorphisms from V onto the modules of the family (V_k) provides a projective representation of V as a direct product of the family (V_k) then the family (η_k) of monomorphisms from (V_k) into V described in our discussion after Theorem 9.1 provides an injective representation of V as a direct sum of the family (V_k).

Let V be a left R-module, W a submodule. Then W is said to be a *direct summand* of V if there exists a submodule W' of V such that V is the internal direct sum of W and W'. Every such submodule W' is called a *supplement* of W, and we say that W and W' are *supplementary submodules* of V. It follows immediately from Theorem 9.3 that the submodules W and W' are supplementary if and only if $W + W' = V$ and $W \cap W' = 0$. In this situation every element x of V can be expressed uniquely in the form $x = w + w'$ with w in W and w' in W'. We notice also that according to Theorem 7.4 the factor module $V/W = (W + W')/W$ is isomorphic to $W'/(W \cap W')$, and hence to W' (since $W \cap W' = 0$); similarly, of course, V/W' is isomorphic to W.

A short exact sequence

$$0 \to V' \xrightarrow{\alpha} V \xrightarrow{\beta} V'' \to 0 \qquad [9.2]$$

is called a *split exact sequence* if the submodule $\mathrm{Im}\ \alpha = \alpha(V')$ is a direct summand of V. We now establish alternative criteria for a short exact sequence to be split.

THEOREM 9.4. *The following conditions for the short exact sequence* [9.2] *are equivalent*:

(a) *the sequence is split*;
(b) *there is an epimorphism* α' *from* V *onto* V' *such that* $\alpha'\alpha = I_{V'}$;
(c) *there is a monomorphism* β'' *from* V'' *to* V *such that* $\beta\beta'' = I_{V''}$.

Proof. (1) Suppose the sequence [9.2] is split. Then there exists a submodule W of V such that V is the internal direct sum of $\mathrm{Im}\ \alpha$ and W, and hence every element x of V can be expressed uniquely in the form $x = v + w$ where $v \in \mathrm{Im}\ \alpha$ and $w \in W$. Since α is a monomorphism, there is a unique element x' of V' such that $v = \alpha(x')$; set $\alpha'(x) = x'$. The mapping α' from V to V' so defined is easily seen to be the epimorphism required in condition (b).

(2) Suppose condition (b) is satisfied.

Let x'' be any element of V''; since [9.2] is exact, β is an epimorphism and hence there is an element x of V such that $\beta(x) = x''$. We claim that the element $x - \alpha\alpha'(x)$ depends only on x'' and not on the choice of x. To see this, let x_1 be another element of V such that $\beta(x_1) = x''$. Then $x - x_1 \in \mathrm{Ker}\ \beta = \mathrm{Im}\ \alpha$; say $x - x_1 = \alpha(x')$ where $x' \in V'$. It follows that $\alpha\alpha'(x - x_1) = \alpha\alpha'\alpha(x') = \alpha(x') = x - x_1$, whence $x - \alpha\alpha'(x) = x_1 - \alpha\alpha'(x_1)$.

It is now easy to see that the mapping β'' from V'' to V defined by setting $\beta''(x'') = x - \alpha\alpha'(x)$ is the monomorphism required for condition (c).

(3) Finally suppose condition (c) is satisfied. We shall show that V is the internal direct sum of $\mathrm{Im}\ \alpha$ and $\mathrm{Im}\ \beta''$. If $x \in \mathrm{Im}\ \alpha \cap \mathrm{Im}\ \beta''$

there exist elements x' of V' and x'' of V'' such that $x = \alpha(x') = \beta''(x'')$. Then $0 = \beta\alpha(x') = \beta\beta''(x'') = x''$; hence $x = 0$ and so $\operatorname{Im} \alpha \cap \operatorname{Im} \beta'' = 0$. Next, if x is any element of V we see at once that $x - \beta''\beta(x) \in \operatorname{Ker} \beta = \operatorname{Im} \alpha$; so $x \in \operatorname{Im} \alpha + \operatorname{Im} \beta''$.

Thus V is the internal direct sum of $\operatorname{Im} \alpha$ and $\operatorname{Im} \beta''$, and condition (a) is satisfied.

It is easy to see that when [9.2] splits we have a commutative diagram

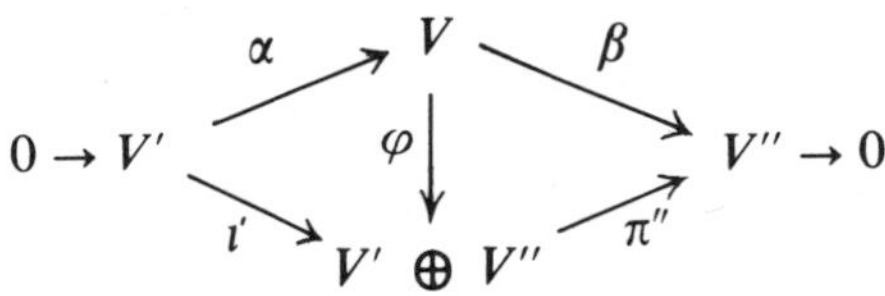

where ι' is the monomorphism given by $\iota'(x') = (x', 0)$ for all elements x' in V', π'' is the epimorphism given by $\pi''(x', x'') = x''$ for all elements (x', x'') in $V' \oplus V''$ and φ is the isomorphism given by $\varphi(x) = (\alpha'(x), \beta(x))$ for all elements x in V.

Now that we have introduced the idea of a split exact sequence we can redeem one of the promises which we made at the end of §8.

THEOREM 9.5. *Let* [9.2] *be a split exact sequence of left R-modules. Then for every left R-module M the sequences*

$$0 \to \operatorname{Hom}(V'', M) \xrightarrow{\beta^*} \operatorname{Hom}(V, M) \xrightarrow{\alpha^*} \operatorname{Hom}(V', M) \to 0 \quad [9.3]$$

and

$$0 \to \operatorname{Hom}(M, V') \xrightarrow{\alpha_*} \operatorname{Hom}(M, V) \xrightarrow{\beta_*} \operatorname{Hom}(M, V'') \to 0 \quad [9.4]$$

are split exact.

Proof. Since [9.2] is split, it follows from Theorem 9.4 that there exists an epimorphism α' from V onto V' such that $\alpha'\alpha = I_{V'}$ and a monomorphism β'' from V'' to V such that $\beta\beta'' = I_{V''}$.

(1) To show that [9.3] is exact all that remains to be shown is that α^* is surjective (cf. Theorem 8.4).

So let θ' be any element of $\operatorname{Hom}(V', M)$. If we set $\theta = \theta'\alpha'$ then $\theta \in \operatorname{Hom}(V, M)$ and we have $\alpha^*(\theta) = \theta\alpha = (\theta'\alpha')\alpha = \theta'(\alpha'\alpha) = \theta'$, i.e. α^* is surjective as required.

Since $\alpha'\alpha = I_{V'}$ we have $\alpha^*(\alpha')^* = (\alpha'\alpha)^* = I^*_{V'}$, from which it follows that [9.3] is split.

(2) To show that [9.4] is exact we have only to show that β_* is surjective.

So let θ'' be any element of Hom (M, V''). If this time we set $\theta = \beta''\theta''$ then $\theta \in \text{Hom}\,(M, V)$ and $\beta_*(\theta) = \beta\theta = \theta''$; so β_* is surjective.

The fact that [9.4] is split follows from the fact that $\beta\beta'' = I_{V''}$.

§10. Direct and Inverse Limits

Let I be a set, ρ an order relation on I; we shall write $i \leqslant j$ instead of $(i, j) \in \rho$. We say that (I, ρ) is a *directed set* if for every pair of elements i, j in I there exists an element k of I such that $i \leqslant k$ and $j \leqslant k$.

Let (I, ρ) be a directed set, R a ring. Then a *direct system* of left R-modules indexed by (I, ρ) is an ordered pair $((V_i)_{i \in I}, (\alpha_{ij})_{(i, j) \in \rho})$ consisting of a family of left R-modules $(V_i)_{i \in I}$ indexed by I and a family of homomorphisms $(\alpha_{ij})_{(i, j) \in \rho}$ indexed by ρ such that

DS1. for each pair (i, j) in ρ, i.e. such that $i \leqslant j$, α_{ij} is a homomorphism from V_i to V_j;

DS2. for all elements i, j, k of I such that $i \leqslant j$ and $j \leqslant k$, $\alpha_{ik} = \alpha_{jk}\alpha_{ij}$.

When the index set I and the order relation ρ are clear from the context we usually denote the direct system $((V_i)_{i \in I}, (\alpha_{ij})_{(i, j) \in \rho})$ simply by (V_i, α_{ij}).

Let (V_i, α_{ij}) be a direct system of left R-modules indexed by (I, ρ). For each index i in I let V'_i be the set $V_i \times \{i\}$; let S be the union $\bigcup_{i \in I} V'_i$. We shall identify each module V_i with the corresponding set V'_i or—what amounts to the same thing—suppose that the modules V_i are pairwise disjoint. Thus for every element x of S there exists a unique element $i(x)$ of I such that $x \in V_{i(x)}$. We define a relation D on S by writing xDy if and only if there is an index i in I such that $i(x) \leqslant i$, $i(y) \leqslant i$ and $\alpha_{i(x)i}(x) = \alpha_{i(y)i}(y)$. It is clear that D is reflexive and symmetric; we claim that it is also transitive. So suppose x, y, z are elements of S such that xDy and yDz; then there are indices i and j in I such that

$$i(x) \leqslant i,\ i(y) \leqslant i,\ \alpha_{i(x)i}(x) = \alpha_{i(y)i}(y)$$

and

$$i(y) \leqslant j,\ i(z) \leqslant j,\ \alpha_{i(y)j}(y) = \alpha_{i(z)j}(z).$$

Since I is a directed set, there exists an index k in I such that $i \leqslant k$

and $j \leqslant k$. Then $i(x) \leqslant k$, $i(z) \leqslant k$ and

$$\alpha_{i(x)k}(x) = \alpha_{ik}(\alpha_{i(x)i}(x)) = \alpha_{ik}(\alpha_{i(y)i}(y)) = \alpha_{i(y)k}(y)$$

while

$$\alpha_{i(z)k}(z) = \alpha_{jk}(\alpha_{i(z)j}(z)) = \alpha_{jk}(\alpha_{i(y)j}(y)) = \alpha_{i(y)k}(y);$$

so $\alpha_{i(x)k}(x) = \alpha_{i(z)k}(z)$ and hence xDz, as asserted. The quotient set $V = S/D$ is called the *direct limit* or *inductive limit* of the direct system (V_i, α_{ij}); we denote it by $\varinjlim V_i$.

Let η be the canonical surjection from S onto V; since each of the modules V_i is a subset of S we may define for each i in I a mapping α_i from V_i to V by setting $\alpha_i(x_i) = \eta(x_i)$ for each element x_i of V_i; we call α_i the *canonical mapping* from V_i to V. We remark that whenever $i \leqslant j$ we have $\alpha_j \alpha_{ij} = \alpha_i$. For if x_i is any element of V_i we have $\alpha_{jj}(\alpha_{ij}(x_i)) = \alpha_{ij}(x_i)$ and hence $\alpha_{ij}(x_i)\, Dx_i$; this implies that $\alpha_j(\alpha_{ij}(x_i)) = \eta(\alpha_{ij}(x_i)) = \eta(x_i) = \alpha_i(x_i)$, i.e. $\alpha_j \alpha_{ij} = \alpha_i$, as required.

We shall now give the set V the structure of a left R-module in such a way that the canonical mappings α_i are all homomorphisms.

To define the addition operation in V, let x and y be any elements of V. Then there are indices i and j in I and elements x_i, y_j of V_i, V_j respectively, such that $x = \alpha_i(x_i)$, $y = \alpha_j(y_j)$. Since I is a directed set there is an index k in I such that $k \geqslant i$ and $k \geqslant j$; we then have $x = \alpha_k \alpha_{ik}(x_i)$ and $y = \alpha_k \alpha_{jk}(y_j)$. We now set

$$x + y = \alpha_k(\alpha_{ik}(x_i) + \alpha_{jk}(y_j)); \qquad [10.1]$$

but before we can assert that this formula defines an addition operation in V we must verify that the right hand side is independent of the various choices involved in its formation. So suppose $x = \alpha_{i'}(x_{i'})$, $y = \alpha_{j'}(y_{j'})$ where $x_{i'} \in V_{i'}$, $y_{j'} \in V_{j'}$, and let k' be an index in I such that $k' \geqslant i'$ and $k' \geqslant j'$. Since $x = \alpha_i(x_i) = \alpha_{i'}(x_{i'})$ there is an index i'' in I such that $\alpha_{ii''}(x_i) = \alpha_{i'i''}(x_{i'})$, and similarly there is an index j'' in I such that $\alpha_{jj''}(y_j) = \alpha_{j'j''}(y_{j'})$. Let k'' be an index such that $k'' \geqslant i'', j'', k, k'$. Then we have

$$\begin{aligned} \alpha_{kk''}(\alpha_{ik}(x_i) + \alpha_{jk}(y_j)) &= \alpha_{kk''}\alpha_{ik}(x_i) + \alpha_{kk''}\alpha_{jk}(y_j) \\ &= \alpha_{ik''}(x_i) + \alpha_{jk''}(y_j) \\ &= \alpha_{i''k''}\alpha_{ii''}(x_i) + \alpha_{j''k''}\alpha_{jj''}(y_j) \\ &= \alpha_{i''k''}\alpha_{i'i''}(x_{i'}) + \alpha_{j''k''}\alpha_{j'j''}(y_{j'}) \\ &= \alpha_{i'k''}(x_{i'}) + \alpha_{j'k''}(y_{j'}) \end{aligned}$$

$$= \alpha_{k'k''}\alpha_{i'k'}(x_{i'}) + \alpha_{k'k''}\alpha_{j'k'}(y_{j'})$$
$$= \alpha_{k'k''}(\alpha_{i'k'}(x_{i'}) + \alpha_{j'k'}(y_{j'})).$$

Hence it follows that

$$\alpha_k(\alpha_{ik}(x_i) + \alpha_{jk}(y_j)) = \alpha_{k'}(\alpha_{i'k'}(x_{i'}) + \alpha_{j'k'}(y_{j'}))$$

as required.

It is now easy to verify that V forms an abelian group under the addition operation we have just defined. To check that this operation is associative, let $x = \alpha_i(x_i)$, $y = \alpha_j(y_j)$, $z = \alpha_k(z_k)$ be elements of V, where $x_i \in V_i$, $y_j \in V_j$, $z_k \in V_k$; let l be an index in I such that $l \geqslant i, j, k$. Then we have

$$\begin{aligned} x + (y + z) &= \alpha_l(\alpha_{il}(x_i) + \alpha_{ll}(\alpha_{jl}(y_j) + \alpha_{kl}(z_k))) \\ &= \alpha_l(\alpha_{ll}\alpha_{il}(x_i) + (\alpha_{ll}\alpha_{jl}(y_j) + \alpha_{ll}\alpha_{kl}(z_k))) \\ &= \alpha_l((\alpha_{ll}\alpha_{il}(x_i) + \alpha_{ll}\alpha_{jl}(y_j)) + \alpha_{ll}\alpha_{kl}(z_k)) \\ &= \alpha_l(\alpha_{ll}(\alpha_{il}(x_i) + \alpha_{jl}(y_j)) + \alpha_{kl}(z_k)) \\ &= (x + y) + z. \end{aligned}$$

The commutativity is established by a similar argument. If i and j are any two indices in I and k is an index such that $k \geqslant i, j$, then we clearly have $\alpha_{ik}(0_i) = 0_k = \alpha_{jk}(0_j)$ where 0_i, 0_j, 0_k are the zero elements of V_i, V_j, V_k respectively. Thus the element $\alpha_i(0_i)$ of V is independent of the index i, and it is obviously the zero element of V. Finally, if $x = \alpha_i(x_i)$ is any element of V, then the element $\alpha_i(-x_i)$ is the negative of x in V.

Next we define a scalar multiplication on the left of V by elements of R. Let $x = \alpha_i(x_i)$ be any element of V, a any element of R. Then we set

$$ax = \alpha_i(ax_i). \qquad [10.2]$$

Again we must verify that the right hand side is independent of the index i. So suppose $x = \alpha_j(x_j)$; then there is an index k such that $k \geqslant i, j$ and $\alpha_{ik}(x_i) = \alpha_{jk}(x_j)$. It follows that $a\alpha_{ik}(x_i) = a\alpha_{jk}(x_j)$ and hence, since α_{ik} and α_{jk} are module homomorphisms, $\alpha_{ik}(ax_i) = \alpha_{jk}(ax_j)$; thus $\alpha_i(ax_i) = \alpha_j(ax_j)$, as required. We readily verify that V forms a left R-module under the addition and scalar multiplication we have defined.

The canonical mappings α_i are homomorphisms; for if x_i and y_i are elements of V_i and a is any element of R the definitions [10.1] and [10.2] give

$$\alpha_i(x_i) + \alpha_i(y_i) = \alpha_i(x_i + y_i) \quad \text{and} \quad a\alpha_i(x_i) = \alpha_i(ax_i).$$

Example 1. Let V be any left R-module; let E be a generating system for V. Let I be the set of finite subsets of E. Then I together with the inclusion relation is a directed set (if i, j are elements of I then $i \subseteq i \cup j$ and $j \subseteq i \cup j$). For each element i in I let V_i be the submodule of V generated by the subset i of E; for each pair of elements i, j in I such that $i \subseteq j$ let α_{ij} be the inclusion monomorphism from V_i to V_j. Then it is clear that (V_i, α_{ij}) is a direct system of modules. We propose to prove that $V_\infty = \varinjlim V_i$ is isomorphic to the original module V.

For each 'index' i in I let α_i be the canonical mapping from V_i to V_∞ and β_i the inclusion monomorphism from V_i to V. Then for each pair of elements i, j of I such that $i \subseteq j$ we have $\alpha_j\alpha_{ij} = \alpha_i$ and $\beta_j\alpha_{ij} = \beta_i$. Now define a mapping β from V_∞ to V as follows. Let x be any element of V_∞; then there is an index i in I and an element x_i of V_i such that $x = \alpha_i(x_i)$. We set $\beta(x) = \beta_i(x_i)$. We must verify that the right hand member depends only on x, not on the choices of i and x_i. So suppose we also have $x = \alpha_j(x_j)$ where $x_j \in V_j$; then there exists an index k such that $k \supseteq i$, $k \supseteq j$ and $\alpha_{ik}(x_i) = \alpha_{jk}(x_j)$. It follows that $\beta_k\alpha_{ik}(x_i) = \beta_k\alpha_{jk}(x_j)$, whence $\beta_i(x_i) = \beta_j(x_j)$, as required. We leave it to the reader to check that β is a homomorphism.

We now show that in fact β is an isomorphism. First let v be any element of V; since E is a generating system for V, there exists a finite subset i of E such that v is expressible as a linear combination of elements of i with coefficients in R. Hence $v \in V_i$ and so $v = \beta_i(v) = \beta(\alpha_i(v))$; thus β is surjective. Next suppose $x = \alpha_i(x_i)$ and $x' = \alpha_j(x'_j)$ are elements of V_∞ such that $\beta(x) = \beta(x')$. Then $\beta_i(x_i) = \beta_j(x'_j)$ and hence, if k is any index such that $k \supseteq i$, $k \supseteq j$, we have $\beta_k\alpha_{ik}(x_i) = \beta_k\alpha_{jk}(x'_j)$. Since β_k is a monomorphism it follows that $\alpha_{ik}(x_i) = \alpha_{jk}(x'_j)$ and hence, using the definition of the direct limit, that $x = \alpha_i(x_i) = \alpha_j(x'_j) = x'$. Thus β is injective.

We often 'identify' V and V_∞ and say informally that the module V is the direct limit of its finitely generated submodules.

Let (V_i, α_{ij}) be a direct system of left R-modules; let M be a left R-module and $(\varphi_i)_{i \in I}$ a family of homomorphisms such that for each index i in I we have $\varphi_i \in \operatorname{Hom}(V_i, M)$. If $\varphi_j\alpha_{ij} = \varphi_i$ for all pairs of indices (i, j) such that $i \leqslant j$ then we say that (φ_i) is a *coherent family of homomorphisms* from the direct system (V_i, α_{ij}) to M. The preceding discussion shows that (α_i) is a coherent family of homomorphisms from (V_i, α_{ij}) to $V = \varinjlim V_i$.

We now say that an ordered pair $(V', (\alpha_i'))$ consisting of a left R-module V' and a coherent family (α_i') of homomorphisms from (V_i, α_{ij}) to V' is *universal for coherent families of homomorphisms* from (V_i, α_{ij}) if for every R-module M and every coherent family (φ_i) of homomorphisms from (V_i, α_{ij}) to M there exists a unique homomorphism φ from V' to M such that for every index i in I the diagram

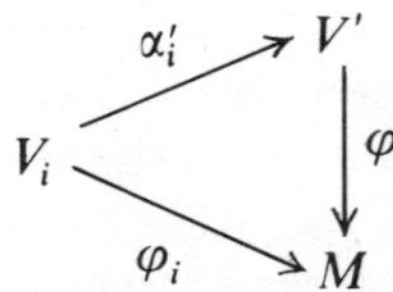

is commutative.

THEOREM 10.1. *Let (V_i, α_{ij}) be a direct system of left R-modules. Then the ordered pair $(V, (\alpha_i))$ consisting of the direct limit $V = \varinjlim V_i$ and the family (α_i) of canonical homomorphisms is universal for coherent families of homomorphisms from (V_i, α_{ij}).*

Proof. Let (φ_i) be a coherent family of homomorphisms from (V_i, α_{ij}) to a left R-module M. We define a mapping φ from V to M as follows. If x is any element of V there is an index i in I and an element x_i of V_i such that $x = \alpha_i(x_i)$; we set $\varphi(x) = \varphi_i(x_i)$. To show that $\varphi(x)$ so defined depends only on x and not on the choices of i and x_i, suppose we also have $x = \alpha_j(x_j)$ where $x_j \in V_j$. Then there is an index k such that $i \leqslant k, j \leqslant k$ and $\alpha_{ik}(x_i) = \alpha_{jk}(x_j)$. Since $\varphi_k\alpha_{ik} = \varphi_i$ and $\varphi_k\alpha_{jk} = \varphi_j$ it follows that $\varphi_i(x_i) = \varphi_j(x_j)$ as desired. Clearly $\varphi\alpha_i = \varphi_i$ for each index i in I.

We claim that φ is a homomorphism. So let $x = \alpha_i(x_i)$ and $y = \alpha_j(y_j)$ be elements of V. There is an index k such that $i \leqslant k$, $j \leqslant k$; then $x = \alpha_k\alpha_{ik}(x_i)$ and $y = \alpha_k\alpha_{jk}(y_j)$ and we have $x + y = \alpha_k(\alpha_{ik}(x_i) + \alpha_{jk}(y_j))$. It follows that

$$\begin{aligned}\varphi(x + y) &= \varphi_k(\alpha_{ik}(x_i) + \alpha_{jk}(y_j))\\ &= \varphi_k\alpha_{ik}(x_i) + \varphi_k\alpha_{jk}(y_j)\\ &= \varphi_i(x_i) + \varphi_j(y_j) = \varphi(x) + \varphi(y);\end{aligned}$$

so φ is a homomorphism.

Finally φ is unique. For suppose we had another homomorphism φ' from V to M such that $\varphi'\alpha_i = \varphi_i$ for every index i in I. Then if $x = \alpha_i(x_i)$ is any element of V we have $\varphi(x) = \varphi\alpha_i(x_i) = \varphi_i(x_i) = \varphi'\alpha_i(x_i) = \varphi'(x)$; hence $\varphi = \varphi'$.

It follows from a general result in §15 (Theorem 15.1*) that $(V, (\alpha_i))$ is essentially the only pair which is universal for coherent families of homomorphisms from (V_i, α_{ij}). (See Example 15 of §15.)

Let now (V_i, α_{ij}) and (W_i, β_{ij}) be direct systems of left R-modules indexed by the same directed set I. Suppose that for each index i in I we have a homomorphism φ_i from V_i to W_i. Then the family $(\varphi_i)_{i \in I}$ is called a *direct system of homomorphisms* from (V_i, α_{ij}) to (W_i, β_{ij}) if all the diagrams

$$\begin{array}{ccc} V_i & \xrightarrow{\varphi_i} & W_i \\ {\scriptstyle \alpha_{ij}}\downarrow & & \downarrow{\scriptstyle \beta_{ij}} \\ V_j & \xrightarrow{\varphi_j} & W_j \end{array}$$

with $i \leqslant j$ are commutative.

Theorem 10.2. *Let (φ_i) be a direct system of homomorphisms from (V_i, α_{ij}) to (W_i, β_{ij}). If $V = \varinjlim V_i$, $W = \varinjlim W_i$ and (α_i), (β_i) are the associated families of canonical homomorphisms, there exists a unique homomorphism φ from V to W such that for every index i the diagram*

$$\begin{array}{ccc} V_i & \xrightarrow{\varphi_i} & W_i \\ {\scriptstyle \alpha_i}\downarrow & & \downarrow{\scriptstyle \beta_i} \\ V & \xrightarrow{\varphi} & W \end{array}$$

is commutative.

Proof. We apply Theorem 10.1 to the family of homomorphisms $(\beta_i\varphi_i)$. First we must show that this is a coherent family of homomorphisms from (V_i, α_{ij}) to W. So suppose $i \leqslant j$; then we have

$$(\beta_j\varphi_j)\alpha_{ij} = \beta_j(\varphi_j\alpha_{ij}) = \beta_j(\beta_{ij}\varphi_i) = (\beta_j\beta_{ij})\varphi_i = \beta_i\varphi_i$$

as required. Hence there exists a unique homomorphism φ from V to W such that for every index i in I we have $\varphi\alpha_i = \beta_i\varphi_i$.

The homomorphism φ constructed in Theorem 10.2 is called the *direct limit* of the family of homomorphisms (φ_i) and we write $\varphi = \varinjlim \varphi_i$.

Let now I be an index set, ρ an arbitrary relation of order on I; thus we no longer require (I, ρ) to be a directed set. An *inverse system of left R-modules* indexed by (I, ρ) is then an ordered pair $((V_i)_{i \in I}, (\alpha_{ji})_{(j,i) \in \rho^{-1}})$—usually abbreviated to (V_i, α_{ji})—consisting of a family of left R-modules $(V_i)_{i \in I}$ indexed by I and a family of homomorphisms $(\alpha_{ji})_{(j,i) \in \rho^{-1}}$ indexed by ρ^{-1} such that

IS1. For each pair (j, i) in ρ^{-1}, i.e. such that $j \geqslant i$, α_{ji} is a homomorphism from V_j to V_i.

IS2. For all elements i, j, k of I such that $i \leqslant j$ and $j \leqslant k$, we have $\alpha_{ki} = \alpha_{ji}\alpha_{kj}$.

Let (V_i, α_{ji}) be an inverse system of left R-modules indexed by (I, ρ). Let P be the direct product module of the family (V_i), and let (π_i) be the family of canonical projection epimorphisms. If V is the subset of P consisting of elements x such that

$$\alpha_{ji}(\pi_j(x)) = \pi_i(x) \qquad [10.3]$$

whenever $i \leqslant j$, it is easy to verify that V is a submodule of P; we call V the *inverse limit* or *projective limit* of the inverse system (V_i, α_{ji}) and we write $V = \varprojlim V_i$. When we write elements of P as $x = (x_i)_{i \in I}$, the condition [10.3] takes the form $\alpha_{ji}(x_j) = x_i$.

For each index i in I we denote the restriction of π_i to V by α_i and call α_i the *canonical homomorphism* from V to V_i. According to [10.3] we have $\alpha_{ji}\alpha_j = \alpha_i$ for all pairs of indices (i, j) such that $i \leqslant j$.

Let M be a left R-module, $(\varphi_i)_{i \in I}$ a family of homomorphisms such that for every index i in I we have $\varphi_i \in \operatorname{Hom}(M, V_i)$. If $\alpha_{ji}\varphi_j = \varphi_i$ whenever $i \leqslant j$ we say that (φ_i) is a *coherent family of homomorphisms* from M to the inverse system (V_i, α_{ji}). According to the preceding paragraph the family of canonical homomorphisms (α_i) is a coherent family of homomorphisms from $\varprojlim V_i$ to (V_i, α_{ji}).

We say that an ordered pair $(V', (\alpha_i'))$ consisting of a left R-module V' and a coherent family of homomorphisms from V' to (V_i, α_{ji}) is *couniversal for coherent families of homomorphisms* to (V_i, α_{ji}) if for every left R-module M and every coherent family (φ_i) of homomorphisms from M to (V_i, α_{ji}) there exists a unique homomorphism φ from M to V' such that for every index i in I the diagram

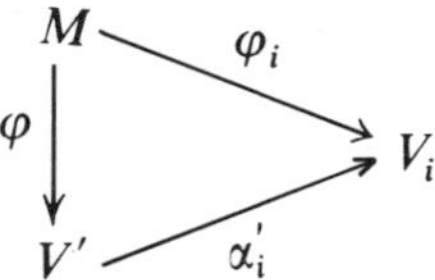

is commutative.

THEOREM 10.3. *Let (V_i,α_{ji}) be an inverse system of left R-modules. Then the ordered pair $(V,(\alpha_i))$ consisting of the inverse limit $V = \varprojlim V_i$ and the family (α_i) of canonical homomorphisms is couniversal for coherent families of homomorphisms to (V_i, α_{ji}).*

Proof. Let M be any left R-module, (φ_i) a coherent family of homomorphisms from M to (V_i,α_{ji}).

Consider the mapping φ from M to $P = \prod_{i \in I} V_i$ defined by setting

$$\varphi(x) = (\varphi_i(x))_{i \in I}$$

for every element x of M. It is clear that φ is a homomorphism; we claim that it actually maps M into V, i.e. that $\varphi(x) \in V$ for every element x of M. To show this we have only to remark that whenever $i \leqslant j$ we have

$$\alpha_{ji}[\pi_j(\varphi(x))] = \alpha_{ji}(\varphi_j(x)) = \varphi_i(x) = \pi_i(\varphi(x)).$$

Thus $\varphi(x) \in V$.

The uniqueness of φ follows from the fact that if $\alpha_i\varphi = \alpha_i\varphi' = \varphi_i$ for every index i in I then for each element x of M and each index i we have

$$\pi_i(\varphi(x)) = \varphi_i(x) = \pi_i(\varphi'(x))$$

and hence $\varphi(x) = \varphi'(x)$. So $\varphi = \varphi'$.

It will follow from Theorem 15.1 that (in a quite precise sense) the pair $(V_i,(\alpha_i))$ is essentially the only one which is couniversal for coherent families of homomorphisms to (V_i,α_{ji}). For the details see Example 16 of §15.

Suppose now that we have two inverse systems (V_i,α_{ji}) and (W_i,β_{ji}) indexed by the same ordered set I, and for each index i in I a homomorphism φ_i from V_i to W_i. The family $(\varphi_i)_{i \in I}$ is called an *inverse system of homomorphisms* if all the diagrams

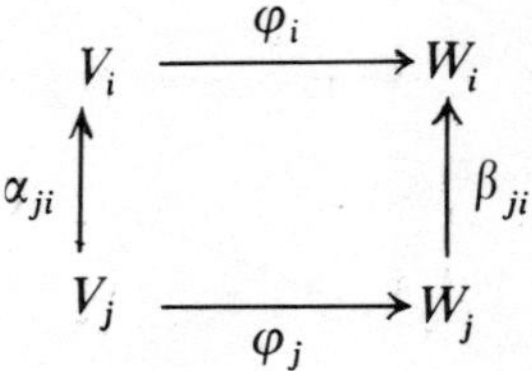

with $i \leqslant j$ are commutative.

THEOREM 10.4. *Let* $(\varphi_i)_{i \in I}$ *be an inverse system of homomorphisms from* (V_i, α_{ji}) *to* (W_i, β_{ji}). *If* $V = \varprojlim V_i$, $W = \varprojlim W_i$ *and* (α_i), (β_i) *are the associated families of canonical homomorphisms, there exists a unique homomorphism* φ *from* V *to* W *such that for every index i the diagram*

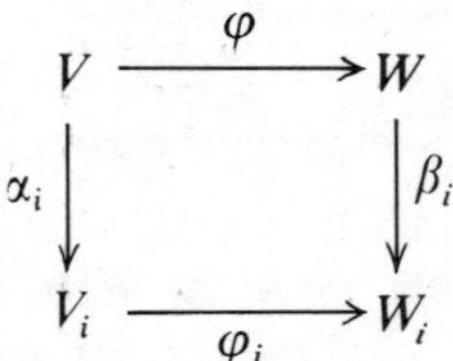

is commutative.

Proof. We consider the family of homomorphisms $(\varphi_i \alpha_i)_{i \in I}$ from V to (W_i). If i and j are indices such that $i \leqslant j$, we have

$$\beta_{ji}(\varphi_j \alpha_j) = (\beta_{ji}\varphi_j)\alpha_j = (\varphi_i \alpha_{ji})\alpha_j = \varphi_i(\alpha_{ji}\alpha_j) = \varphi_i \alpha_i.$$

So $(\varphi_i \alpha_i)$ is a coherent family of homomorphisms from V to (W_i). It follows from Theorem 10.3 that there exists a unique homomorphism φ from V to W such that $\beta_i \varphi = \varphi_i \alpha_i$ for every index i in I.

The homomorphism φ constructed in Theorem 10.4 is called the *inverse limit* of the inverse system (φ_i) and we write $\varphi = \varprojlim \varphi_i$.

§11. Free, Projective and Injective Modules

Let R be a ring with identity element e, E any set. Consider the family $(V_x)_{x \in E}$ of left R-modules indexed by E in which each member V_x is the module R_l, i.e. the ring itself considered as left R-module.

We form the external direct sum of this family and denote it by $F(E)$. According to the definition of the external direct sum of a family of modules, $F(E)$ is the subset of Map (E, R) consisting of those mappings α from E to R such that $\alpha(x)$ is non-zero for at most finitely many elements x of E; of course if E is a finite set we have $F(E) = \text{Map}\,(E, R)$. We recall the definitions of the addition and scalar multiplication in $F(E)$. If α_1 and α_2 are elements of $F(E)$, we define $\alpha_1 + \alpha_2$ by setting

$$(\alpha_1 + \alpha_2)(x) = \alpha_1(x) + \alpha_2(x) \qquad [11.1]$$

for each element x of E; if a is any element of R and α any element of $F(E)$, we set

$$(a\alpha)(x) = a\alpha(x)$$

for each element x of E.

The module $F(E)$ is called the *free R-module based on E*.

Consider next the mapping φ from E to $F(E)$ defined by setting $\varphi(x) = \iota_x(e)$ for every element x of E, where ι_x is the canonical injection monomorphism from $V_x = R_l$ to $F(E)$. We call φ the *natural mapping* from E to $F(E)$.

If α is any element of $F(E)$ then for each element x of E we write a_x instead of $\pi_x(\alpha) = \alpha(x)$. Then we have

$$\begin{aligned}\alpha &= \sum_{x \in E} \iota_x \pi_x(\alpha) \\ &= \sum_{x \in E} \iota_x(\pi_x(\alpha) \,.\, e) \\ &= \sum_{x \in E} \pi_x(\alpha)\, \iota_x(e) = \sum_{x \in E} a_x \varphi(x).\end{aligned}$$

If we identify E with its image under φ in $F(E)$ we have expressed α as a 'formal sum'

$$\alpha = \sum_{x \in E} a_x x \qquad [11.2]$$

of elements of E with coefficients in R. The expression [11.2] for α is plainly unique.

We say that a pair (G, θ) consisting of a left R-module G and a mapping θ from E to G is *universal* for mappings from E to left R-modules if for every mapping ψ from E to a left R-module V

there exists a unique R-module homomorphism ν from G to V such that the diagram

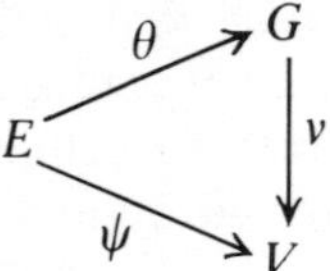

is commutative.

THEOREM 11.1. *Let E be any set. The pair $(F(E), \varphi)$ consisting of the free R-module based on E and the natural mapping from E to $F(E)$ is universal for mappings from E to left R-modules.*

Proof. Throughout the proof we use F as an abbreviation for $F(E)$.

Let V be any left R-module, ψ a mapping from E to V. Consider the mapping ν from F to V defined by setting for each element α of F

$$\nu(\alpha) = \sum_{x \in E} \alpha(x)\psi(x).$$

(If we express α as a formal sum, $\alpha = \sum_{x \in E} \alpha_x x$, this definition becomes

$$\nu(\sum a_x x) = \sum a_x \psi(x).)$$

Then it is easy to verify that ν is a homomorphism. Further, $\nu\varphi = \psi$; for if t is any element of E, we have

$$\nu(\varphi(t)) = \sum_{x \in E} [\varphi(t)](x)\psi(x) = \psi(t).$$

The homomorphism ν is plainly unique.

So $(F(E), \varphi)$ is universal for mappings from E to left R-modules.

As usual in these 'universal' situations we remark that an 'essential uniqueness' property will follow from the general result of Theorem 15.1*. For the details see § 15, Example 17.

A left R-module G is said to be *free* if there exists a set E such that G is isomorphic to $F(E)$, the free left R-module based on E. The next theorem gives a useful criterion for a module to be free.

THEOREM 11.2. *A left R-module is free if and only if it has a basis.*

Proof. (1) Let G be a left R-module with basis X. We shall show that G is isomorphic to $F(X)$.

We define a mapping λ from G to $F(X)$ as follows. If g is any element of G, then, according to Theorem 6.4, g can be uniquely expressed in the form $g = \sum_{x \in X} a_x x$ where $(a_x)_{x \in X}$ is a quasi-finite family of elements of R. The element $\lambda(g)$ of $F(X)$ is defined by setting for each element t of X,

$$[\lambda(g)]\,(t) = a_t.$$

It is easily verified that λ is an isomorphism. (If we identify X with its image in $F(X)$ under the natural mapping φ and write the elements of $F(X)$ as formal sums then of course $F(X) = G$ and λ is the identity mapping.)

(2) Conversely, suppose G is free. Then there is a set E and an isomorphism μ from $F(E)$ onto G. We shall show that $\mu(E)$ is a basis for G.

Let us write the elements of $F(E)$ as formal sums. Since μ is surjective, if g is any element of G there is an element $\sum_{x \in E} a_x x$ of $F(E)$ such that

$$g = \mu(\textstyle\sum a_x x) = \sum a_x \mu(x);$$

this shows that $\mu(E)$ is a generating system for G. To show that $\mu(E)$ is linearly independent, suppose $(b_x)_{x \in E}$ is a quasi-finite family of elements of R such that $\sum b_x \mu(x) = 0$. Then $\mu(\sum b_x x) = 0$ and hence, since μ is injective, $\sum b_x x = 0$ in $F(E)$, i.e. all the elements b_x are zero.

Thus $\mu(E)$ is indeed a basis for G.

COROLLARY 1. *For every ring R the left R-module R_l is free.*

Proof. The set consisting of the identity alone is a basis.

COROLLARY 2. *If R is a division ring every left R-module (left vector space over R) is free.*

Proof. The Corollary to Theorem 6.5 shows that every such module has a basis.

In the next two theorems we establish two basic properties of free modules.

THEOREM 11.3. *If V is any left R-module there exists an epimorphism from a free R-module onto V.*

Proof. Let $F = F(V)$ be the free left R-module based on V, φ the natural mapping from V to $F(V)$.

Since (F, φ) is universal for mappings from V to left R-modules (Theorem 11.1) and the identity mapping I_V is a mapping from V to the left R-module V there exists a homomorphism v from F to V such that $v\varphi = I_V$. Then, if x is any element of V, we have $x = I_V(x) = v(\varphi(x))$; so v is an epimorphism.

The construction we have just given is by no means unique, and in fact it is an extremely uneconomical one. The reader may care to verify that if X is any generating system for V then V is an epimorphic image of the free module $F(X)$—instead of the identity mapping I_V in the argument above we use the canonical injection of X into V.

THEOREM 11.4. *Let V and V'' be left R-modules, η an epimorphism from V onto V'', F a free left R-module and α a homomorphism from F to V''. Then there exists a homomorphism β from F to V such that $\eta\beta = \alpha$.*

Proof. Let X be a basis for F. Since η is an epimorphism it follows that for every element x of X there is an element v_x of V such that $\eta(v_x) = \alpha(x)$. We now define a mapping β from F to V as follows. If t is any element of F it may be expressed uniquely in the form $t = \sum_{x \in X} a_x x$ where $(a_x)_{x \in X}$ is as usual a quasi-finite family of elements of R. Set $\beta(t) = \sum_{x \in X} a_x v_x$. Then it is easy to verify that β is a homomorphism and $\eta\beta = \alpha$.

The property established for free modules in Theorem 11.4 is of such importance that we give a special name to those modules which enjoy it. In introducing this we restate the property in the more transparent language of diagrams: a left R-module P is said to be *projective* if in every diagram

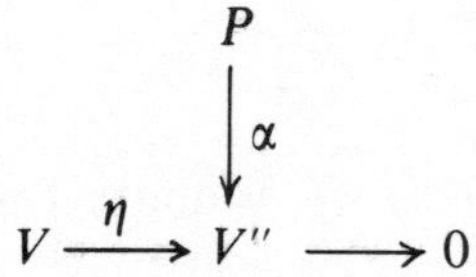

of modules and homomorphisms in which the row is exact we can insert a homomorphism $\beta:P \to V$ such that the diagram

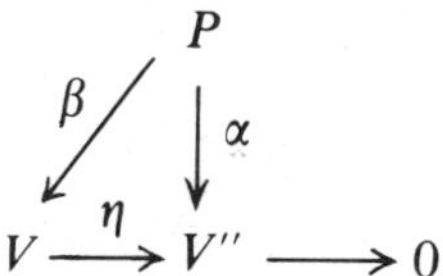

is commutative. (We say in this situation that α has been *lifted* to β.) We may thus restate Theorem 11.4 briefly in the following way. *Every free left R-module is projective.* Theorem 11.3 then allows us to say that *every left R-module is an epimorphic image of a projective module* (though Theorem 11.3 is of course a stronger result than this, this is what we shall use later). Our next theorem gives us two useful criteria for a left R-module to be projective.

THEOREM 11.5. *Let P be a left R-module; then the following conditions are equivalent:*

(a) *P is projective;*

(b) *every short exact sequence of left R-modules of the form*

$$0 \to V' \to V \to P \to 0$$

splits;

(c) *P is a direct summand of a free left R-module.*

Proof. (1) Suppose P is a projective left R-module. Let

$$0 \to V' \to V \xrightarrow{\eta} P \to 0 \qquad [11.3]$$

be a short exact sequence, so that η is an epimorphism. Consider the identity mapping I_P from P to P. Since P is projective this mapping can be lifted to a homomorphism η' from P to V, i.e. there is a homomorphism η' from P to V such that $\eta\eta' = I_P$. According to Theorem 9.4, the sequence [11.3] splits (η' is clearly injective).

Thus (a) implies (b).

(2) Suppose condition (b) is satisfied.

In the proof of Theorem 11.3 we showed how to construct a short exact sequence

$$0 \to \operatorname{Ker} \nu \to F(P) \xrightarrow{\nu} P \to 0.$$

According to condition (b) this sequence splits and hence, as we remarked after proving Theorem 9.4, the direct sum $\operatorname{Ker}\nu \oplus P$ is isomorphic to the free module $F(P)$—and hence is free. So P is a direct summand of a free left R-module.

Thus (b) implies (c).

(3) Finally suppose condition (c) is satisfied. Then there exists a free left R-module F and a submodule P' of F such that F is the internal direct sum of P and P'. It follows that the external direct sum $S = P \oplus P'$ is free, and hence projective. Let π be the canonical projection epimorphism from S onto P, ι the canonical injection monomorphism from P to S; we recall that $\pi\iota = I_P$.

Now let V and V'' be any two modules, η an epimorphism from V onto V'', α a homomorphism from P to V''. Then $\alpha\pi$ is a homomorphism from S to V'', and since S is projective there is therefore a homomorphism β_1 from S to V such that $\eta\beta_1 = \alpha\pi$. Set $\beta = \beta_1\iota$; then β is a homomorphism from P to V and we have $\eta\beta = \eta\beta_1\iota = \alpha\pi\iota = \alpha I_P = \alpha$. So P is projective.

Hence (c) implies (a) and the proof is complete.

We return to one of the questions raised at the end of §8 which we are now in a position to answer.

THEOREM 11.6. *Let M be a left R-module and let*

$$0 \to V' \xrightarrow{\alpha} V \xrightarrow{\beta} V'' \to 0 \qquad [11.4]$$

be a short exact sequence of left R-modules. If M is projective then the sequence

$$0 \to \operatorname{Hom}(M, V') \xrightarrow{\alpha_*} \operatorname{Hom}(M, V) \xrightarrow{\beta_*} \operatorname{Hom}(M, V'') \to 0 \qquad [11.5]$$

is exact. Conversely, if [11.5] *is exact for every short exact sequence* [11.4], *then M is projective.*

Proof. It follows from Theorem 8.4 that the sequence [11.5] is exact if and only if β_* is an epimorphism.

(1) Suppose M is projective and let θ'' be any element of $\operatorname{Hom}(M, V'')$. Then we have a diagram

$$\begin{array}{ccc} & M & \\ & \downarrow{\scriptstyle \theta''} & \\ V \xrightarrow{\beta} & V'' & \to 0 \end{array} \qquad [11.6]$$

in which the row is exact. Since M is projective, there exists a homomorphism θ from M to V such that $\beta\theta = \theta''$. But, by definition,

$\beta_*(\theta) = \beta\theta$, so we have indeed shown that $\theta'' \in \text{Im}\ \beta_*$, i.e. that β_* is an epimorphism and so [11.5] is exact.

(2) Conversely, suppose that for every short exact sequence [11.4] the induced homomorphism β_* is an epimorphism. Let [11.6] be a diagram in which the row is exact. Since β_* is an epimorphism from Hom (M, V) onto Hom (M, V'') and $\theta'' \in$ Hom (M, V'') it follows that there is a homomorphism θ from M to V such that $\beta_*(\theta) = \theta''$, i.e. such that $\beta\theta = \theta''$. Thus M is projective.

We next introduce a property of modules which, in a technical sense we shall discuss in §15, is dual to the projective property. A left R-module Q is said to be *injective* if in every diagram

$$\begin{array}{ccccc} 0 & \longrightarrow & V' & \xrightarrow{\kappa} & V \\ & & \alpha \big\downarrow & & \\ & & Q & & \end{array}$$

of modules and homomorphisms in which the row is exact we can insert a homomorphism $\beta: V \to Q$ such that the diagram

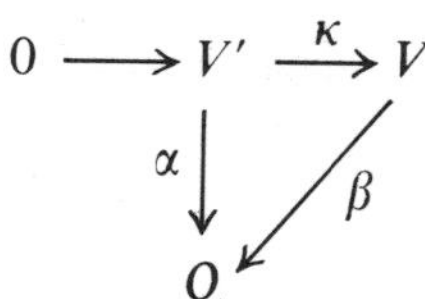

is commutative; β is called an *extension* of α to V.

Example 1. Let D be a division ring. We shall show that every left D-module (left vector space over D) is injective. So let Q be any left D-module and suppose we have a diagram

$$\begin{array}{ccccc} 0 & \longrightarrow & V' & \xrightarrow{\kappa} & V \\ & & \alpha \big\downarrow & & \\ & & Q & & \end{array}$$

with exact row, i.e. a monomorphism κ from V' to V and a homomorphism α from V' to Q. According to the Corollary of Theorem

6.5 there exists a basis B' for V'; since κ is a monomorphism, $\kappa(B')$ is a free subset of V. Theorem 6.5 itself now shows that there exists a subset B'' of V, disjoint from $\kappa(B')$, such that $\kappa(B') \cup B''$ is a basis for V; then every element x of V can be expressed uniquely in the form

$$x = \sum_{s \in B'} a_s \kappa(s) + \sum_{t \in B''} a_t t,$$

where $(a_s)_{s \in B'}$ and $(a_t)_{t \in B''}$ are quasi-finite families of elements of R. Define a mapping β from V to Q by setting

$$\beta(x) = \sum_{s \in B'} a_s \alpha(s)$$

for each element x of V. Then it is clear that β is a homomorphism and that $\beta\kappa = \alpha$. So Q is injective, as we claimed.

We shall obtain another example once we have established the following criterion for a module to be injective.

THEOREM 11.7. *Let Q be a left R-module. Then Q is injective if and only if for every left ideal L of R (considered as left R-module) and every homomorphism α from L to Q there exists an element a of Q such that $\alpha(l) = la$ for all elements l of L.*

Proof. (1) Suppose Q is injective.

Consider the diagram

$$\begin{array}{ccccc} 0 & \longrightarrow & L & \xrightarrow{\iota} & R_l \\ & & \Big\downarrow{\scriptstyle\alpha} & & \\ & & Q & & \end{array}$$

in which L is a left ideal of R, ι is the inclusion monomorphism from L to R and α is a homomorphism from L to Q. Since, by hypothesis, Q is injective, there exists a homomorphism β from R_l to Q such that $\beta\iota = \alpha$. Then for each element l of L we have $\alpha(l) = \beta\iota(l) = l\beta\iota(e)$; so $\beta\iota(e)$ is the required element a of Q.

(2) Conversely, suppose Q satisfies the condition of the theorem.

Let κ be a monomorphism from a left R-module V' to another left R-module V and let α be a homomorphism from V' to Q.

Making use of Zorn's Lemma we shall show that there exists a homomorphism β from V to Q such that $\beta\kappa = \alpha$.

We consider the set E of all ordered pairs (W, φ) consisting of a submodule W of V which includes $\kappa(V')$ and a homomorphism φ from W to Q such that $\varphi\kappa = \alpha$. Write $(W, \varphi) \leqslant (W', \varphi')$ if W is a submodule of W' and $\varphi'(w) = \varphi(w)$ for every element w of W. The reader is invited to verify that E is inductively ordered by the relation $\leqslant$; according to Zorn's Lemma, E has a maximal element, say (V_0, φ_0). We claim that $V_0 = V$; once we have established this, φ_0 will serve as the required homomorphism β. Suppose, to the contrary, that $V_0 \neq V$ and let x be an element of V which does not belong to V_0. If V_1 is the submodule of V generated by V_0 and x, every element y of V_1 can be expressed in the form

$$y = y_0 + rx, \qquad [11.7]$$

where $y_0 \in V_0$ and $r \in R$. We propose to define a homomorphism φ_1 from V_1 to Q such that $\varphi_1(y_0) = \varphi_0(y_0)$ for all elements y_0 of V_0; we shall then have $(V_1, \varphi_1) > (V_0, \varphi_0)$, contradicting the maximal property of (V_0, φ_0).

To this end, let L be the set of elements r of the ring R such that $rx \in V_0$; we see at once that L is a left ideal of R. Consider the mapping α from L to Q defined by setting $\alpha(l) = \varphi_0(lx)$ for all elements l of L. This mapping is clearly a homomorphism and hence, by hypothesis, there is an element a of Q such that $\varphi_0(lx) = \alpha(l) = la$ for all elements l of L. We now define a mapping φ_1 from V_1 to Q by setting, for each element $y = y_0 + rx$ of V_1,

$$\varphi_1(y) = \varphi_0(y_0) + ra.$$

This is independent of the expression [11.7] for y; if we have $y = y_0 + rx = y_0' + r'x$, then $(r - r')x = y_0' - y_0 \in V_0$ and hence

$$\begin{aligned}\varphi_0(y_0') - \varphi_0(y_0) = \varphi_0(y_0' - y_0) = \varphi_0((r - r')x) &= (r - r')a \\ &= ra - r'a,\end{aligned}$$

so that

$$\varphi_0(y_0) + ra = \varphi_0(y_0') + r'a.$$

It is now a trivial matter to check that φ_1 is a homomorphism satisfying the required property. Thus our contradiction is established.

Example 2. An additive abelian group A is said to be *divisible* if for every element x of A and every non-zero integer m there exists an element x' of A such that $x = mx'$. We shall show that a **Z**-module A is injective if and only if it is divisible. This will follow from Theorem 11.7 if we can show that the property of divisibility is equivalent to that described in the theorem.

First, suppose that A is divisible and let L be an ideal of **Z**, α a homomorphism from L to A. Since all the ideals of **Z** are principal there is an integer m such that every element l of L has the form $l = km$ where $k \in \mathbf{Z}$. Let $\alpha(m) = b$; since A is divisible, there is an element a of A such that $b = ma$. (If $m = 0$ we have $b = 0$ and we take $a = 0$.) Now we have, for each element l of L, $\alpha(l) = \alpha(km) = k\alpha(m) = kb = kma = la$. Thus A satisfies the condition of Theorem 11.7 and hence is injective.

Conversely, suppose that A is injective. Let b be any element of A, m any non-zero integer. If L is the principal ideal of **Z** generated by m, we may define a homomorphism α from L to A by setting $\alpha(km) = kb$ for each element km of L. According to Theorem 11.7 there is an element a of A such that $\alpha(l) = la$ for every element l of L. We have in particular $b = \alpha(m) = ma$; so A is divisible.

We can now deal with the second of the questions raised at the end of §8.

THEOREM 11.8. *Let M be a left R-module and let*

$$0 \to V' \xrightarrow{\alpha} V \xrightarrow{\beta} V'' \to 0 \qquad [11.8]$$

be a short exact sequence of left R-modules. If M is injective then the sequence

$$0 \to \mathrm{Hom}(V'', M) \xrightarrow{\beta^*} \mathrm{Hom}(V, M) \xrightarrow{\alpha^*} \mathrm{Hom}(V', M) \to 0 \qquad [11.9]$$

is exact. Conversely, if [11.9] *is exact for every short exact sequence* [11.8] *then M is injective.*

Proof. It follows from Theorem 8.4 that the sequence [11.9] is exact if and only if α^* is an epimorphism.

(1) Suppose M is injective and let θ' be any element of $\mathrm{Hom}(V', M)$. Then we have a diagram

$$\begin{array}{ccccc} 0 & \longrightarrow & V' & \xrightarrow{\alpha} & V \\ & & \theta' \big\downarrow & & \\ & & M & & \end{array} \qquad [11.10]$$

in which the row is exact. According to the injective property of M there exists a homomorphism θ from V to M such that $\theta\alpha = \theta'$. But, by definition of α^*, we have $\alpha^*(\theta) = \theta\alpha$; so $\theta' = \alpha^*(\theta)$. Hence α^* is an epimorphism and [11.9] is exact.

(2) Conversely, suppose that for every short exact sequence [11.8] the induced homomorphism α^* from Hom (V, M) to Hom (V', M) is an epimorphism. Let [11.10] be a diagram in which the row is exact. Then, since α^* is an epimorphism and $\theta' \in$ Hom (V', M), there is a homomorphism θ in Hom (V, M) such that $\alpha^*(\theta) = \theta'$, i.e. such that $\theta\alpha = \theta'$. Hence M is injective.

We proved earlier that for every left R-module V there exists a projective module P and an epimorphism from P onto V. The dual of this statement (see §15) is then that for every left R-module V there exist an injective module Q and a monomorphism from V to Q; if we were to identify V with its image under this monomorphism we could say that *every left R-module is a submodule of an injective module.* We now set out to prove this dual result.

We begin by considering the factor group $\mathbf{T} = \mathbf{Q}/\mathbf{Z}$ of the additive group of rational numbers modulo the additive group of integers. First of all it is clear that $\mathbf{T}$ is a divisible abelian group: if A is any element of $\mathbf{T}$, let r be any rational number in the coset A; then if m is any non-zero integer we have $A = mA'$, where A' is the coset containing the rational number r/m. It follows from Example 2 that $\mathbf{T}$ is an injective $\mathbf{Z}$-module. Next we remark that if C is any non-zero cyclic group, generated by x say, there exists a non-zero abelian group homomorphism α from C to $\mathbf{T}$. Namely, if x is free (so that C is an infinite cyclic group) we may define $\alpha(x)$ to be any non-zero element of $\mathbf{T}$; while if x has order m we define $\alpha(x)$ to be the coset which contains $1/m$.

Now, let M be any left R-module; since M is of course an abelian group (i.e. a $\mathbf{Z}$-module) we may form the abelian group

$$\hat{M} = \mathrm{Hom}_{\mathbf{Z}}(M, \mathbf{T}).$$

For each element a of R and each element χ of $\hat{M}$ we define a mapping χa from M to $\mathbf{T}$ by setting $(\chi a)(x) = \chi(ax)$ for every element x of M. It is easy to verify first that this mapping χa is a homomorphism, so that χa belongs to $\hat{M}$, and then that under the right scalar multiplication so defined $\hat{M}$ is a right R-module. We call $\hat{M}$ the *character module* of M.

We may of course also form the character module $\hat{\hat{M}}$ of $\hat{M}$; this is again a *left* R-module, under the scalar multiplication defined by setting $(a\hat{\chi})(\chi) = \hat{\chi}(\chi a)$ for all elements a of R, $\hat{\chi}$ of $\hat{\hat{M}}$, χ of $\hat{M}$ respectively. There is an obvious device by which we may map the original module M into $\hat{\hat{M}}$. Namely, for each element x of M let $\iota(x)$ be the mapping from $\hat{M}$ to $\mathbf{T}$ defined by setting

$$[\iota(x)](\chi) = \chi(x)$$

for every element χ of M. Then $\iota(x)$ is actually a homomorphism from $\hat{M}$ to $\mathbf{T}$ and the mapping ι so defined is a homomorphism from M to $\hat{\hat{M}}$.

We claim that ι is in fact a monomorphism. So suppose x is a non-zero element of M such that $\iota(x)$ is the zero element of M. This means that for every element χ of M we have $\chi(x) = [\iota(x)](\chi) = 0$. According to our remarks above, if C is the cyclic subgroup ($\mathbf{Z}$-submodule) of M generated by x, there exists a $\mathbf{Z}$-homomorphism α from C to $\mathbf{T}$ such that $\alpha(x) \neq 0$. Since $\mathbf{T}$ is an injective $\mathbf{Z}$-module there exists an extension χ of α to M, i.e. there is an element χ of $\hat{M}$ such that $\chi(x) = \alpha(x) \neq 0$. This is a contradiction; hence ι is a monomorphism, as asserted. We call it the *canonical monomorphism* from M to $\hat{\hat{M}}$.

We are now in a position to prove the result announced above.

THEOREM 11.9. *Let V be any left R-module. Then there exist an injective module Q and a monomorphism from V to Q.*

Proof. According to the analogue of Theorem 11.3 for right R-modules there exists a free right R-module F and an R-epimorphism η from F onto the character module $\hat{V}$ of V. Then the sequence of right R-modules

$$F \xrightarrow{\eta} \hat{V} \to 0$$

is exact. It follows from Theorem 8.4 that the sequence of left R-modules

$$0 \to \mathrm{Hom}_{\mathbf{Z}}(\hat{V}, \mathbf{T}) \xrightarrow{\eta^*} \mathrm{Hom}_{\mathbf{Z}}(F, \mathbf{T})$$

is exact. (Here η^* is the homomorphism induced by η: we easily verify that it is in fact an R-homomorphism.) Thus we have a monomorphism η^* from $\hat{\hat{V}}$ to $\hat{F}$; composing this with the canonical monomorphism ι_V from V to $\hat{\hat{V}}$ we have a monomorphism $\eta^*\iota_V$ from V to $\hat{F}$.

Our theorem will be established if we can show that $\hat{F} = \mathrm{Hom}_Z(F, \mathbf{T})$ is an injective left R-module. So suppose we have a diagram

$$\begin{array}{ccccc} 0 & \longrightarrow & X' & \xrightarrow{\kappa} & X \\ & & {\scriptstyle\alpha}\downarrow & & \\ & & \hat{F} & & \end{array}$$

of left R-modules and R-homomorphisms in which the row is exact. We may construct from this the following diagram of right R-modules and R-homomorphisms

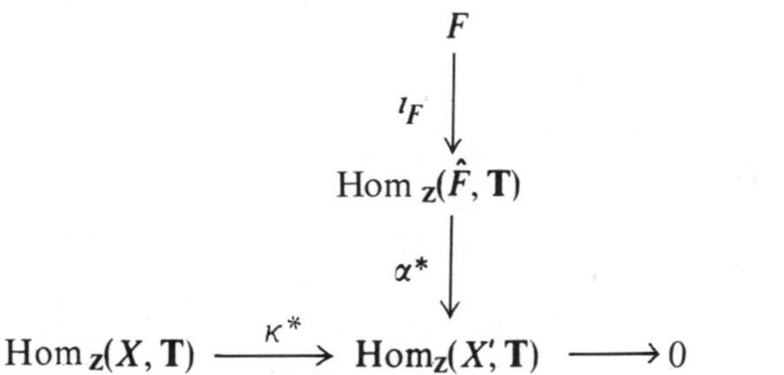

in which ι_F is the canonical monomorphism from F to $\hat{\hat{F}} = \mathrm{Hom}_Z(\hat{F}, \mathbf{T})$ and α^* and κ^* are the homomorphisms induced by α and κ respectively (easily shown to be R-homomorphisms).

Since $\mathbf{T}$ is injective it follows from Theorem 11.8 that the row in the last diagram is exact. Now F, being free, is a projective right R-module. Hence there exists a homomorphism θ from F to $\hat{X} = \mathrm{Hom}_Z(X, \mathbf{T})$ such that $\kappa^*\theta = \alpha^*\iota_F$.

We now contend that $(\theta^*\iota_X)\kappa = \alpha$, where θ^* is the homomorphism from $\hat{\hat{X}} = \mathrm{Hom}_Z(\hat{X}, \mathbf{T})$ to $\hat{F} = \mathrm{Hom}_Z(F, \mathbf{T})$ induced by θ and ι_X is the canonical monomorphism from X to $\hat{\hat{X}}$. So let x' be any element of X'; we must show that $\theta^*\iota_X\kappa(x') = \alpha(x')$. Since these are both elements of $\hat{F}$, let t be any element of F and consider

$$\begin{aligned}[\theta^*\iota_X\kappa(x')]\,(t) = [\iota_X(\kappa(x'))]\,(\theta(t)) &= [\theta(t)]\,(\kappa(x')) \\ &= [\kappa^*\theta(t)]\,(x') \\ &= [\alpha^*\iota_F(t)]\,(x') \\ &= [\iota_F(t)]\,(\alpha(x')) \\ &= [\alpha(x')]\,(t).\end{aligned}$$

Hence $\theta^* \iota_X \kappa = \alpha$, and $\hat{F}$ is injective as required.

This completes the proof.

As a consequence of this theorem we obtain another criterion for a module to be injective, analogous to one of those in Theorem 11.5 for projective modules.

Theorem 11.10. *A left R-module Q is injective if and only if every short exact sequence of the form*

$$0 \to Q \to V \to V'' \to 0$$

splits.

Proof. (1) Suppose Q is injective. Let

$$0 \to Q \xrightarrow{\kappa} V \to V'' \to 0 \qquad [11.11]$$

be a short exact sequence, so that κ is a monomorphism. Consider the identity mapping I_Q from Q onto Q. Since Q is injective there is an extension κ' of this mapping to V, i.e. a homomorphism κ' from V to Q such that $\kappa'\kappa = I_Q$. Then, according to Theorem 9.4, the sequence [11.11] splits.

(2) Conversely, suppose every such sequence splits.

According to Theorem 11.9 there exists an injective module Q_1 and a monomorphism κ from Q to Q_1. Then the sequence

$$0 \to Q \xrightarrow{\kappa} Q_1 \xrightarrow{\eta} Q_1/\kappa(Q) \to 0$$

(where η is the canonical epimorphism from Q_1 onto $Q_1/\kappa(Q)$) is exact and therefore splits. Hence there is a homomorphism κ' from Q_1 to Q such that $\kappa'\kappa = I_Q$.

Now suppose we have a diagram

$$\begin{array}{ccccc} 0 & \longrightarrow & V' & \xrightarrow{\lambda} & V \\ & & \downarrow{\scriptstyle\alpha} & & \\ & & Q & & \end{array}$$

in which the row is exact. Then $\kappa\alpha$ is a homomorphism from V' to the injective module Q_1 and so there exists a homomorphism β from V to Q_1 such that $\beta\lambda = \kappa\alpha$. It follows that $(\kappa'\beta)\,\lambda = (\kappa'\kappa)\,\alpha = I_Q\alpha = \alpha$. Thus Q is injective.

§12. Tensor Products

Let R be a ring with identity, V a right R-module and W a left R-module. We form the free $\mathbf{Z}$-module $F = F(V \times W)$ based on the Cartesian product of V and W; as in §11, we shall write the elements of F as 'formal sums' $\sum_{(x,y)\in V\times W} n_{(x,y)}(x, y)$, where the family of integers $(n_{(x,y)})$ is quasi-finite, i.e., only finitely many of the integers $n_{(x,y)}$ are non-zero. In particular, if $n_{(x_1,y_1)} = 1$ and $n_{(x,y)} = 0$ for $(x, y) \neq (x_1, y_1)$ we shall abbreviate the formal sum simply to (x_1, y_1).

Consider the subgroup K of F generated by the set of all elements of the forms

$$\begin{gathered}(x_1 + x_2, y) - (x_1, y) - (x_2, y),\\ (x, y_1 + y_2) - (x, y_1) - (x, y_2),\\ (xa, y) - (x, ay),\end{gathered} \qquad [12.1]$$

where $x, x_1, x_2 \in V$, $y, y_1, y_2 \in W$ and $a \in R$. The factor group F/K is called the *tensor product* of V and W and is denoted by $V \otimes_R W$, or simply by $V \otimes W$ if the ring R is clear from the context. We read $V \otimes_R W$ as 'V-tensor-R-W'. Comparison between this construction and others (such as the homomorphism groups discussed in § 8) would be facilitated by the introduction of the notation $\mathrm{Ten}_R(V, W)$ instead of $V \otimes_R W$; but the latter is too well established for there to be any hope of changing it.

Let η be the canonical epimorphism from F onto F/K. For each ordered pair of elements (x, y) in $V \times W$ we denote $\eta(x, y)$ by $x \otimes y$ (read 'x-tensor-y'). Then every element of the tensor product has the form

$$\eta\left(\sum n_{(x,y)}(x, y)\right) = \sum n_{(x,y)} x \otimes y$$

where $(n_{(x,y)})$ is a quasi-finite family of integers; clearly we may write this alternatively in the form

$$\sum_{i\in I} x_i \otimes y_i$$

where (x_i) and (y_i) are families of elements of V and W respectively indexed by the same finite index set I.

Since all elements of the forms [12.1] belong to the kernel of η

it follows that we have

$$\begin{aligned}(x_1 + x_2) \otimes y &= x_1 \otimes y + x_2 \otimes y,\\ x \otimes (y_1 + y_2) &= x \otimes y_1 + x \otimes y_2,\\ xa \otimes y &= x \otimes ay,\end{aligned} \qquad [12.2]$$

for all elements x, x_1, x_2 of V, y, y_1, y_2 of W and a of R. We deduce that for every element y of W we have $0 \otimes y = 0$; namely

$$0 \otimes y = (0 + 0) \otimes y = 0 \otimes y + 0 \otimes y$$

and hence $0 \otimes y = 0$ as asserted. Similarly we have $x \otimes 0 = 0$ for every element x of V.

Still let V and W be right and left R-modules respectively, and let A be any additive abelian group. A mapping φ from the Cartesian product $V \times W$ to A is said to be *balanced* if for all elements x, x_1, x_2 of V, y, y_1, y_2 of W and a of R we have

$$\begin{aligned}\varphi(x_1 + x_2, y) &= \varphi(x_1, y) + \varphi(x_2, y),\\ \varphi(x, y_1 + y_2) &= \varphi(x, y_1) + \varphi(x, y_2),\\ \varphi(xa, y) &= \varphi(x, ay).\end{aligned}$$

For example, the equations [12.2] show that the mapping v from $V \times W$ to $V \otimes W$ defined by setting $v(x, y) = x \otimes y$ for every ordered pair (x, y) in $V \times W$ is a balanced mapping; we call it the *natural mapping* from $V \times W$ to $V \otimes W$.

A pair (B, β) consisting of an additive abelian group B and a balanced mapping β from $V \times W$ to B is said to be *universal* for balanced mappings from $V \times W$ if for every balanced mapping φ from $V \times W$ to an additive abelian group A there exists a unique abelian group homomorphism α from B to A such that the diagram

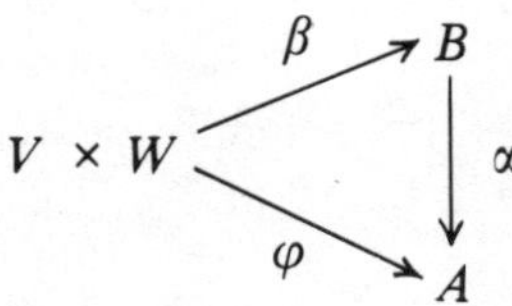

is commutative.

THEOREM 12.1. *Let V and W be right and left R-modules respectively. Then the pair $(V \otimes W, v)$ consisting of the tensor product of V and W*

and the natural mapping from $V \times W$ to $V \otimes W$ is universal for balanced mappings from $V \times W$.

Proof. Let φ be a balanced mapping from $V \times W$ to an additive abelian group A. According to Theorem 11.1 there is a unique **Z**-module (abelian group) homomorphism φ' from the free **Z**-module $F(V \times W)$ to A such that $\varphi'(x, y) = \varphi(x, y)$ for every pair (x, y) in $V \times W$. (As usual we are identifying each pair in $V \times W$ with its image under the natural mapping in $F(V \times W)$.) From the hypothesis that φ is balanced, it follows easily that all the elements of K are contained in the kernel of the homomorphism φ': for instance, we have

$$\begin{aligned}\varphi'((x_1 + x_2, y) - (x_1, y) - (x_2, y))\\ &= \varphi'(x_1 + x_2, y) - \varphi'(x_1, y) - \varphi'(x_2, y)\\ &= \varphi(x_1 + x_2, y) - \varphi(x_1, y) - \varphi(x_2, y) = 0.\end{aligned}$$

We now define a mapping α from $V \otimes W$ to A as follows. If C is any element of $V \otimes W$, i.e. a coset of $F(V \times W)$ modulo K, choose any element c from C and set $\alpha(C) = \varphi'(c)$; this clearly depends only on C, not on the choice of the element c, for if c' is another element in C we have $c - c' \in K \subseteq \operatorname{Ker} \varphi'$ and so $\varphi'(c) = \varphi'(c')$. This mapping α is easily seen to be a homomorphism, and we clearly have $\alpha\nu = \varphi$.

To see that α is the unique homomorphism satisfying this condition, let α' be another homomorphism from $V \otimes W$ to A such that $\alpha'\nu = \varphi$. Let C be any element of $V \otimes W$; then, as we have seen, C can be expressed in the form $C = \sum_{i \in I} x_i \otimes y_i = \sum_{i \in I} \nu(x_i, y_i)$. Then we have

$$\alpha(C) = \alpha\left(\sum \nu(x_i, y_i)\right) = \sum \alpha\nu(x_i, y_i) = \sum \varphi(x_i, y_i)$$

and similarly

$$\alpha'(C) = \sum \varphi(x_i, y_i).$$

Thus $\alpha = \alpha'$, as asserted.

It follows from a general result in §15 (Theorem 15.1*) that $(V \otimes_R W, \nu)$ is essentially the only pair which is universal for balanced mappings from $V \times W$. The precise meaning of 'essentially' is discussed in Example 18 of §15.

Suppose for a moment that R is a commutative ring, and let V and W be right and left R-modules respectively. Let r be any element of R and consider the mapping φ_r from $V \times W$ to $V \otimes W$ defined by setting $\varphi_r(x, y) = xr \otimes y = x \otimes ry$ for each ordered pair

(x, y) in $V \times W$; we contend that φ_r is balanced. So let x, x_1, x_2 be elements of V, y, y_1, y_2 elements of W and a an element of R. Then

$$\begin{aligned}\varphi_r(x_1 + x_2, y) = ((x_1 + x_2)r) \otimes y &= (x_1 r + x_2 r) \otimes y \\ &= (x_1 r) \otimes y + (x_2 r) \otimes y \\ &= \varphi_r(x_1, y) + \varphi_r(x_2, y);\end{aligned}$$

similarly

$$\varphi_r(x, y_1 + y_2) = \varphi_r(x, y_1) + \varphi_r(x, y_2);$$

further

$$\begin{aligned}\varphi_r(xa, y) = (xa)r \otimes y = x(ar) \otimes y \overset{*}{=} x(ra) \otimes y &= (xr)a \otimes y \\ &= xr \otimes ay = \varphi_r(x, ay).\end{aligned}$$

The commutativity of R is used at the step marked with an asterisk. It follows that there is a homomorphism α_r from $V \otimes W$ to $V \otimes W$ such that $\alpha_r(x \otimes y) = \varphi_r(x, y) = xr \otimes y$ for every ordered pair (x, y) in $V \times W$.

We now define a scalar multiplication on the left of $V \otimes W$ by elements of R by setting, for each element t of $V \otimes W$ and each element r of R,

$$rt = \alpha_r(t)$$

and claim that under this scalar multiplication the abelian group $V \otimes W$ is turned into a left R-module.

To establish this result, we remark that since α_r is a homomorphism we have

$$r(t_1 + t_2) = \alpha_r(t_1 + t_2) = \alpha_r(t_1) + \alpha_r(t_2) = rt_1 + rt_2$$

for every pair of elements t_1, t_2 of $V \otimes W$. It now remains to show that for every element t of $V \otimes W$ and every pair of elements r_1, r_2 of R we have

$$(r_1 + r_2)t = r_1 t + r_2 t, \quad (r_1 r_2)t = r_1(r_2 t), \quad et = t.$$

Since, as we have seen, every element t of $V \otimes W$ can be expressed as the sum of a finite family of elements of the form $x \otimes y$, it is sufficient to establish these relations for elements of this form—and this is trivial.

If R is once more a general ring (i.e. not necessarily commutative) then tensor products $V \otimes W$ are in general only abelian groups, not R-modules. But there are two special cases to consider even in the general situation, where tensor products can be given an R-module structure. These are the cases $V = R_r$ (R itself, considered as a right R-module) and $W = R_l$ (R as left R-module). We consider the first of these in some detail. Let b be any element of R and consider the mapping φ_b from $R \times W$ to $R_r \otimes W$ defined by setting $\varphi_b(a, y) = ba \otimes y$ for each ordered pair (a, y) in $R \times W$. Routine computations show that φ_b is a balanced mapping; hence there is a homomorphism α_b from $R_r \otimes W$ to itself such that $\alpha_b \nu = \varphi_b$, i.e. such that $\alpha_b(a \otimes y) = ba \otimes y$. We define a scalar multiplication on the left of $R_r \otimes W$ by elements of R by setting $bt = \alpha_b(t)$ for each element b of R and each element t of $R_r \otimes W$; under this scalar multiplication $R_r \otimes W$ becomes a left R-module. By a similar device we turn $V \otimes R_l$ into a right R-module.

THEOREM 12.2. *Let R be a ring with identity, V and W right and left R-modules respectively. Then the left R-module $R_r \otimes W$ is isomorphic to W and the right R-module $V \otimes R_l$ is isomorphic to V.*

Proof. Consider the mapping φ from $R \times W$ to W defined by setting $\varphi(a, y) = ay$ for each ordered pair (a, y) in $R \times W$. Since W is a left R-module this mapping φ is balanced, and hence there exists an abelian group homomorphism α from $R_r \otimes W$ to W such that $\alpha \nu = \varphi$. We claim that when $R_r \otimes W$ is considered as a left R-module in the manner described above this mapping α is actually an R-module homomorphism. To see this, first let a, b be elements of R, y an element of W; then we have

$$\alpha(b(a \otimes y)) = \alpha((ba) \otimes y) = \alpha(\nu(ba, y)) = \varphi(ba, y) = (ba)\,y$$

while

$$b\alpha(a \otimes y) = b\alpha(\nu(a, y)) = b\varphi(a, y) = b(ay) = (ba)\,y.$$

So $\alpha(b(a \otimes y)) = b\alpha(a \otimes y)$. Then, since every element t of $R_r \otimes W$ can be expressed as a sum of a finite family of elements of the form $a \otimes y$, it follows easily than $\alpha(bt) = b\alpha(t)$ for all elements b of R and all elements t of $R_r \otimes W$. Thus α is indeed an R-module homomorphism.

To show that α is an epimorphism, let y be any element of W. Then $y = ey = \varphi(e, y) = \alpha(\nu(e, y)) = \alpha(e \otimes y)$.

Next we show that α is a monomorphism. Let t be any element of Ker α. Then t can be expressed in the form $t = \sum_{i=1}^{n} a_i \otimes y_i$ $= \sum_{i=1}^{n} ea_i \otimes y_i = \sum_{i=1}^{n} e \otimes a_i y_i = e \otimes \sum_{i=1}^{n} a_i y_i = e \otimes y$ say, and we have $0 = \alpha(t) = ey = y$. Thus $t = e \otimes 0 = 0$, as required.

Hence α is an isomorphism from the left R-module $R_r \otimes W$ onto W. Similarly we set up an isomorphism from the right R-module $V \otimes R_l$ onto V.

Let V_1 and V_2 be right R-modules, W_1 and W_2 left R-modules, α and β homomorphisms from V_1 to V_2 and W_1 to W_2 respectively. Consider the mapping φ from $V_1 \times W_1$ to $V_2 \otimes W_2$ defined by setting $\varphi(x_1, y_1) = \alpha(x_1) \otimes \beta(y_1)$ for every ordered pair (x_1, y_1) in $V_1 \times W_1$. We easily verify that φ is a balanced mapping; hence there exists an abelian group homomorphism γ from $V_1 \otimes W_1$ to $V_2 \otimes W_2$ such that $\gamma\nu = \varphi$, i.e. such that $\gamma(x_1 \otimes y_1) = \alpha(x_1) \otimes \beta(y_1)$ for every ordered pair (x_1, y_1) in $V_1 \times W_1$. We denote this homomorphism γ by $\alpha \otimes \beta$ and call it the *tensor product* of α and β.

In the next two theorems we establish fundamental properties of the tensor product of homomorphisms.

THEOREM 12.3. *Let V_1, V_2, V_3 be right R-modules, W_1, W_2, W_3 be left R-modules and let $\alpha, \alpha', \beta, \beta'$ be homomorphisms from V_1 to V_2, V_2 to V_3, W_1 to W_2 and W_2 to W_3 respectively. Then $(\alpha' \otimes \beta')(\alpha \otimes \beta) = (\alpha'\alpha) \otimes (\beta'\beta)$.*

Proof. This follows at once from the consideration that for every ordered pair (x_1, y_1) in $V_1 \times W_1$ we have

$$\begin{aligned}(\alpha' \otimes \beta')[(\alpha \otimes \beta)(x_1 \otimes y_1)] &= (\alpha' \otimes \beta')(\alpha(x_1) \otimes \beta(y_1)) \\ &= \alpha'(\alpha(x_1)) \otimes \beta'(\beta(y_1)) \\ &= \alpha'\alpha(x_1) \otimes \beta'\beta(y_1).\end{aligned}$$

THEOREM 12.4. *Let V_1, V_2 be right R-modules, W_1, W_2 left R-modules; let $\alpha, \alpha_1, \alpha_2$ be homomorphisms from V_1 to V_2 and β, β_1, β_2 homomorphisms from W_1 to W_2. Then*

$$(\alpha_1 + \alpha_2) \otimes \beta = \alpha_1 \otimes \beta + \alpha_2 \otimes \beta$$

and

$$\alpha \otimes (\beta_1 + \beta_2) = \alpha \otimes \beta_1 + \alpha \otimes \beta_2.$$

Proof. For every ordered pair (x_1, y_1) in $V_1 \times W_1$ we have

$$\begin{aligned}[(\alpha_1 + \alpha_2) \otimes \beta]\,(x_1 \otimes y_1) &= (\alpha_1 + \alpha_2)(x_1) \otimes \beta(y_1) \\ &= (\alpha_1(x_1) + \alpha_2(x_1)) \otimes \beta(y_1) \\ &= \alpha_1(x_1) \otimes \beta(y_1) + \alpha_2(x_1) \otimes \beta(y_1) \\ &= (\alpha_1 \otimes \beta)(x_1 \otimes y_1) + (\alpha_2 \otimes \beta)(x_1 \otimes y_1) \\ &= [\alpha_1 \otimes \beta + \alpha_2 \otimes \beta]\,(x_1 \otimes y_1).\end{aligned}$$

Thus $(\alpha_1 + \alpha_2) \otimes \beta = (\alpha_1 \otimes \beta) + (\alpha_2 \otimes \beta)$ and the other result follows similarly.

We now examine the effect on a short exact sequence of right R-modules of forming the tensor products of its terms with a fixed left R-module.

THEOREM 12.5. *Let*

$$0 \to V' \xrightarrow{\alpha} V \xrightarrow{\beta} V'' \to 0$$

be an exact sequence of right R-modules; let M be a left R-module. Then the sequence

$$V' \otimes M \xrightarrow{\alpha \otimes I_M} V \otimes M \xrightarrow{\beta \otimes I_M} V'' \otimes M \to 0$$

is exact.

Proof. Let us abbreviate $\alpha \otimes I_M$ and $\beta \otimes I_M$ to α_* and β_* respectively. We must show that (a) $\operatorname{Im} \alpha_* \subseteq \operatorname{Ker} \beta_*$, (b) $\operatorname{Ker} \beta_* \subseteq \operatorname{Im} \alpha_*$, and (c) β_* is an epimorphism.

(a) Let t be any element of $\operatorname{Im} \alpha_*$; then there is an element t' of $V' \otimes M$ such that $t = \alpha_*(t')$. This element t' can be expressed in the form $t' = \sum_{i \in I} x'_i \otimes m_i$ where $(x'_i)_{i \in I}$, $(m_i)_{i \in I}$ are finite families of elements of V' and M respectively. For each index i in I we have

$$\beta_*(\alpha_*(x'_i \otimes m_i)) = \beta_*(\alpha(x'_i) \otimes m_i) = \beta(\alpha(x'_i)) \otimes m_i = 0 \otimes m_i = 0.$$

It follows that $\beta_*(t) = \beta_*(\alpha_*(t')) = 0$ and so $\operatorname{Im} \alpha_* \subseteq \operatorname{Ker} \beta_*$ as required.

(b) To show that $\operatorname{Ker} \beta_* \subseteq \operatorname{Im} \alpha_*$ we begin by defining an abelian group homomorphism γ from $\operatorname{Coker} \alpha_* = (V \otimes M)/\operatorname{Im} \alpha_*$ to $V'' \otimes M$ as follows. If C is any coset of $V \otimes M$ modulo $\operatorname{Im} \alpha_*$, choose an element t from C and define $\gamma(C)$ to be $\beta_*(t)$; then $\gamma(C)$ is clearly independent of the choice of t, for if t_1 is another element in C we have $t - t_1 \in \operatorname{Im} \alpha_* \subseteq \operatorname{Ker} \beta_*$ and hence $\beta_*(t_1) = \beta_*(t)$.

Next consider the mapping φ from $V'' \times M$ to $\operatorname{Coker} \alpha_*$ defined by setting $\varphi(t'', m)$ = coset of $t \otimes m$ modulo Im α_* where t is any element of $\beta^{-1}(t'')$. Then $\varphi(t'', m)$ so defined is independent of the choice of t; for if t_1 is another element of $\beta^{-1}(t'')$ we have $t - t_1 \in \operatorname{Ker} \beta = \operatorname{Im} \alpha$ and hence $(t - t_1) \otimes m \in \operatorname{Im} \alpha_*$. It is easy to verify that φ is a balanced mapping. So there exists a homomorphism γ' from $V'' \otimes M$ to Coker α_* such that $\gamma'(t'' \otimes m) = \varphi(t'', m)$ for every pair (t'', m) in $V'' \times M$.

If C is any element of Coker α_* we clearly have $\gamma'\gamma(C) = C$. So γ is a monomorphism. But it follows at once from the definition that $\operatorname{Ker} \gamma = \operatorname{Ker} \beta_*/\operatorname{Im} \alpha_*$. Hence $\operatorname{Ker} \beta_* = \operatorname{Im} \alpha_*$, as required.

(c) Let t'' be any element of $V'' \otimes M$; t'' can be expressed in the form $t'' = \sum_{i \in I} x_i'' \otimes m_i$ where (x_i'') and (m_i) are finite families of elements of V'' and M respectively. Since β is an epimorphism it follows that for each index i in I there is an element x_i of V such that $x_i'' = \beta(x_i)$. Then we have

$$t'' = \sum \beta(x_i) \otimes m_i = \sum \beta_*(x_i \otimes m_i) = \beta_*(\sum x_i \otimes m_i).$$

Thus β_* is an epimorphism, as required.

This completes the proof.

By an exactly similar argument we can prove the analogue of Theorem 12.5 for short exact sequences of left R-modules.

THEOREM 12.6. *Let*

$$0 \to V' \xrightarrow{\alpha} V \xrightarrow{\beta} V'' \to 0$$

be an exact sequence of left R-modules, M a right R-module. Then the sequence

$$M \otimes V' \xrightarrow{I_M \otimes \alpha} M \otimes V \xrightarrow{I_M \otimes \beta} M \otimes V'' \to 0$$

is exact.

To show that the results of Theorems 12.5 and 12.6 cannot be improved in general we give an example of a right R-module monomorphism α from V' to V say and a left R-module M such that the homomorphism $\alpha \otimes I_M$ from $V' \otimes M$ to $V \otimes M$ is not a monomorphism. Namely, we take $R = \mathbf{Z}$, $V' = \mathbf{Z}_r$, $V = \mathbf{Q}$ (the additive group of rational numbers considered as a $\mathbf{Z}$-module), α the inclusion monomorphism from V' to V, and M an abelian group of order 2; M consists of two elements 0 and m, say, and $2m = 0$. According to Theorem 12.2, the tensor product $V' \otimes M$ is isomorphic to M. But the tensor product $V \otimes M$ is a zero module; for if x is any rational number there exists a rational number x' such that $x = x'2$ (we write it like this since we are thinking of V as a right **Z**-module); hence we have $x \otimes m = x'2 \otimes m = x' \otimes 2m = x' \otimes 0 = 0$, and of course $x \otimes 0 = 0$. Thus $\alpha \otimes I_M$ is not a monomorphism.

Let V be a right R-module, $(V_k)_{k \in K}$ a family of right R-modules and $(\eta_k)_{k \in K}$ a family of homomorphisms from (V_k) to V yielding an injective representation of V as a direct sum of the family (V_k); for each index k in K let φ_k be the associated homomorphism from V to V_k such that for each index k in K we have $\varphi_k \eta_k = I_k$, for each pair of distinct indices k, k' in K $\varphi_k \eta_{k'} = \zeta_{kk'}$ and for each element x of V, $\sum_{k \in K} \eta_k \varphi_k(x) = x$. (This situation was discussed just after Theorem 9.2) Let W be a left R-module, $(W_l)_{l \in L}$ a family of left R-modules and $(\theta_l)_{l \in L}$ a family of homomorphisms from (W_l) to W yielding an injective representation of W as a direct sum of the family (W_l); let (ψ_l) be the associated family of mappings from W to (W_l).

THEOREM 12.7. *In the notation just described, the family* $(\eta_k \otimes \theta_l)_{(k,l) \in K \times L}$ *yields an injective representation of* $V \otimes W$ *as a direct sum of the family* $(V_k \otimes W_l)$.

Proof. Let S be the external direct sum of the family $(V_k \otimes W_l)$ with (ι_{kl}) the family of canonical injection monomorphisms. From the universal property of $(S, (\iota_{kl}))$ there exists a homomorphism β from S to $V \otimes W$ such that $\beta \iota_{kl} = \eta_k \otimes \theta_l$ for all pairs (k, l) in $K \times L$. (We are here applying Theorem 9.2 to abelian groups, i.e. **Z**-modules.)

Consider now the mapping φ from $V \times W$ to S defined by setting

$$\varphi(x, y) = \sum_{k,l} \iota_{kl}(\varphi_k(x) \otimes \psi_l(y)),$$

the sum being essentially finite since $\varphi_k(x)$ and $\psi_l(y)$ are non-zero for only finitely many indices k and l respectively. It is easy to check that φ is a balanced mapping; hence there is a unique homomorphism β' from $V \otimes W$ to S such that for each pair (x, y) in $V \times W$ we have $\beta'(x \otimes y) = \varphi(x, y)$. Then routine computations show that $\beta\beta'$ and $\beta'\beta$ are the identity mappings of $V \otimes W$ and S respectively.

Hence β is an isomorphism and so the family $(\eta_k \otimes \theta_l)$ gives an injective representation of $V \otimes W$ as a direct sum of the family $(V_k \otimes W_l)$.

COROLLARY 1. *Let V be a right R-module, W a free left R-module with basis Y. Then every element of $V \otimes W$ can be expressed uniquely in the form $\sum_{y \in Y} x_y \otimes y$ where $(x_y)_{y \in Y}$ is a family of elements of V only finitely many of which are non-zero.*

Proof. Since Y is a basis for W we see easily that the family of inclusion monomorphisms from the submodules Ry for all elements y of the basis Y gives an injective representation of W as a direct sum of the family $(Ry)_{y \in Y}$. The identity automorphism I_V may also be regarded as giving an injective representation of V as direct sum of the family whose only member is V. It then follows from the theorem that $V \otimes W$ is the internal direct sum of the family $(V \otimes Ry)_{y \in Y}$. Hence every element t of $V \otimes W$ can be expressed uniquely in the form $t = \sum_{y \in Y} t_y$ where for each element y of Y we have $t_y \in V \otimes Ry$ and only finitely many of these elements are non-zero. Now t_y can be expressed as a finite sum $t_y = \sum_{i \in I} v_i \otimes a_i y$; so it follows that we have

$$t_y = \sum (v_i a_i \otimes y) = (\sum v_i a_i) \otimes y.$$

Thus, setting $x_y = \sum v_i a_i$, we have

$$t = \sum_{y \in Y} x_y \otimes y$$

as required.

COROLLARY 2. *Let R be a commutative ring; let V be a free right R-module with basis X, W a free left R-module with basis Y. Then the tensor product $V \otimes W$, considered as a left R-module, is isomorphic to the free left R-module with basis $X \times Y$.*

Proof. Consider the families $(V_x)_{x \in X}$ and $(W_y)_{y \in Y}$ of right and left R-modules respectively defined by setting $V_x = R_r$ for every element x of X and $W_y = R_l$ for every element y of Y. Then one checks easily that the family of mappings $(\eta_x)_{x \in X}$ from (V_x) to V defined by setting $\eta_x(r) = xr$ for each element x of X and r of $V_x = R$ provides an injective representation of V as direct sum of the family (V_x). Similarly the family $(\theta_y)_{y \in Y}$ of mappings from (W_y) to W defined by setting $\theta_y(r) = ry$ for each element y of Y and r of R gives an injective representation of W as a direct sum of the family (W_y).

It follows from the theorem that the family $(\eta_x \otimes \theta_y)$ gives an injective representation of $V \otimes W$ as a direct sum of the family $(V_x \otimes W_y)$. For each ordered pair of indices (x, y) in $X \otimes Y$ the tensor product $V_x \otimes W_y = R_r \otimes R_l$, which (according to Theorem 12.2) is isomorphic, as left R-module, to R_l. Hence $V \otimes W$ is isomorphic to the free left R-module with basis $X \times Y$.

We return now to the situation with which we were concerned in Theorems 12.5 and 12.6 and we say that a right R-module M is *flat* if for every short exact sequence

$$0 \to V' \xrightarrow{\alpha} V \xrightarrow{\beta} V'' \to 0 \qquad [12.3]$$

of left R-modules the sequence of abelian groups

$$0 \to M \otimes V' \xrightarrow{I_M \otimes \alpha} M \otimes V \xrightarrow{I_M \otimes \beta} M \otimes V'' \to 0 \qquad [12.4]$$

is also exact. Flat left modules are defined in a similar way.

Example 1. The right R-module R_r is flat. To prove this assertion let [12.3] be any short exact sequence of left R-modules. To show that the sequence [12.4] is exact, all that remains to be done (by virtue of Theorem 12.6) is to prove that $I_M \otimes \alpha$ is a monomorphism. To do this we refer to the proof of Theorem 12.2 and deduce that there are isomorphisms γ' and γ from $R_r \otimes V'$ onto V' and $R_r \otimes V$ onto V respectively such that $\gamma'(a \otimes x') = ax'$ and $\gamma(a \otimes x) = ax$ for all elements a of R, x' of V', x of V. It is easy to show that the square

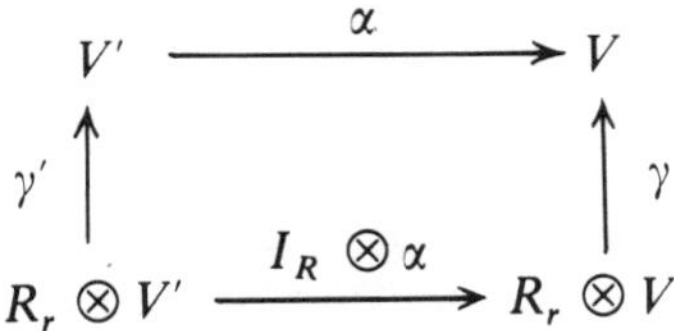

is commutative. Thus $I_R \otimes \alpha = \gamma^{-1}\alpha\gamma'$ and hence is a monomorphism, since γ^{-1}, α and γ' are all monomorphisms.

The next theorem will enable us to give further examples of flat modules.

Theorem 12.8. *Let M be a right R-module, $(M_i)_{i \in I}$ a family of right R-modules and $(\mu_i)_{i \in I}$ a family of mappings from (M_i) to M which yields an injective representation of M as a direct sum of the family*

(M_i). Then M is a flat module if and only if all the modules of the family (M_i) are flat.

Proof. Let [12.3] be any short exact sequence of left R-modules.

According to Theorem 12.7 the families $(\mu_i \otimes I_{V'})$ and $(\mu_i \otimes I_V)$ yield injective representations of $M \otimes V'$ and $M \otimes V$ as direct sums of the families $(M_i \otimes V')$ and $(M_i \otimes V)$ respectively. Let (π_i') and (π_i) be the families of epimorphisms from $M \otimes V'$ and $M \otimes V$ onto the families $(M_i \otimes V')$ and $(M_i \otimes V)$ respectively associated with the monomorphisms $(\mu_i \otimes I_{V'})$ and $(\mu_i \otimes I_V)$. Then for each index i the diagrams

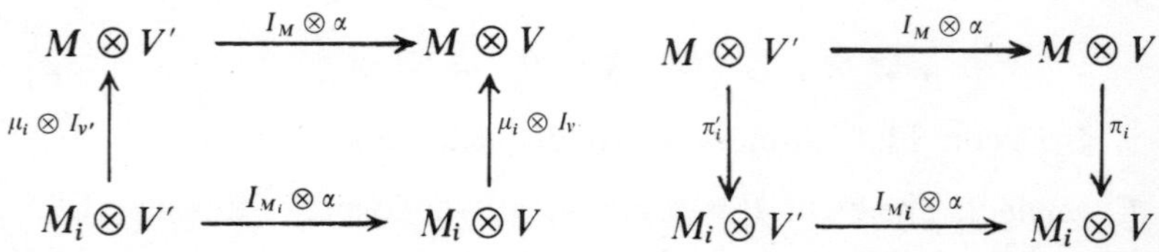

$$
\begin{array}{ccc}
M \otimes V' & \xrightarrow{I_M \otimes \alpha} & M \otimes V \\
{\scriptstyle \mu_i \otimes I_{V'}} \uparrow & & \uparrow {\scriptstyle \mu_i \otimes I_V} \\
M_i \otimes V' & \xrightarrow{I_{M_i} \otimes \alpha} & M_i \otimes V
\end{array}
\qquad
\begin{array}{ccc}
M \otimes V' & \xrightarrow{I_M \otimes \alpha} & M \otimes V \\
{\scriptstyle \pi_i'} \downarrow & & \downarrow {\scriptstyle \pi_i} \\
M_i \otimes V' & \xrightarrow{I_{M_i} \otimes \alpha} & M_i \otimes V
\end{array}
$$

are commutative, as the reader will easily verify.

Suppose first that M is flat; then $I_M \otimes \alpha$ is a monomorphism. To establish that the modules M_i are flat all that remains to show (by virtue of Theorem 12.6) is that the homomorphisms $I_{M_i} \otimes \alpha$ are injective.

So let t_i' be an element of Ker $(I_{M_i} \otimes \alpha)$. Then we have

$$[(I_M \otimes \alpha)(\mu_i \otimes I_{V'})]\ (t_i') = [(\mu_i \otimes I_V)(I_{M_i} \otimes \alpha)]\ (t_i') = 0.$$

Since $I_M \otimes \alpha$ and $\mu_i \otimes I_{V'}$ are monomorphisms it follows that $t_i' = 0$; hence $I_{M_i} \otimes \alpha$ is injective, as required.

Conversely, suppose that each of the modules M_i is flat; then the mappings $I_{M_i} \otimes \alpha$ are all monomorphisms. As usual, in order to show that M is flat all that remains is to show that $I_M \otimes \alpha$ is a monomorphism.

So let t be any element of Ker $(I_M \otimes \alpha)$. Then for each index i we have

$$[(I_{M_i} \otimes \alpha)\ \pi_i']\ (t) = [\pi_i(I_M \otimes \alpha)]\ (t) = 0.$$

Since $I_{M_i} \otimes \alpha$ is a monomorphism, it follows that $\pi_i'(t) = 0$. Hence we have

$$t = \sum_{i \in I} [(\mu_i \otimes I_V) \pi_i'] (t) = 0$$

and so $I_M \otimes \alpha$ is a monomorphism as required.

This completes the proof.

COROLLARY. *Every free right R-module is flat and every projective right R-module is flat.*

Proof. (1) Let F be a free right R-module with basis X. Then there is an injective representation of F as a direct sum of a family $(V_x)_{x \in X}$ of right R-modules such that $V_x = R_r$ for every element x of X. Example 1 above shows that each of these modules V_x is flat; so it follows from the theorem that F is flat.

(2) Let P be a projective right R-module. According to Theorem 11.5, P is a direct summand of a free (and hence flat) module. It now follows from Theorem 12.8 that P itself is flat.

We conclude this section by establishing a connexion between the groups of homomorphisms which we studied in §8 and the tensor products we have just been discussing.

Let R and S be two rings with identity. Let A be a right R-module, C a right S-module and B a left R-module which is also a right S-module; suppose that the left R-module and right S-module structures on B are linked by the requirement that for all elements r, s, b of R, S, B respectively we have $(rb)s = r(bs)$. In this situation we shall show that the abelian groups $A \otimes_R B$ and $\mathrm{Hom}_S(B, C)$ can be given the structures of a right S-module and a right R-module respectively.

Consider first the case of $A \otimes_R B$. For each element s of S consider the mapping φ_s from $A \times B$ to $A \otimes_R B$ defined by setting $\varphi_s(a, b) = a \otimes bs$ for every ordered pair (a, b) in $A \times B$. For all elements a, a_1, a_2 of A, b, b_1, b_2 of B and r of R we have

$$\begin{aligned}
\varphi_s(a_1 + a_2, b) &= (a_1 + a_2) \otimes bs = a_1 \otimes bs + a_2 \otimes bs \\
&= \varphi_s(a_1, b) + \varphi_s(a_2, b), \\
\varphi_s(a, b_1 + b_2) &= a \otimes (b_1 + b_2)s = a \otimes (b_1 s + b_2 s) \\
&= a \otimes b_1 s + a \otimes b_2 s \\
&= \varphi_s(a, b_1) + \varphi_s(a, b_2), \\
\varphi_s(ar, b) &= ar \otimes bs = a \otimes r(bs) = a \otimes (rb)s = \varphi_s(a, rb).
\end{aligned}$$

That is to say φ_s is a balanced mapping; so there is a homomorphism α_s from $A \otimes_R B$ to $A \otimes_R B$ such that $\alpha_s(a \otimes b) = a \otimes bs$ for all ordered pairs (a, b) of $A \times B$. We now define a right scalar multiplication of $A \otimes_R B$ by elements of S by setting $xs = \alpha_s(x)$ for every element x of $A \otimes_R B$ and every element s of S. It is now a routine matter to check that under this scalar multiplication $A \otimes_R B$ becomes a right S-module.

Next we consider $\mathrm{Hom}_S(B, C)$. For each element α of $\mathrm{Hom}_S(B, C)$ and each element r of R we define the mapping αr from B to C by setting $(\alpha r)(b) = \alpha(rb)$ for every element b of B. This mapping is in fact an S-homomorphism from B to C; for if b, b_1, b_2 are any elements of B, s any element of S we have

$$\begin{aligned}(\alpha r)(b_1 + b_2) &= \alpha(r(b_1 + b_2)) = \alpha(rb_1 + rb_2)\\ &= \alpha(rb_1) + \alpha(rb_2) = (\alpha r)(b_1) + (\alpha r)(b_2),\end{aligned}$$

$$(\alpha r)(bs) = \alpha(r(bs)) = \alpha((rb)s) = (\alpha(rb))s = ((\alpha r)(b))\, s.$$

Thus we have defined a right scalar multiplication of $\mathrm{Hom}_S(B, C)$ by elements of R. Once again it is a routine matter to check that under this scalar multiplication $\mathrm{Hom}_S(B, C)$ becomes a right R-module.

From this discussion it follows that we may form the homomorphism groups $\mathrm{Hom}_S(A \otimes_R B, C)$ and $\mathrm{Hom}_R(A, \mathrm{Hom}_S(B, C))$. We proceed to define mappings from these groups to one another.

First, let α be any element of $\mathrm{Hom}_S(A \otimes_R B, C)$; we define a mapping α' from A to $\mathrm{Hom}_S(B, C)$ as follows. For each element a of A and b of B we set

$$[\alpha'(a)]\,(b) = \alpha(a \otimes b).$$

We have to verify that the mapping $\alpha'(a)$ so defined is in fact an S-homomorphism from B to C, but this is easily done. It is also easy to check that α' is an R-homomorphism from A to $\mathrm{Hom}_S(B, C)$ and then finally that the mapping Φ from $\mathrm{Hom}_S(A \otimes_R B, C)$ to $\mathrm{Hom}_R(A, \mathrm{Hom}_S(B, C))$ defined by setting $\Phi(\alpha) = \alpha'$ for every element α of the first group is an abelian group homomorphism.

Next let β be any element of $\mathrm{Hom}_R(A, \mathrm{Hom}_S(B, C))$. Consider the mapping φ from $A \times B$ to C defined by setting

$$\varphi(a, b) = [\beta(a)]\,(b)$$

for every ordered pair (a, b) in $A \times B$. We check that φ is R-balanced (in the obvious sense of that term); hence there exists an abelian group homomorphism β' from $A \otimes_R B$ to C such that $\beta'(a \otimes b) = [\beta(a)](b)$ for all ordered pairs (a, b) in $A \times B$. Easy calculations show that β' is actually an S-module homomorphism from $A \otimes_R B$ to C and that the mapping Ψ from $\mathrm{Hom}_R(A, \mathrm{Hom}_S(B, C))$ to $\mathrm{Hom}_S(A \otimes_R B, C)$ defined by setting $\Psi(\beta) = \beta'$ for every element β of the first group is an abelian group homomorphism.

THEOREM 12.9. *In the situation just described, the mapping Φ is an isomorphism from* $\mathrm{Hom}_S(A \otimes_R B, C)$ *onto* $\mathrm{Hom}_R(A, \mathrm{Hom}_S(B, C))$ *and the mapping Ψ is its inverse.*

Proof. Let α be any element of $\mathrm{Hom}_S(A \otimes_R B, C)$; we shall show that $\Psi\Phi(\alpha) = \alpha$. So let a and b be elements of A and B respectively; then

$$[\Psi\Phi(\alpha)]\,(a \otimes b) = [(\Phi(\alpha))\,(a)]\,(b) = \alpha(a \otimes b).$$

Since every element of $A \otimes_R B$ can be expressed as a sum of elements of the form $a \otimes b$, it follows that $[\Psi\Phi(\alpha)]\,(t) = \alpha(t)$ for every element t of $A \otimes_R B$. Thus $\Psi\Phi(\alpha) = \alpha$.

Next let β be any element of $\mathrm{Hom}_R(A, \mathrm{Hom}_S(B, C))$; we shall prove that $\Phi\Psi(\beta) = \beta$. Thus let a be any element of A, b any element of B; then

$$\{[\Phi\Psi(\beta)]\,(a)\}\,(b) = [\Psi(\beta)]\,(a \otimes b) = [\beta(a)]\,(b).$$

Hence $\Phi\Psi(\beta) = \beta$.

It follows that Φ and Ψ are mutually inverse bijections, and since we have already checked that they are homomorphisms it follows that they are isomorphisms, as asserted.

§13. Composition Series

Let V be a left R-module, l a natural number. A *normal series of length l* for V is a sequence $(V_i)_{i \in [0, l]}$ of submodules of V such that $V_0 = V$, $V_l = 0$ and $V_{i-1} \supseteq V_i$ for $i = 1, \ldots, l$. We usually write such a sequence in the form

$$V = V_0 \supseteq V_1 \supseteq \ldots \supseteq V_l = 0. \qquad [13.1]$$

Clearly V has a normal series of length 0 if and only if it is a zero module. If all the inclusions in [13.1] are proper, i.e. if $V_{i-1} \supset V_i$ for $i = 1, \ldots, l$, we say that [13.1] is a *non-repeating* normal series.

If $l > 0$ then the factor modules V_{i-1}/V_i $(i = 1, \ldots, l)$ are called the *factors* of the normal series [13.1]; the normal series of length 0 for a zero module has no factors.

Suppose now we have a second normal series for V,

$$V = V'_0 \supseteq V'_1 \supseteq \ldots \supseteq V'_m = 0. \qquad [13.2]$$

We say that [13.2] is a *refinement* of [13.1] if there is an injection κ from $[0, l]$ to $[0, m]$ such that for $i = 0, \ldots, l$ we have $V'_{\kappa(i)} = V_i$; so [13.2] is a refinement of [13.1] if and only if all the modules V_i occur among the modules V'_j. Again, we say that the normal series [13.2] is *equivalent* or *isomorphic* to [13.1] if $l = m$ and there is a bijection π from $[1, l]$ onto itself such that for $i = 1, \ldots, l$ the factor V_{i-1}/V_i is isomorphic to the factor $V'_{\pi(i)-1}/V'_{\pi(i)}$. It is easy to verify that the relation between normal series which we have just defined is indeed an equivalence relation on the set of normal series for V.

If V is a non-zero module the normal series [13.1] is called a *composition series* for V if all the factors V_{i-1}/V_i are simple modules. Since by definition a simple module is non-zero, this implies that all the inclusions in [13.1] are proper, so a composition series is non-repeating; and since a simple module has no submodules except itself and the zero submodule, we deduce (from Theorem 7.5) that for $i = 1, \ldots, l$ it is impossible to find a submodule V' of V such that $V_{i-1} \supset V' \supset V_i$. If V is a zero module we shall say that the normal series $V = V_0 = 0$ of length 0 is a composition series for V, and of course it is the only one.

In the next theorem we take a first step towards the proof of the Jordan-Hölder Theorem.

THEOREM 13.1. *Let V be a left R-module, l a natural number. Then V has a composition series of length l if and only if l is the least upper bound of the set of lengths of all non-repeating normal series for V.*

Proof. (1) Let l be this least upper bound. Then there is a non-repeating normal series

$$V = V_0 \supset V_1 \supset \ldots \supset V_l = 0 \qquad [13.3]$$

of length l for V. Suppose the factor V_{i-1}/V_i of [13.3] is not simple. Since this module is non-zero, it must have a proper non-zero submodule and hence, according to Theorem 7.5, there is a submodule V' of V such that $V_{i-1} \supset V' \supset V_i$. It follows that we have a non-repeating normal series

$$V = V_0 \supset V_1 \supset \ldots \supset V_{i-1} \supset V' \supset V_i \supset \ldots \supset V_l = 0$$

of length $l + 1$ for V. This contradicts the definition of l, and so [13.3] is a composition series for V.

(2) Conversely, suppose that V has a composition series [13.3] of length l.

If $l = 0$ then V is a zero module. In this case the only non-repeating normal series for V is V itself; so certainly $l = 0$ is the least upper bound of the set of lengths of all such series.

If $l > 0$ we make the inductive hypothesis that for every module W with a composition series of length r less than l the least upper bound of the set of lengths of all non-repeating normal series for W is r. Under this hypothesis we shall show that if

$$V = V'_0 \supset V'_1 \supset \ldots \supset V'_m = 0 \qquad [13.4]$$

is any non-repeating normal series for V then $m \leqslant l$.

First we remark that [13.3] gives rise to a composition series

$$V_1 \supset V_2 \supset \ldots \supset V_l = 0$$

of length $l - 1$ for V_1, and that [13.4] gives a non-repeating normal series

$$V'_1 \supset V'_2 \supset \ldots \supset V'_m = 0 \qquad [13.5]$$

of length $m - 1$ for V'_1.

If $V'_1 = V_1$ then, according to the inductive hypothesis, we have $m - 1 \leqslant l - 1$, and so $m \leqslant l$.

If $V'_1 \subset V_1$ then we have a non-repeating normal series

$$V_1 \supset V'_1 \supset V'_2 \supset \ldots \supset V'_m = 0$$

of length m for V_1. By the inductive hypothesis, $m \leqslant l - 1$ and so of course we again have $m \leqslant l$.

Since V/V_1 is simple we cannot have $V \supset V'_1 \supset V_1$; so the only possibility which remains is that neither of the two modules V_1, V'_1 is included in the other. In this case $V_1 + V'_1 \supset V_1$ and hence, since V/V_1 is simple, we have $V_1 + V'_1 = V$; furthermore, $V_1 \cap V'_1 \subset V_1$ and $V_1 \cap V'_1 \subset V'_1$. If now

$$V_1 \cap V'_1 = W_0 \supset W_1 \supset \ldots \supset W_s = 0$$

is a non-repeating normal series of length s for $V_1 \cap V'_1$, then

$$V_1 \supset W_0 \supset W_1 \supset \ldots \supset W_s = 0$$

is a non-repeating normal series of length $s + 1$ for V_1. According tc the inductive hypothesis we have $s + 1 \leqslant l - 1$, whence $s \leqslant l - 2$. Hence the set of lengths of non-repeating normal series for $V_1 \cap V'_1$ is bounded above by $l - 2$; this set has therefore a least upper bound t and $t \leqslant l - 2$. It follows from the first part of the theorem that $V_1 \cap V'_1$ has a composition series

$$V_1 \cap V'_1 = W'_0 \supset W'_1 \supset \ldots \supset W'_t = 0.$$

We claim that

$$V'_1 \supset W'_0 \supset W'_1 \supset \ldots . \supset W'_t = 0 \qquad [13.6]$$

is a composition series for V'_1. To prove this we have only to show that V'_1/W'_0 is simple. But, according to Theorem 7.4, $V'_1/W'_0 = V'_1/V_1 \cap V'_1$ is isomorphic to $(V_1 + V'_1)/V_1 = V/V_1$, which is a simple module. Since [13.6] has length $t + 1 \leqslant l - 1$ we may apply the inductive hypothesis to the composition series [13.6] and the non-repeating normal series [13.5]; we obtain $m - 1 \leqslant t + 1 \leqslant l - 1$, whence $m \leqslant l$.

This completes the proof.

COROLLARY 1. *If the left R-module V has one composition series of length l, then a non-repeating normal series for V is a composition series for V if and only if it has length l.*

COROLLARY 2. *If the left R-module V has a composition series, then so does every submodule of V.*

Proof. Suppose V has a composition series of length l. Let W be a submodule of V. If $W = V$, then of course W has a composition series; if $W \neq V$, let

$$W = W_0 \supset W_1 \supset \ldots \supset W_m = 0$$

be a non-repeating normal series of length m for W. Then

$$V \supset W_0 \supset W_1 \supset \ldots \supset W_m = 0$$

is a non-repeating normal series of length $m + 1$ for V. So $m + 1 \leqslant l$ and hence $m \leqslant l - 1$. It follows that the set of lengths of non-repeating normal series for W has a least upper bound m_0, where $m_0 \leqslant l - 1$. Thus W has a composition series.

COROLLARY 3. *If the left R-module V has a composition series of length l then every non-repeating normal series for V has a refinement which is a composition series for V.*

Proof. Let

$$V = V_0 \supset V_1 \supset \ldots \supset V_m = 0 \qquad [13.7]$$

be a non-repeating normal series for V of length m; by the theorem, $m \leqslant l$.

If $m = l$ then, by Corollary 1, [13.7] is a composition series for V and hence is itself the required refinement.

If $m < l$, [13.7] is not a composition series; hence at least one of its factors is not simple, say V_{i-1}/V_i. As in the first part of the proof of the theorem we obtain a submodule V' of V such that $V_{i-1} \supset V' \supset V_i$; thus we have a non-repeating normal series

$$V = V_0 \supset V_1 \supset \ldots \supset V_{i-1} \supset V' \supset V_i \supset \ldots \supset V_m = 0$$

of length $m + 1$ which is a refinement of [13.7]. Clearly after $l - m$ applications of this procedure we reach a composition series for V which is a refinement of [13.7].

We can now establish the central result on composition series.

THEOREM 13.2. (*Jordan–Hölder Theorem*). *Let V be a left R-module which has at least one composition series. Then all composition series for V are equivalent.*

Proof. According to Corollary 1 above, all composition series for V have the same length, l say. Let

$$V = V_0 \supset V_1 \supset \ldots \supset V_l = 0 \qquad [13.8]$$

and

$$V = V'_0 \supset V'_1 \supset \ldots \supset V'_l = 0 \qquad [13.9]$$

be two composition series for V.

If $l = 0$, these series are certainly equivalent.

We now proceed by induction, making the inductive hypothesis that for every module with a composition series of length less than l all composition series are equivalent. There are two cases to consider.

(1) If $V_1 = V'_1$ the series [13.8] and [13.9] give rise to composition series

$$V_1 \supset V_2 \supset \ldots \supset V_l = 0$$

and

$$V_1 = V'_1 \supset V'_2 \supset \ldots \supset V'_l = 0$$

for V_1. According to the inductive hypothesis, these series are equivalent; so there is a bijection π' from $[2, l]$ onto itself such that for $i = 2, \ldots, l$ the factor V_{i-1}/V_i is isomorphic to the factor $V'_{\pi'(i)-1}/V'_{\pi'(i)}$. If we now define the bijection π from $[1, l]$ onto itself by setting $\pi(1) = 1$ and $\pi(i) = \pi'(i)$ for $i = 2, \ldots, l$, we see that for $i = 1, \ldots, l$ the factors V_{i-1}/V_i and $V'_{\pi(i)-1}/V'_{\pi(i)}$ are isomorphic.

So in this case [13.8] and [13.9] are equivalent.

(2) If $V_1 \neq V'_1$ then an argument used in the course of proving Theorem 13.1 shows that $V_1 + V'_1 = V$ and that $V_1 \cap V'_1$ is properly included in both V_1 and V'_1. Further, $V_1/V_1 \cap V'_1$ is isomorphic to the simple module V/V'_1 and $V'_1/V_1 \cap V'_1$ is isomorphic to the simple module V/V_1.

According to Corollary 2, $V_1 \cap V'_1$ has a composition series, say

$$V_1 \cap V'_1 = W_0 \supset W_1 \supset \ldots \supset W_m = 0.$$

Then it follows from the preceding paragraph that the normal series

$$V = V_0 \supset V_1 \supset W_0 \supset W_1 \supset \ldots \supset W_m = 0 \qquad [13.10]$$

and

$$V = V'_0 \supset V'_1 \supset W_0 \supset W_1 \supset \ldots \supset W_m = 0 \qquad [13.11]$$

are equivalent composition series for V. (Incidentally, this shows that $m = l - 2$.) By (1) above, [13.10] and [13.8] are equivalent and [13.11] and [13.9] are equivalent.

Hence in this case also [13.8] and [13.9] are equivalent.

The results we have obtained in this section have all depended on the existence of composition series for the modules under consideration. In §14 we shall give a criterion for a module to have a composition series, but meanwhile we simply remark that while every finite abelian group, considered as **Z**-module, has a composition series, the **Z**-module $\mathbf{Z}_l$ does not.

According to Corollary 1 of Theorem 13.1, if a left R-module V does have a composition series then all its composition series have the same length. We denote this common length by $l(V)$ and call it the *length* of the module V. Clearly, $l(V) = 0$ if and only if $V = 0$;

and $l(V) = 1$ if and only if V is simple. It is convenient to write $l(V) = +\infty$ if the module V has no composition series. If V is a vector space over a division ring D the length of V is usually called the *dimension* of V and denoted by $\dim V$ or $\dim_D V$.

Example. Let V be a vector space over a division ring D with a finite basis $\{x_1, \ldots, x_n\}$ consisting of n elements. We shall show that $\dim V = n$. Set $V_n = 0$ and for $i = 0, 1, \ldots, n-1$ let V_i be the submodule of V generated by $\{x_1, \ldots, x_{n-i}\}$. Thus we have a normal series

$$V = V_0 \supseteq V_1 \supseteq \ldots \supseteq V_n = 0. \qquad [13.12]$$

We claim that this is in fact a composition series for V.

To this end we show first that all the inclusion relations are proper; we show in fact that for $i = 1, \ldots, n$ the element x_{n-i+1} does not belong to V_i. This is clear when $i = n$; for $V_n = 0$ while x_n is non-zero, being an element of a linearly independent set. So suppose $i < n$ and assume that $x_{n-i+1} \in V_i$. Then there exist elements $a_1, \ldots, a_{n-i}$ of D such that $x_{n-i+1} = a_1x_1 + \ldots + a_{n-i}x_{n-i}$; hence we have

$$a_1x_1 + \ldots + a_{n-i}x_{n-i} + (-e)x_{n-i+1} = 0,$$

which contradicts the linear independence of $\{x_1, \ldots, x_n\}$. Hence $x_{n-i+1} \notin V_i$ and so $V_{i-1} \supset V_i (i = 1, \ldots, n)$.

Now, let V' be any submodule of V such that $V_{i-1} \supseteq V' \supset V_i$ $(i = 1, \ldots, n)$; we shall show that $V' = V_{i-1}$, so completing the proof that V_{i-1}/V_i is simple. Let x be any element of V' which does not belong to V_i. Since $x \in V_{i-1}$, there are elements $a_1, \ldots, a_{n-i+1}$ of D such that

$$x = a_1x_1 + \ldots + a_{n-i}x_{n-i} + a_{n-i+1}x_{n-i+1}.$$

The last coefficient a_{n-i+1} must be non-zero, since otherwise x would lie in V_i. It follows that

$$x_{n-i+1} = a_{n-i+1}^{-1}(x - a_1x_1 - \ldots - a_{n-i}x_{n-i}) \in V'$$

and hence that $V_{i-1} \subseteq V'$. Thus $V' = V_{i-1}$, as asserted, and [13.12] is a composition series for V; hence $\dim V = n$.

We notice that we have shown that for a finite-dimensional vector space all bases have the same cardinal (equal to the dimension).

We have now a result connecting the lengths of a module, a submodule and the corresponding factor module.

THEOREM 13.3. *Let W be a submodule of a left R-module V. Then $l(V) = l(W) + l(V/W)$.*

Note. The equation $l(V) = l(W) + l(V/W)$ is intended to include the statement that if $l(V) = +\infty$ then either $l(W) = +\infty$ or $l(V/W) = +\infty$ (or both) and conversely.

Proof. Let

$$W = W_0 \supset W_1 \supset \ldots \supset W_m = 0 \qquad [13.13]$$

and

$$V/W = \bar{V}_0 \supset \bar{V}_1 \supset \ldots \supset \bar{V}_n = 0 \qquad [13.14]$$

be non-repeating normal series of lengths m and n for W and V/W respectively. Let η be the canonical epimorphism from V onto V/W. Then

$$V = \eta^{-1}(\bar{V}_0) \supset \eta^{-1}(\bar{V}_1) \supset \ldots \supset \eta^{-1}(\bar{V}_n) = W_0 \supset W_1 \supset \ldots W_m = 0 \qquad [13.15]$$

is a non-repeating normal series of length $m + n$ for V.

If $l(W) = +\infty$ we may take m as large as we please, and if $l(V/W) = +\infty$ we may take n as large as we please; so in either case we may obtain non-repeating normal series for V of arbitrary length. Hence $l(V) = +\infty$.

Suppose now that $l(W)$ and $l(V/W)$ are both finite. Then we may take $m = l(W)$ and $n = l(V/W)$, so that [13.13] and [13.14] are composition series. For $i = 1, \ldots, n$ we deduce from Theorem 7.5 that $\eta^{-1}(\bar{V}_{i-1})/\eta^{-1}(\bar{V}_i)$ is isomorphic to $\bar{V}_{i-1}/\bar{V}_i$ and hence is simple. It follows that [13.15] is a composition series for V. Hence $l(V) = m + n = l(W) + l(V/W)$.

§14. Artinian and Noetherian Modules and Rings

Let R be a ring with identity. An infinite sequence $(V_k)_{k\in\mathbf{N}}$ of left R-modules is said to be an *infinite descending chain* if $V_k \supseteq V_{k+1}$ for all natural numbers k; such a chain is said to be *strictly descending* if all the inclusions are proper, i.e. if $V_k \supset V_{k+1}$ for all natural numbers k. A finite sequence $(V_k)_{k\in I}$ of left R-modules, indexed by an interval $I = [p, q]$ of natural numbers is called a *finite descending chain* if $V_k \supseteq V_{k+1}$ for $k = p, \ldots, q-1$; again the chain is said to be *strictly descending* if all the inclusions are proper. By reversing all the inclusions we obtain definitions for *ascending* and *strictly ascending chains.*

We say that an ascending or descending chain (V_k) *levels off,* or *is eventually constant* if there is an index k_0 such that $V_k = V_{k_0}$ for all indices $k \geqslant k_0$; we say that the chain levels off at V_{k_0}. Clearly every finite ascending or descending chain levels off at its last term.

A left R-module V is said to satisfy the *descending chain condition* if every descending chain of submodules of V (infinite as well as finite) levels off; clearly V satisfies this condition if and only if there are no infinite strictly descending chains of submodules of V. Similarly V is said to satisfy the *ascending chain condition* if every ascending chain of submodules of V levels off; this is equivalent to requiring that there be no infinite strictly ascending chains of submodules of V.

A left R-module V is said to satisfy the *minimum condition* if every non-empty set of submodules of V has a minimal element with respect to the inclusion relation. Thus V satisfies the minimum condition if and only if in every non-empty set E of submodules of V there is a submodule V_0 such that no submodule in E is properly included in V_0. Similarly V is said to satisfy the *maximum condition* if every non-empty set of submodules of V has a maximal element with respect to the inclusion relation.

THEOREM 14.1. *A left R-module satisfies the minimum condition if and only if it satisfies the descending chain condition; it satisfies the maximum condition if and only if it satisfies the ascending chain condition.*

Proof. We shall prove the first statement of the theorem; the proof of the second part can then be easily obtained by making the obvious modifications.

(1) Suppose V satisfies the minimum condition.

If $(V_k)_{k \in \mathbf{N}}$ were an infinite strictly descending chain of submodules of V the set of all the submodules V_k in this chain would be a non-empty set of submodules of V without a minimal element. Hence there can be no infinite strictly descending chain of submodules of V.

So V satisfies the descending chain condition.

(2) Conversely, suppose V satisfies the descending chain condition.

Let us assume that V does not satisfy the minimum condition and deduce a contradiction; so we suppose that there is a non-empty set E of submodules of V which has no minimal element.

Let V_0 be any element of E. Now suppose that for some natural number k we have constructed a strictly descending chain

$$V_0 \supset V_1 \supset \ldots \supset V_k$$

of submodules in E. Since E has no minimal element there is a submodule V_{k+1} in E such that $V_k \supset V_{k+1}$; so we have a strictly descending chain

$$V_0 \supset V_1 \supset \ldots \supset V_k \supset V_{k+1}$$

of submodules in E. Proceeding in this way, we obtain an infinite strictly descending chain of submodules of V. This, however, contradicts the descending chain condition in V.

Hence V satisfies the minimum condition.

A left R-module which satisfies the minimum condition (and hence the descending chain condition) is called an *Artinian module*; a left R-module which satisfies the maximum condition (and hence the ascending chain condition) is called a *Noetherian module.* These names are given in honour of Emil Artin and Emmy Noether, two of the great pioneers of modern algebra.

We now give an alternative criterion for a module to be Noetherian.

THEOREM 14.2. *A left R-module V is Noetherian if and only if every submodule of V is finitely generated.*

Proof. (1) Suppose every submodule of V is finitely generated.

Let $(V_k)_{k \in \mathbf{N}}$ be an infinite ascending chain of submodules of V. Let $W = \bigcup_{k \in \mathbf{N}} V_k$. Then W is a submodule of V—for if x and y are elements of W, there are natural numbers n_1, n_2 such that $x \in V_{n_1}$, $y \in V_{n_2}$ and hence $x + y \in V_n \subseteq W$ where $n = \sup(n_1, n_2)$; further, if a is any element of R, we have $ax \in V_{n_1} \subseteq W$.

By hypothesis, W is generated by a finite subset $\{x_1, \ldots, x_r\}$ say. Since the elements $x_1, \ldots, x_r$ belong to the union $\bigcup_{k \in \mathbf{N}} V_k$ there exist natural numbers $n_1, \ldots, n_r$ such that $x_i \in V_{n_i}$ $(i = 1, \ldots, r)$. Let $k_0 = \sup(n_1, \ldots, n_r)$. Then $\{x_1, \ldots, x_r\} \subseteq V_{k_0}$ and hence $W \subseteq V_{k_0}$; but of course $V_{k_0} \subseteq W$. So $V_{k_0} = W$ and for every natural number $k \geqslant k_0$ we have $W = V_{k_0} \subseteq V_k \subseteq W$, whence $V_k = V_{k_0}$.

Thus (V_k) levels off. Hence V is Noetherian.

(2) Conversely, suppose V is a Noetherian module.

Let W be any submodule of V and let E be the set of finitely generated submodules of W; E is clearly non-empty since the zero submodule belongs to it. Since V satisfies the maximum condition, E has a maximal element, W_0 say. If we can show that $W_0 = W$ it will follow that W is finitely generated.

Suppose, to the contrary, that $W_0 \neq W$; then there is an element x of W which does not belong to W_0. The submodule W_1 generated by $W_0 \cup \{x\}$ is clearly finitely generated and so belongs to E; but $W_1 \supset W_0$ and this contradicts the maximal property of W_0.

Hence $W = W_0$ and so W is finitely generated, as required.

We show next that the Artinian and Noetherian properties carry over to submodules and factor modules.

THEOREM 14.3. *Let V be an Artinian (Noetherian) left R-module, W a submodule of V. Then W and V/W are Artinian (Noetherian).*

Proof. We shall give the proof in the case where V is an Artinian module; the proof for the Noetherian case is obtained by making the obvious modifications of setting 'maximal' for 'minimal' and 'ascending' for 'descending'.

(1) Let E be a non-empty set of submodules of W. Then of course E is a non-empty set of submodules of V and so, by hypothesis, has a minimal element. Thus W is Artinian.

(2) Let $(\overline{V}_k)_{k \in \mathbf{N}}$ be an infinite descending chain of submodules of V/W. Then $(\eta^{-1}(\overline{V}_k))_{k \in \mathbf{N}}$ is a descending chain of submodules of V (where η is, as usual, the canonical epimorphism from V onto V/W). Since V is Artinian, there is a natural number k_0 such that $\eta^{-1}(\overline{V}_k) = \eta^{-1}(\overline{V}_{k_0})$ for all natural numbers $k \geqslant k_0$. So, since η is an epimorphism, it follows that $\overline{V}_k = \overline{V}_{k_0}$ for all natural numbers $k \geqslant k_0$. Thus V/W is Artinian.

The converse of Theorem 14.3 is true, but the next theorem shows that in order to prove that a module is Artinian (Noetherian) it is unnecessary to assume that all submodules and factor modules are Artinian (Noetherian)—it is enough to have this for one submodule and the corresponding factor module.

THEOREM 14.4. *Let V be a left R-module, W a submodule of V. If W and V/W are Artinian (Noetherian) then V is Artinian (Noetherian).*

Proof. We give the proof this time in the Noetherian case.

Let $(V_k)_{k \in \mathbf{N}}$ be an ascending chain of submodules of V. Then $(W \cap V_k)_{k \in \mathbf{N}}$ and $(\eta(V_k))_{k \in \mathbf{N}}$ are ascending chains of submodules of W and V/W (again η is the canonical epimorphism from V onto

V/W). By hypothesis there exist natural numbers k_1, k_2 such that $W \cap V_k = W \cap V_{k_1}$ for all natural numbers $k \geqslant k_1$ and $\eta(V_k) = \eta(V_{k_2})$ for all natural numbers $k \geqslant k_2$. Let $k_0 = \sup(k_1, k_2)$; we claim that $V_k = V_{k_0}$ for all natural numbers $k \geqslant k_0$.

So we let k be any natural number such that $k \geqslant k_0$. We certainly have $V_k \supseteq V_{k_0}$. Suppose next that x is any element of V_k. Then $\eta(x) \in \eta(V_k) = \eta(V_{k_0})$; hence there is an element x_0 of V_{k_0} such that $\eta(x) = \eta(x_0)$, whence $x - x_0 \in W$. But since $V_{k_0} \subseteq V_k$ we also have $x - x_0 \in V_k$; so $x - x_0 \in W \cap V_k = W \cap V_{k_0}$. It follows that $x \in V_{k_0}$ and so $V_k = V_{k_0}$, as required.

Hence V is Noetherian.

COROLLARY 1. *The external direct sum of a finite family of Artinian (Noetherian) left R-modules is Artinian (Noetherian).*

Proof. Let V_1 and V_2 be Artinian left R-modules, and let $V = V_1 \oplus V_2$. If ι_1 and π_2 have the obvious meanings, we have $\operatorname{Ker} \pi_2 = \operatorname{Im} \iota_1$ and hence $V/\operatorname{Im} \iota_1 = V/\operatorname{Ker} \pi_2$ is isomorphic to $\operatorname{Im} \pi_2 = V_2$. Thus $V/\operatorname{Im} \iota_1$ is Artinian; and $\operatorname{Im} \iota_1$, which is isomorphic to V_1, is also Artinian. According to the Theorem it follows that V is Artinian. To obtain the result for finite families with more than two members, we proceed by induction using the easily established fact that $V_1 \oplus \ldots \oplus V_r$ is isomorphic to $(V_1 \oplus \ldots \oplus V_{r-1}) \oplus V_r$ $(r \geqslant 3)$.

COROLLARY 2. *The sum of a finite family of Artinian (Noetherian) submodules of a left R-module is Artinian (Noetherian).*

Proof. The sum of the family of submodules is an epimorphic image of the external direct sum of the family and hence is isomorphic to a factor module of this external direct sum. But the external direct sum is Artinian (Noetherian) by Corollary 1 and so the desired result follows from Theorem 14.3.

We now revert to the problem raised in § 13 of giving a criterion for a left R-module to have a composition series.

THEOREM 14.5. *A left R-module has a composition series if and only if it is both Artinian and Noetherian.*

Proof. (1) Let V be a left R-module which is both Artinian and Noetherian.

We set $V_0 = V$; now suppose that for some natural number k

we have defined submodules $V_0, V_1, \ldots, V_k$ of V such that

$$V_0 \supset V_1 \supset \ldots \supset V_{k-1} \supset V_k$$

and all the factor modules V_i/V_{i+1} $(i = 0, \ldots, k-1)$ are simple. If V_k is not the zero submodule, the set of proper submodules of V_k is non-empty and hence, by the Noetherian condition, has a maximal element V_{k+1} say. Then we have

$$V_0 \supset V_1 \supset \ldots \supset V_{k-1} \supset V_k \supset V_{k+1}$$

and all the factor modules V_i/V_{i+1} $(i = 0, \ldots, k)$ are simple.

If none of the submodules V_n obtained in this way is the zero submodule we should obtain an infinite strictly descending chain of submodules of V, contradicting our hypothesis that V is Artinian. Hence there is a natural number l such that $V_l = 0$, and

$$V = V_0 \supset V_1 \supset \ldots \supset V_l = 0$$

is a composition series for V.

(2) Conversely, suppose V has a composition series of length l. If we had an infinite strictly descending chain $(V_k)_{k \in \mathrm{N}}$ of submodules of V we may assume without loss of generality that $V_0 = V$. Then

$$V = V_0 \supset V_1 \supset \ldots \supset V_l \supset 0$$

is a non-repeating normal series for V of length $l + 1$; but this contradicts Theorem 13.1. So there are no infinite strictly descending chains of submodules of V; thus V is Artinian.

Similarly, if we had an infinite strictly ascending chain $(V_k)_{k \in \mathbf{N}}$ of submodules of V we may assume that $V_0 = 0$. Then

$$V \supset V_l \supset V_{l-1} \supset \ldots \supset V_0 = 0$$

is a non-repeating normal series for V of length $l + 1$, again contradicting Theorem 13.1. Thus V is Noetherian.

Example 1. According to the Example of § 13, if D is a division ring then every left vector space over D (left D-module) with a finite basis has a composition series and hence is both Artinian and Noetherian.

The ring R is said to be a *left Artinian ring* if the left R-module R_l is an Artinian module; R is said to be a *left Noetherian ring* if R_l is a Noetherian module. Since the submodules of the left R-module R_l are the left ideals of R (§ 6, Example 8), we see that R is a left Artinian ring if and only if every non-empty set of left ideals of R has a minimal element with respect to the inclusion relation; an equivalent condition is that every infinite descending chain of left ideals of R should level off. We have similar conditions for R to be a left Noetherian ring; referring to Theorem 14.2 we see also that R is a left Noetherian ring if and only if every left ideal is finitely generated.

Example 2. Every division ring D is both left Artinian and left Noetherian, since the only left ideals of D are D itself and the zero ideal.

Example 3. Since every ideal of $\mathbf{Z}$ is a principal ideal, i.e. is generated by a single element (§ 3, Example 1), $\mathbf{Z}$ is a left Noetherian ring. But $\mathbf{Z}$ is not a left Artinian ring; if for every natural number n we let I_n be the principal ideal generated by 2^n, the sequence $(I_n)_{n \in \mathbf{N}}$ is an infinite strictly descending chain of ideals of $\mathbf{Z}$.

Example 4. Let R be any ring, $P_\infty(R)$ the infinite order polynomial ring with coefficients in R(§ 5, Example 8). Then $P_\infty(R)$ is neither left Artinian nor left Noetherian: for each natural number n we let I_n be the principal left ideal generated by X_1^n and let J_n be the left ideal generated by the set $\{X_1, \ldots, X_n\}$. Then the sequence $(I_n)_{n \in \mathbf{N}}$ is an infinite strictly descending chain of left ideals of $P_\infty(R)$ and the sequence $(J_n)_{n \in \mathbf{N}}$ is an infinite strictly ascending chain of left ideals of $P_\infty(R)$.

It is not by chance that we fail to give an example of a ring which is Artinian but not Noetherian: we shall prove later that there are no such rings (Theorem 23.10).

The following simple result will be needed later.

THEOREM 14.6. *Let R be an Artinian (Noetherian) ring, I a two-sided ideal of R. Then R/I is an Artinian (Noetherian) ring.*

Proof. We have only to remark that the (R/I)-submodules of R/I are precisely the same as the R-submodules of the R-module R/I.

The next theorem, one of Hilbert's many seminal contributions to algebra, shows (among many other important consequences) the existence of a large class of Noetherian rings.

THEOREM 14.7. (*Hilbert's Basis Theorem*). *If R is a commutative Noetherian ring (with identity) then so is the polynomial ring $P(R)$.*

Proof. As usual we shall identify R with its image in $P(R)$ under the canonical monomorphism κ.

Let A be any ideal in $P(R)$. (Since $P(R)$ is commutative, left, right and two-sided ideals all coincide, so we shall just talk of 'ideals'.)

Let $L(A)$ be the set of elements of R each of which occurs as the leading coefficient of at least one polynomial in A. We claim that $L(A)$ is an ideal of R. So suppose a and b are elements of $L(A)$, r any element of R. Then there exist polynomials α and β in A which have a and b respectively as leading coefficients, say $\alpha = aX^m + \ldots$, $\beta = bX^n + \ldots$ If $m \leqslant n$ then $X^{n-m}\alpha - \beta = (a-b)X^n + \ldots$ is an element of A and so $a - b \in L(A)$. Further, $r\alpha = raX^m + \ldots \in A$ and hence $ra \in L(A)$. Thus $L(A)$ is an ideal of R, as asserted.

Since R is Noetherian, $L(A)$ is finitely generated, say by the set of elements $\{a_1, \ldots, a_s\}$ of R. Let $\alpha_1, \ldots, \alpha_s$ be polynomials in the ideal A which have $a_1, \ldots, a_s$ respectively as leading coefficients. Let $n = \sup(\partial\alpha_1, \ldots, \partial\alpha_s)$.

Next, for $i = 0, 1, \ldots, n-1$ let $L_i(A)$ be the subset of R consisting of the zero element together with the leading coefficients of all polynomials in A of degree i. These sets $L_i(A)$ are easily seen to be ideals of R and hence are all finitely generated. Suppose $L_i(A)$ is generated by the subset $\{a_{i1}, \ldots, a_{is_i}\}$ $(i = 0, \ldots, n-1)$; let α_{ij} $(i = 0, \ldots, n-1; j = 1, \ldots, s_i)$ be a polynomial of degree i in A with leading coefficient a_{ij}.

We shall show that A is generated by the finite set of polynomials

$$P = \{\alpha_1, \ldots, \alpha_s; \alpha_{01}, \ldots, \alpha_{0s_0}; \ldots; \alpha_{n-1,1}, \ldots, \alpha_{n-1,s_{n-1}}\}.$$

To this end let $\varphi = bX^k + \ldots$ be any polynomial in A, with leading coefficient b.

Suppose first that $\partial\varphi = k \geqslant n$. Since b is a leading coefficient of a polynomial in A, $b \in L(A)$; hence, since $L(A)$ is generated by $\{a_1, \ldots, a_s\}$ there are elements $b_1, \ldots, b_s$ of R such that $b = b_1a_1 + \ldots + b_sa_s$. Then the polynomial

$$\varphi_1 = \varphi - b_1X^{k-\partial\alpha_1}\alpha_1 - b_2X^{k-\partial\alpha_2}\alpha_2 - \ldots - b_sX^{k-\partial\alpha_s}\alpha_s$$

belongs to A and $\partial\varphi_1 < \partial\varphi$. Repeating this procedure a finite number of times we obtain polynomials $\beta_1, \ldots, \beta_s$ in $P(R)$ such that the polynomial

$$\psi = \varphi - \beta_1\alpha_1 - \ldots - \beta_s\alpha_s \qquad [14.1]$$

belongs to A and $\partial\psi < n$.

So suppose now that we have a polynomial $\psi = cX^l + \ldots$ in A with leading coefficient c and $\partial\psi = l < n$. Since c is a leading coefficient of a polynomial of degree l in A, $c \in L_l(A)$; hence there exist elements $c_{l1}, \ldots, c_{ls_l}$ of R such that $c = c_{l1}a_{l1} + \ldots + c_{ls_l}a_{ls_l}$. Then the polynomial

$$\psi_1 = \psi - c_{l1}\alpha_{l1} - \ldots - c_{ls_l}\alpha_{ls_l}$$

belongs to A and $\partial\psi_1 < \partial\psi$. Repeating this procedure a finite number of times we obtain elements c_{ij} of R such that the polynomial

$$\psi - \sum_{i=0}^{n-1} \sum_{j=1}^{s_i} c_{ij}\alpha_{ij} \qquad [14.2]$$

is the zero polynomial.

Our assertion that A is generated by the finite set P now follows from [14.1] and [14.2]. This completes the proof.

An obvious inductive argument gives the following result.

COROLLARY. *Let R be a commutative Noetherian ring with identity. Then for each natural number $n \geqslant 1$ the nth order polynomial ring $P_n(R)$ is also Noetherian.*

Chapter 3

CATEGORIES

§ 15 Categories

We begin with some elementary remarks about the class $\boldsymbol{E}$ of all sets. First we notice that for every ordered pair of elements (A, B) of $\boldsymbol{E}$ (i.e. every ordered pair of sets) there exists a set Map (A, B), possibly empty, consisting of all mappings from A to B, such that if $(A, B) \neq (A', B')$ then Map (A, B) and Map (A', B') are disjoint. Next, for each triple of elements A, B, C of $\boldsymbol{E}$, each element α of Map (A, B) and each element β of Map (B, C) there exists an element of Map (A, C) which we denote by $\beta\alpha$—the composition of A and B. If A, B, C, D are elements of $\boldsymbol{E}$ and α, β, γ are elements of Map (A, B), Map (B, C), Map (C, D) respectively, then $\gamma(\beta\alpha) = (\gamma\beta)\alpha$. Finally, for each element A of $\boldsymbol{E}$ there exists an element I_A of Map (A, A)—this is the identity mapping of A—such that if B is any element of $\boldsymbol{E}$, α and β any elements of Map (A, B) and Map (B, A) respectively, then $\alpha I_A = \alpha$ and $I_A\beta = \beta$.

Guided by this fundamental example we introduce the notion of a category. A *category* $\boldsymbol{C}$ is a class, whose elements are called the *objects* of the category, together with

(1) for each ordered pair of objects (A, B) a set, possibly empty, which we denote by $\boldsymbol{C}(A, B)$ and call the set of *morphisms* of $\boldsymbol{C}$ with *domain* A and *codomain* B (or morphisms from A to B), and

(2) for every triple of objects A, B, C a mapping $\kappa_{A,B,C}$ from $\boldsymbol{C}(A, B) \times \boldsymbol{C}(B, C)$ to $\boldsymbol{C}(A, C)$.

If $\alpha \in \boldsymbol{C}(A, B)$ and $\beta \in \boldsymbol{C}(B, C)$ we write $\beta\boldsymbol{C}\alpha$ or simply $\beta\alpha$ instead of $\kappa_{A,B,C}(\alpha, \beta)$ and call it the *composition* of the morphisms α and β. We require further that the following conditions be satisfied:

*C*1. If A, A', B, B' are objects of the category $\boldsymbol{C}$ such that $(A, B) \neq (A', B')$ then $\boldsymbol{C}(A, B) \cap \boldsymbol{C}(A', B') = \phi$;

*C*2. If A, B, C, D are objects of $\boldsymbol{C}$ and α, β, γ are morphisms from A to B, B to C, C to D respectively, then $\gamma(\beta\alpha) = (\gamma\beta)\alpha$;

*C*3. Corresponding to each object A of $\boldsymbol{C}$ there exists a morphism I_A from A to A such that if B is any object of $\boldsymbol{C}$ and α and β are morphisms from A to B, B to A then $\alpha I_A = \alpha$ and $I_A\beta = \beta$.

The morphism I_A is called the *identity morphism* of A. It is immediate that the identity morphism of each object A is unique. For suppose that ι and ι' are identity morphisms of A; then we have $\iota = \iota\iota' = \iota'$.

We notice that if α and β are morphisms of the category $\boldsymbol{C}$ then the composition $\beta C \alpha$ is defined if and only if the codomain of α coincides with the domain of β. It is customary to write $\alpha : A \to B$ as an alternative to $\alpha \in \boldsymbol{C}(A, B)$; but we must not let this notation deceive us into thinking that the morphisms of a category must be mappings between sets. Keeping this warning in mind, we shall also allow ourselves to use diagrams of objects and morphisms.

A category is said to be *small* if the objects of the category constitute a set.

Example 1. Our first example is the one which we discussed in detail before giving the formal definition; this is the *category of sets*. Its objects are sets; its morphisms are mappings of sets; and the composition of morphisms is the usual composition of mappings.

Example 2. If R is any ring with identity we may form the *category of left R-modules*. The objects of this category are left R-modules; if A and B are left R-modules then the morphisms from A to B are the R-homomorphisms from A to B; the composition of morphisms is the usual composition of mappings. We denote this category by ${}_R\boldsymbol{M}$, or simply $\boldsymbol{M}$ if it is clear from the context that we are dealing with left modules. Clearly we may also form the *category of right R-modules*, $\boldsymbol{M}_R$.

Example 3. If in Example 2 we consider the special case in which the ring R is the ring of integers $\mathbf{Z}$, we obtain the *category of abelian groups*.

Example 4. Again let R be any ring with identity. We may form the *category of short exact sequences* of left R-modules. The objects of this category are short exact sequences. To describe the morphisms, let E_i $(i = 1, 2)$ be the short exact sequence

$$0 \to V_i' \xrightarrow{\alpha_i} V_i \xrightarrow{\beta_i} V_i'' \to 0;$$

then a morphism from E_1 to E_2 is a triple of R-homomorphisms $(\varphi', \varphi, \varphi'')$ from V_1' to V_2', V_1 to V_2, V_1'' to V_2'' respectively, such that the diagram

$$\begin{array}{ccccccccc} 0 & \longrightarrow & V_1' & \xrightarrow{\alpha_1} & V_1 & \xrightarrow{\beta_1} & V_1'' & \longrightarrow & 0 \\ & & \varphi' \downarrow & & \varphi \downarrow & & \varphi'' \downarrow & & \\ 0 & \longrightarrow & V_2' & \xrightarrow{\alpha_2} & V_2 & \xrightarrow{\beta_2} & V_2'' & \longrightarrow & 0 \end{array}$$

is commutative. The composition of two morphisms $(\varphi', \varphi, \varphi'')$ and (ψ', ψ, ψ'') is the triple $(\psi'\varphi', \psi\varphi, \psi''\varphi'')$.

Example 5. As in §10 let (I, ρ) be a directed set. We may then form a category whose objects are the direct systems of left R-modules indexed by (I, ρ). If (A_i, α_{ij}) and (B_i, β_{ij}) are objects of this category, a morphism from the first to the second is a direct system of homomorphisms from (A_i, α_{ij}) to (B_i, β_{ij}). The composition of two such morphisms (φ_i) and (ψ_i) is the direct system $(\psi_i\varphi_i)$.

In an analogous way we may form for each ordered set (I, ρ) the category of inverse systems of left R-modules indexed by (I, ρ).

Example 6. To show that categories occur in other than algebraic contexts we mention the *category of topological spaces*. The objects of this category are topological spaces, its morphisms are continuous mappings and the composition of these morphisms is once again the ordinary composition of mappings.

Example 7. To illustrate our earlier remark that the morphisms of a category need not be mappings between sets, consider the following situation. Let (E, ρ) be an ordered set, and construct a category O as follows. The objects of O are the elements of the set E; for each ordered pair (a, b) of objects of O the set of morphisms $O(a, b)$ is defined by setting

$$O(a, b) = \begin{cases} (a, b) & \text{if } a\rho b \\ \phi & \text{otherwise.} \end{cases}$$

We have next to define composition of morphisms: if $\alpha \in O(a, b)$ and $\beta \in O(b, c)$, then these sets of morphisms are non-empty, and this can occur only if $a\rho b$ and $b\rho c$; hence, since ρ is an order relation, we have $a\rho c$, and so $O(a, c)$ is non-empty. Thus we define the composition $\beta O\alpha$ to be the unique element (a, c) of $O(a, c)$. It is now easy to verify that conditions *C1*, *C2*, *C3* are satisfied, i.e. that O is a category.

Let C be any category; then the *dual* or *opposite category* of C is the category C^* defined as follows. The objects of C^* are the same as the objects of C. For each ordered pair (A, B) of objects of C^* the set of C^*-morphisms, $C^*(A, B)$, is defined to be the set of C-morphisms $C(B, A)$. If A, B, C are objects of C^* and α and β are C^*-morphisms from A to B and B to C respectively, then the C^*-composition $\beta C^* \alpha$ is defined to be the C-composition $\alpha C \beta$. We readily verify that C^* satisfies conditions *C1*, *C2*, *C3*. Clearly the dual of C^* is the original category C itself.

Let S be any statement which can be made about a category, involving reference to objects, morphisms and their domains and codomains, and composition of morphisms. The *dual statement* of S is the statement S^* obtained from S when we replace each occurrence of 'domain' in S by 'codomain' and each occurrence of 'codomain' by 'domain' and replace each composition $\beta\alpha$ of morphisms α and β by the composition $\alpha\beta$. It is clear that to make the statement S^* about a category C is equivalent to making the statement S about the dual category C^*; thus if the statement S is true for every category C, so also is the statement S^*. This discussion should not, however, lead us to suppose that if a statement S of the type described above is true for some particular category then the dual statement S^* is also true for that category.

Next let P be a property expressible in terms of objects, morphisms, domains, codomains and compositions which may be enjoyed by an object or a morphism of a category. Then we define the property P^* *dual* to P by postulating that an object or morphism of a category C shall be said to enjoy the property P^* in C if and only if it enjoys the property P in the dual category C^*. Clearly the property dual to P^* is P itself.

Although we have tried to emphasise that the morphisms of a category need not be mappings between sets we can nevertheless define in general categories certain properties of their morphisms which are originally defined for mappings of sets by reference to the elements of the domain and range sets.

For example, let A and B be objects of a category C, α a morphism from A to B. Then α is said to be *epic* if for every pair of morphisms β_1, β_2 from B to an object Y of C such that $\beta_1\alpha = \beta_2\alpha$ it follows that $\beta_1 = \beta_2$. Again, α is said to be *monic* if for every pair of morphisms φ_1, φ_2 from an object X of C to A such that $\alpha\varphi_1 = \alpha\varphi_2$ it follows that $\varphi_1 = \varphi_2$.

Example 8. The epic and monic morphisms of the category of sets are the surjective and injective mappings respectively.

Example 9. Although the result will hardly surprise him the reader should nevertheless verify that in the category of left R-modules the epic and monic morphisms are the epimorphisms and monomorphisms respectively.

We remark that the properties of being epic and being monic are dual in the sense described above. To see this, let α be a morphism of a category $\mathbf{C}$; say $\alpha \in \mathbf{C}(A, B)$. Then we see that the morphism $\alpha \in \mathbf{C}(A, B) = \mathbf{C}^*(B, A)$ is epic in the dual category $\mathbf{C}^*$ if and only if for every object X of $\mathbf{C}^*$ and every pair of morphisms φ_1, φ_2 in $\mathbf{C}^*(A, X)$ such that $\varphi_1 \mathbf{C}^* \alpha = \varphi_2 \mathbf{C}^* \alpha$ we have $\varphi_1 = \varphi_2$. That is to say, α is epic in $\mathbf{C}^*$ if and only if for every object X of $\mathbf{C}$ and every pair of morphisms φ_1, φ_2 in $\mathbf{C}(X, A)$ such that $\alpha \mathbf{C} \varphi_1 = \alpha \mathbf{C} \varphi_2$ we have $\varphi_1 = \varphi_2$. So α is epic in $\mathbf{C}^*$ if and only if it is monic in $\mathbf{C}$.

A morphism α from A to B is said to be an *isomorphism* if there exists a morphism α' from B to A such that $\alpha'\alpha = I_A$ and $\alpha\alpha' = I_B$. If α is an isomorphism, the morphism α' is easily seen to be unique. It is also an isomorphism, which we call the *inverse* of α and denote by α^{-1}. Clearly an isomorphism is both monic and epic, and the composition of two isomorphisms is again an isomorphism; the property of being an isomorphism is self-dual. If there exists an isomorphism from A to B, we say that A is *isomorphic* to B.

Example 10. In the category of sets the isomorphisms are just the bijective mappings. In the category of left R-modules the isomorphisms are the R-isomorphisms.

As another example of dual properties we consider the notions of projective and injective objects in a category. An object P of a category $\mathbf{C}$ is said to be *projective* if for every diagram

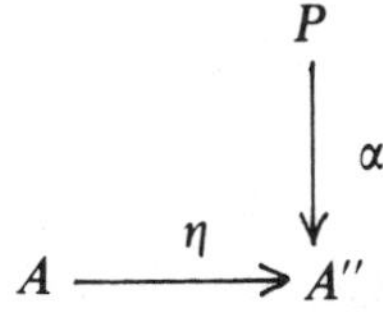

in which the morphism η is epic there exists a morphism β from P to A such that $\eta\beta = \alpha$. Dually an object Q of $\mathbf{C}$ is said to be *injective*

if for every diagram

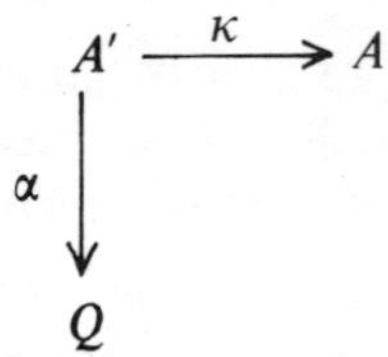

in which κ is monic there exists a morphism β from A to Q such that $\beta\kappa = \alpha$.

Example 11. In the category of sets every object is both projective and injective. Suppose for example that η is a surjection from A onto A''. Then there exists a mapping θ from A'' to A such that $\eta\theta = I_{A''}$. Let P be any set, α any mapping from P to A''; if we set $\beta = \theta\alpha$ we have $\eta\beta = \eta(\theta\alpha) = (\eta\theta)\alpha = \alpha$. Thus P is projective. A dual argument shows that every set is injective.

Example 12. In the category of left R-modules the projective and injective objects are the projective and injective modules respectively.

An object T of a category $\mathbf{C}$ is said to be *terminal* if for each object A of $\mathbf{C}$ there is precisely one morphism of $\mathbf{C}$ with domain A and codomain T; an object J of $\mathbf{C}$ is said to be *initial* if there is exactly one morphism of $\mathbf{C}$ with domain J and codomain A.

Example 13. In the category of sets every set consisting of a single element is a terminal object; the empty set ϕ is an initial object.

Clearly the properties of being a terminal object and being an initial object are dual. Although we cannot assert that every category has terminal or initial objects, if we know that one such object exists we can characterize them all.

THEOREM 15.1. *If a category $\mathbf{C}$ has a terminal object T then an object T' of $\mathbf{C}$ is terminal if and only if there is a unique isomorphism from T' to T.*

Proof. Since T is terminal, the set $\mathbf{C}(T', T)$ consists of a single element, α say.

(1) Suppose T' is terminal. Then $\mathbf{C}(T, T')$ also consists of a single element, β say. Then $\alpha\beta \in \mathbf{C}(T, T)$ and $\beta\alpha \in \mathbf{C}(T', T')$. But each of

these sets consists of a single element, which must be the appropriate identity morphism. So $\alpha\beta = I_T$ and $\beta\alpha = I_{T'}$, i.e. α is an isomorphism.

(2) Conversely, suppose α is an isomorphism; let β be its inverse. Let A be any object in $\mathbf{C}$. Since T is terminal, there exists precisely one morphism θ in the set $\mathbf{C}(A, T)$. Then $\beta\theta$ is an element of $\mathbf{C}(A, T')$ and we claim it is the only one; for if $\varphi \in \mathbf{C}(A, T')$ then we have $\alpha\varphi \in \mathbf{C}(A, T)$, whence $\alpha\varphi = \theta$ and so $\varphi = (\beta\alpha)\varphi = \beta(\alpha\varphi) = \beta\theta$. So T' is terminal, as asserted.

It follows from our general remarks about duality that the dual of this result also holds.

THEOREM 15.1*. *If a category* $\mathbf{C}$ *has an initial object* J *then an object* J' *of* $\mathbf{C}$ *is initial if and only if there is a unique isomorphism from* J *to* J'.

An object Z of a category $\mathbf{C}$ which is both initial and terminal is called a *zero object* of $\mathbf{C}$. It follows immediately from Theorems 15·1 and 15·1* that if $\mathbf{C}$ has a zero object Z then an object Z' of $\mathbf{C}$ is a zero object if and only if Z' is isomorphic to Z.

Example 14. There are no zero objects in the category of sets. In the category of left R-modules every module which consists of a zero element alone is a zero object.

Theorems 15.1 and 15.1* can now be applied to deal with the question of 'essential uniqueness' which we raised every time we considered 'universal' and 'couniversal' situations.

Example 15. Let (I, ρ) be a directed set; let (V_i, α_{ij}) be a direct system of left R-modules indexed by (I, ρ). We construct a category $\mathbf{C}$ as follows: the objects of $\mathbf{C}$ are pairs $(M,(\varphi_i))$ consisting of a left R-module M and a coherent family of homomorphisms (φ_i) from (V_i, α_{ij}) to M (see §10); if $(M,(\varphi_i))$ and $(M',(\varphi_i'))$ are two such pairs, the $\mathbf{C}$-morphisms from $(M,(\varphi_i))$ to $(M',(\varphi_i'))$ are the homomorphisms θ from M to M' such that $\theta\varphi_i = \varphi_i'$ for every index i in I; composition of morphisms in $\mathbf{C}$ is defined to be ordinary composition of mappings.

It is easily verified that a pair $(V',(\alpha_i'))$ is initial in $\mathbf{C}$ if and only if it is universal for coherent families of homomorphisms from (V_i, α_{ij}) in the sense described in §10. Theorem 10.1 thus shows that the pair $(V,(\alpha_i))$ consisting of the direct limit $V = \varinjlim V_i$ and the family (α_i) of canonical homomorphisms from (V_i) to V is initial in $\mathbf{C}$. It now follows easily from Theorem 15.1* that $(V',(\alpha_i'))$ is universal for coherent families of homomorphisms from (V_i, α_{ij}), i.e. is initial in $\mathbf{C}$, if and only if there is a unique isomorphism α from V onto V' such that $\alpha\alpha_i = \alpha'_i$ for each index i in I.

Example 16. Let (I, ρ) be an ordered set, (V_i, α_{ji}) an inverse system of left R-modules indexed by (I, ρ). As objects of a category C we take the pairs $(M,(\varphi_i))$ consisting of a left R-module M and a coherent family (φ_i) of homomorphisms from M to (V_i, α_{ji}); the C-morphisms from $(M,(\varphi_i))$ to $(M',(\varphi'_i))$ are defined to be the homomorphisms θ from M to M' such that $\varphi'_i\theta = \varphi_i$; composition of morphisms in C is defined to be ordinary composition of mappings.

We easily verify that a pair $(V',(\alpha'_i))$ is terminal in the category C if and only if it is couniversal for coherent families of homomorphisms to (V_i, α_{ji}) in the sense of §10. According to Theorem 10.3 the pair $(V,(\alpha_i))$ consisting of the inverse limit $V = \varprojlim V_i$ and the family (α_i) of canonical homomorphisms from V to (V_i) is terminal in C. It follows now from Theorem 15.1 that $(V',(\alpha'_i))$ is couniversal for coherent families of homomorphisms to (V_i, α_{ji}), i.e. terminal in C, if and only if there exists a unique isomorphism α' from V' onto V such that $\alpha_i\alpha' = \alpha'_i$ for every index i in I.

Example 17. Let R be a ring with identity, E a set. Let C be the category whose objects are pairs (G, θ) consisting of a left R-module G and a mapping θ from E to G, and whose morphisms and composition operation are defined in the obvious way. Then an object of C is initial if and only if it is universal for mappings from E to left R-modules in the sense of §11.

According to Theorem 11.1 the pair $(F(E), \varphi)$ consisting of the free left R-module $F(E)$ based on E and the natural mapping φ from E to $F(E)$ is initial in C. It follows now from Theorem 15.1* that a pair (G, θ) is universal for mappings from E to left R-modules, i.e. is initial in C, if and only if there exists a unique isomorphism λ from $F(E)$ onto G such that $\lambda\varphi = \theta$.

Example 18. Let R be a ring with identity, V and W right and left R-modules respectively. Then similar arguments show that a pair (B, β) consisting of an abelian group B and a balanced mapping β from $V \times W$ to B is universal for balanced mappings from $V \times W$ if and only if there exists a unique isomorphism θ from $V \otimes_R W$ onto B such that $\beta = \theta\nu$. (Cf. Theorem 12.1.)

We now use the notions of terminal and initial objects in a category to define products and coproducts of families of objects in an arbitrary category. The reader may find the procedure easier to accept if he has another glance at the earlier part of § 9.

Let C be a category, $(A_k)_{k\in K}$ a family of objects of C. We construct a new category C' as follows. The objects of C' are pairs $(E,(\varphi_k))$ consisting of an object E of C and a family (φ_k) of morphisms from E to (A_k); the C'-morphisms from $(E,(\varphi_k))$ to $(E',(\varphi'_k))$ are defined to be the C-morphisms θ from E to E' such that $\varphi'_k\theta = \varphi_k$ for all indices k in K; composition of morphisms in C' is defined to be the same as composition of morphisms in C.

The terminal objects of the category C', if any, are called *products* of the family (A_k). Thus a product of (A_k) is a pair $(P,(\pi_k))$ consisting of an object P and a family (π_k) of morphisms from P to (A_k) such that for every object E and every family (φ_k) of morphisms from E to (A_k) there exists a unique morphism φ from E to P such that $\pi_k\varphi = \varphi_k$ for every index k in K. Thus, in a terminology adapted from §9, $(P,(\pi_k))$ is a product of the family (A_k) if and only if it is couniversal for families of morphisms with codomains (A_k).

We cannot assert in a general category that every family of objects has a product. But as soon as we know that there exists one product for a given family. Theorem 15.1 gives us a characterization of all the products of that family.

Example 19. If (V_k) is a family of left R-modules it follows from Theorem 9.1 that the pair $(P,(\pi_k))$ consisting of the direct product module $P = \prod V_k$ and the family (π_k) of canonical projection epimorphisms is a product of the family (V_k) in the category of left R-modules. Appealing in the usual way to Theorem 15.1 we see that a pair $(P',(\pi'_k))$ is couniversal for homomorphisms to (V_k) if and only if there exists a unique isomorphism θ' from P' onto P such that $\pi_k\theta' = \pi'_k$ for every index k in K.

The notion of coproduct is dual to that of product. Let (A_k) be a family of objects in a category C. Then a *coproduct* of (A_k) is a pair $(S,(\iota_k))$ consisting of an object S of C and a family of morphisms (ι_k) from (A_k) to S such that for every object E of C and every family (φ_k) of morphisms from (A_k) to E there exists a unique morphism φ from S to E such that $\varphi\iota_k = \varphi_k$ for every index k in K. Adapting again the terminology of §9 we may say that a pair $(S,(\iota_k))$ is a coproduct for (A_k) if and only if it is universal for morphisms from (A_k).

Example 20. Let (V_k) be a family of left R-modules. Then, according to Theorem 9.2, the pair $(S,(i_k))$ consisting of the direct sum $S = \oplus V_k$

and the family (ι_k) of canonical injection monomorphisms is a coproduct of (V_k) in the category of left R-modules. Using Theorem 15.1* we deduce that $(S',(\iota'_k))$ is universal for homomorphisms from (V_k) if and only if there is a unique isomorphism θ from S onto S' such that $\theta\iota_k = \iota'_k$ for every index k in K.

The following result is sometimes useful.

THEOREM 15.2. *Let* $(P_k)_{k\in K}$ *be a family of projective objects of a category* $\mathbf{C}$. *If this family has a coproduct* $(P,(\iota_k))$ *then* P *is a projective object of* $\mathbf{C}$.

Proof. Let η be an epic morphism from A to A'' and let α be a morphism from P to A''.

For each index k in K we have a morphism $\alpha\iota_k$ from the projective object P_k to A''. Since η is epic, it follows that for each index k there exists a morphism φ_k from P_k to A such that $\eta\varphi_k = \alpha\iota_k$.

Since $(P,(\iota_k))$ is universal for morphisms from (P_k) it follows that there exists a unique morphism φ from P to A such that $\varphi\iota_k = \varphi_k$ for each index k in K. Thus we have $\eta\varphi\iota_k = \eta\varphi_k = \alpha\iota_k$ for every index k, and hence (since $(P,(\iota_k))$ is universal) $\eta\varphi = \alpha$. So P is projective.

A dual argument yields the dual result.

THEOREM 15.2*. *Let* $(Q_k)_{k\in K}$ *be a family of injective objects of a category* $\mathbf{C}$. *If this family has a product* $(Q,(\pi_k))$ *then* Q *is an injective object of* $\mathbf{C}$.

We now describe a notion which will be of considerable importance to us in setting up the machinery of §18. Let A, B, C be objects of a category $\mathbf{C}$, α and β morphisms from A and B respectively to C. Then a *pullback* for $(A, B, C; \alpha, \beta)$ is a triple $(X; \xi, \eta)$ consisting of an object X of $\mathbf{C}$ and morphisms ξ, η from X to A and B respectively such that $\alpha\xi = \beta\eta$ and such that for every commutative square

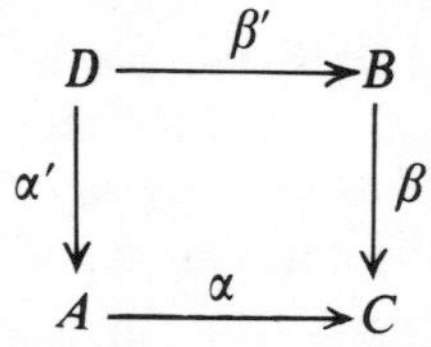

$$\begin{array}{ccc} D & \xrightarrow{\beta'} & B \\ {\scriptstyle \alpha'}\downarrow & & \downarrow{\scriptstyle \beta} \\ A & \xrightarrow{\alpha} & C \end{array} \qquad [15.1]$$

of objects and morphisms of $\mathbf{C}$ there exists a unique morphism γ

from D to X such that the diagram

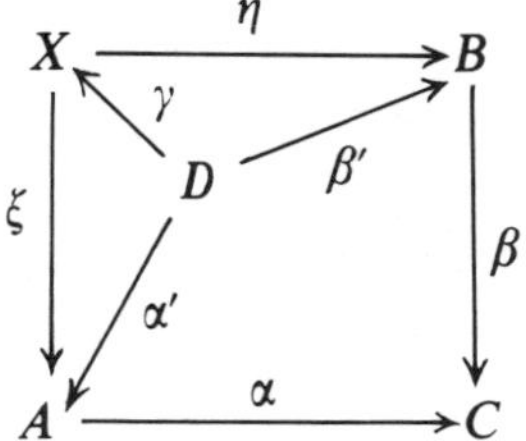

is commutative. (The last diagram should explain the name 'pullback'.) In the situation we have described we also say that the commutative square

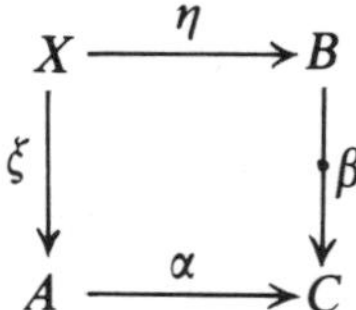

is couniversal for commutative diagrams of the form [15.1]. Applying Theorem 15.1 in the usual way we deduce easily that if $(X; \xi, \eta)$ is a pullback for $(A, B, C; \alpha, \beta)$ then $(X'; \xi', \eta')$ is another pullback if and only if there exists an isomorphism θ from X to X' such that $\xi'\theta = \xi$ and $\eta'\theta = \eta$.

Pullbacks are also known as *Cartesian squares*; the dual notion, with which we shall not be concerned, is known as a *pushout* or a *co-Cartesian square*. Of course there is no guarantee in an arbitrary category that pullbacks or pushouts exist.

The following result on pullbacks will be crucial in §18.

Theorem 15.3. *Consider the diagram*

$$\begin{array}{ccccc} X & \xrightarrow{\varphi} & Y & \xrightarrow{\eta} & C \\ {\scriptstyle\theta}\downarrow & & \downarrow{\scriptstyle\xi} & & \downarrow{\scriptstyle\gamma} \\ A & \xrightarrow{\alpha} & B & \xrightarrow{\beta} & D \end{array}$$

of objects and morphisms of a category **C**. *If* $(Y; \xi, \eta)$ *is a pullback for*

$(B, C, D; \beta, \gamma)$ and $(X; \theta, \varphi)$ is a pullback for $(A, Y, B; \alpha, \xi)$, then $(X; \theta, \eta\varphi)$ is a pullback for $(A, C, D; \beta\alpha, \gamma)$.

Proof. Let E be any object of the category $\mathbf{C}$ and let α' and γ' be morphisms from E to A and C respectively such that $(\beta\alpha)\,\alpha' = \gamma\gamma'$.

Then $\alpha\alpha'$ and γ' are morphisms from E to B and C respectively and $\beta(\alpha\alpha') = \gamma\gamma'$. So, since the right-hand square is a pullback, there is a unique morphism ψ from E to Y such that $\xi\psi = \alpha\alpha'$ and $\eta\psi = \gamma'$. Hence α' and ψ are morphisms from E to A and Y respectively and $\xi\psi = \alpha\alpha'$. It follows, since the left-hand square is a pullback, that there is a unique morphism ε from E to X such that $\theta\varepsilon = \alpha'$ and $(\eta\varphi)\,\varepsilon = \eta(\varphi\varepsilon) = \eta\psi = \gamma'$.

This completes the proof.

§16 Abelian Categories

Let $\mathbf{C}$ be a category which has at least one zero object Z. If A and B are any two objects of $\mathbf{C}$ there is a unique morphism ζ_{AZ} from A to Z and a unique morphism ζ_{ZB} from Z to B. Consider the composition $\zeta_{ZB}\zeta_{AZ}$ of these morphisms; we claim that this composite morphism depends only on A and B, not on the zero object Z. For suppose Z' is another zero object of $\mathbf{C}$; it follows from Theorem 15.1 that there exist unique isomorphisms η and η' in $\mathbf{C}(Z, Z')$ and $\mathbf{C}(Z', Z)$ respectively. Since $\eta'\eta \in \mathbf{C}(Z, Z)$ and $\mathbf{C}(Z, Z)$ consists of a single morphism, which must of course be the identity morphism I_Z, we have $\eta'\eta = I_Z$. Now $\eta\zeta_{AZ}$ and $\zeta_{AZ'}$ both belong to $\mathbf{C}(A, Z')$; hence since $\mathbf{C}(A, Z')$ consists of a single morphism, we must have $\eta\zeta_{AZ} = \zeta_{AZ'}$. Similarly $\zeta_{ZB}\eta' = \zeta_{Z'B}$, and hence we have

$$\zeta_{Z'B}\zeta_{AZ'} = (\zeta_{ZB}\eta')(\eta\zeta_{AZ}) = \zeta_{ZB}I_Z\zeta_{AZ} = \zeta_{ZB}\zeta_{AZ}$$

as we asserted. We write $\zeta_{AB} = \zeta_{ZB}\zeta_{AZ}$ and call it the *zero morphism* from A to B. We shall often abbreviate ζ_{AB} simply to ζ when the domain and codomain are clear from the context. In our next theorem we show that zero morphisms behave under composition as we should expect zeros to behave.

THEOREM 16.1. *Let A and B be objects of a category $\mathbf{C}$ with a zero object Z. If α is any morphism from an object X to A and β any morphism from B to an object Y we have $\zeta_{AB}\alpha = \zeta_{XB}$ and $\beta\zeta_{AB} = \zeta_{AY}$.*

Proof. By definition we have $\zeta_{AB} = \zeta_{ZB}\zeta_{AZ}$. Hence $\zeta_{AB}\alpha = (\zeta_{ZB}\zeta_{AZ})\,\alpha$

$= \zeta_{ZB}(\zeta_{AZ}\alpha)$. Now $\zeta_{AZ}\alpha \in C(X, Z)$, which consists of a single element ζ_{XZ}. Hence $\zeta_{AZ}\alpha = \zeta_{XZ}$ and so $\zeta_{AB}\alpha = \zeta_{ZB}\zeta_{XZ} = \zeta_{XB}$ as asserted.

Similarly $\beta\zeta_{AB} = \zeta_{AY}$.

We now propose to show that in a category C with at least one zero object we can define the notions of kernel and cokernel of a morphism. In order to see how we ought to proceed, let us consider the category of left R-modules, where R is a ring with identity. Let $\alpha \in \mathrm{Hom}_R(V, W)$ be a morphism of this category and let K be the kernel of α as defined in §7. Then if κ is the inclusion monomorphism from K into V we have of course $\alpha\kappa = \zeta$. Furthermore, if β is any homomorphism from a module V' to V such that $\alpha\beta = \zeta$, we have $\beta(V') \subseteq K$; so we may write $\beta = \kappa\beta'$ where β' is the homomorphism from V' to K defined by setting $\beta'(x') = \beta(x')$ for every element x' of V'. Since κ is a monomorphism, the homomorphism β' such that $\beta = \kappa\beta'$ is uniquely determined. Thus we might say that κ is universal for homomorphisms which annihilate α on the right; this is evidently a notion which we can carry over into arbitrary categories with zero objects.

So let C be a category with zero objects, α a morphism of C. Then a morphism κ of C is called a *kernel* of α if it is universal for morphisms which annihilate α on the right; that is to say κ is a kernel of α if and only if (1) $\alpha\kappa = \zeta$ and (2) for every morphism β of C such that $\alpha\beta = \zeta$ there exists a unique morphism β' such that $\beta = \kappa\beta'$. It follows at once from the definition that, if κ is a kernel of α, then κ is monic; for if $\kappa\varphi_1 = \kappa\varphi_2 = \beta$ say, then (by condition (1)) $\alpha\beta = \alpha(\kappa\varphi_1) = (\alpha\kappa)\varphi_1 = \zeta\varphi_1 = \zeta$ and so (by condition (2)) β has a unique factorisation in the form $\beta = \kappa\beta'$, whence $\beta' = \varphi_1 = \varphi_2$ and κ is monic as asserted. It follows also by arguments we have used frequently in dealing with 'universal' situations that if κ is a kernel of α then a morphism κ' is a kernel of α if and only if there is an isomorphism θ from the domain of κ' to that of κ such that $\kappa\theta = \kappa'$.

Although we have defined the notion of kernel in an arbitrary category C with a zero object, we are not asserting that every morphism of C possesses a kernel. We show, however, in the next two examples, that zero morphisms and isomorphisms do have kernels—and in fact the kernels we should expect from our study of module homomorphisms.

Example 1. Let $\alpha = \zeta_{AB}$. Then $\alpha I_A = \zeta_{AB}$ and if β is a morphism from an object A' to A such that $\alpha\beta = \zeta_{A'B}$ we can express β uniquely in the form $\beta = I_A\beta'$ simply by taking $\beta' = \beta$. Thus I_A is a kernel of α; hence it easily follows that a morphism κ from an object A' to A is a kernel of ζ_{AB} if and only if it is an isomorphism from A' to A.

Example 2. Let α be an isomorphism from A to B. If Z is any zero object then $\alpha\zeta_{ZA} = \zeta_{ZB}$, and if β is a morphism from an object A' to A such that $\alpha\beta = \zeta_{A'B}$ then, since α is an isomorphism, we have $\beta = (\alpha^{-1}\alpha)\beta = \alpha^{-1}(\alpha\beta) = \alpha^{-1}\zeta_{A'B} = \zeta_{A'A} = \zeta_{ZA}\zeta_{A'Z}$ and this expression is plainly unique since Z is a zero object. Hence ζ_{ZA} is a kernel of α and it follows that the kernels of α are precisely the zero morphisms from zero objects to A.

Consider again the case of a homomorphism α from V to W where V and W are left R-modules; the cokernel of α was defined in §7 to be the factor module $W/\text{Im }\alpha$. If η is the canonical epimorphism from W onto $W/\text{Im }\alpha$ we have $\eta\alpha = \zeta$. Further, if γ is any homomorphism with domain W such that $\gamma\alpha = \zeta$ then $\text{Im }\alpha \subseteq \text{Ker }\gamma$ and we may define a homomorphism γ' from $W/\text{Im }\alpha$ to the codomain of γ such that $\gamma = \gamma'\eta$; namely, if C is any element of $W/\text{Im }\alpha$, choose any element y from C and set $\gamma'(C) = \gamma(y)$, which depends only on C, not on the choice of y. Since η is an epimorphism the homomorphism γ' such that $\gamma = \gamma'\eta$ is uniquely determined.

We turn once more to the consideration of a morphism α in an arbitrary category $\mathbf{C}$ with zero objects. Guided by the preceding paragraph we say that a morphism η is a *cokernel* for α if it is universal for morphisms which annihilate α on the left; in other words, η is a cokernel for α if and only if (1) $\eta\alpha = \zeta$ and (2) for every morphism γ of $\mathbf{C}$ such that $\gamma\alpha = \zeta$ there exists a unique morphism γ' such that $\gamma = \gamma'\eta$. Thus the notion of cokernel is dual to that of kernel. Arguments dual to those for kernels show that a cokernel is epic and that if η is a cokernel of α then a morphism η' is a cokernel of α if and only if there is an isomorphism θ from the codomain of η to that of η' such that $\theta\eta = \eta'$.

It is not the case in an arbitrary category with zero objects that every morphism has cokernels; but we shall show that isomorphisms and zero morphisms in a general category have the cokernels which we should expect them to have from an examination of the module case.

Example 3. Let $\alpha = \zeta_{AB}$. Then $I_B\alpha = \zeta_{AB}$ and, if β is a morphism from B to an object B' such that $\beta\alpha = \zeta_{AB'}$ we can express β uniquely in the form $\beta = \beta' I_B$ by taking $\beta' = \beta$. Thus I_B is a cokernel of ζ_{AB}; hence the cokernels of ζ_{AB} are the isomorphisms with domain B.

Example 4. Let α be an isomorphism from A to B. Then if Z is any zero object we have $\zeta_{BZ}\alpha = \zeta_{AZ}$; and if β is a morphism from B to B' such that $\beta\alpha = \zeta_{AB'}$ we deduce as in Example 2 that $\beta = \zeta_{BB'}$ and so is uniquely expressible as $\beta = \zeta_{ZB'}\zeta_{BZ}$. Hence ζ_{BZ} is a cokernel of α and so the cokernels of α are precisely the zero morphisms from B to zero objects.

A category C is said to be *exact* if it has at least one zero object (so that it makes sense to talk of kernels and cokernels of morphisms of C) and for every morphism α of C, with domain A and codomain B say, there exist objects K, I, L and morphisms κ, α_1, α_2, η of C related as in the diagram

$$K \xrightarrow{\kappa} A \xrightarrow{\alpha_1} I \xrightarrow{\alpha_2} B \xrightarrow{\eta} L \qquad [16.1]$$

such that

(1) $\alpha = \alpha_2\alpha_1$,

(2) κ is a kernel of α, and η is a cokernel of α,

(3) α_1 is a cokernel of κ, and α_2 is a kernel of η.

The diagram [16.1] is called an *analysis* of the morphism α; the morphism α_2 is called an *image* and α_1 is called a *coimage* of α. We showed earlier that kernels are monic and cokernels are epic; so if [16.1] is an analysis of α, the morphisms κ and α_2 are monic, while α_1 and η are epic.

We have already seen that for a given morphism in a category with a zero object its kernels and cokernels are 'unique up to an isomorphism'. The analogous result for the images and coimages of a morphism in an exact category follows from our next theorem.

THEOREM 16.2. *Let C be an exact category. Then every morphism α of C can be expressed as the composition of an epic morphism followed by a monic morphism, and if $\alpha = \mu_1\theta_1 = \mu_2\theta_2$ where μ_1, μ_2 are monic and θ_1, θ_2 are epic then there exists an isomorphism φ such that $\varphi\theta_1 = \theta_2$ and $\mu_2\varphi = \mu_1$.*

Proof. The first statement follows at once from the existence of an analysis for α.

Suppose now that [16.1] is an analysis for α and that $\alpha = \mu_1\theta_1$ where θ_1 is an epic morphism from A to J_1 and μ_1 is a monic morphism from J_1 to B. We consider the diagram

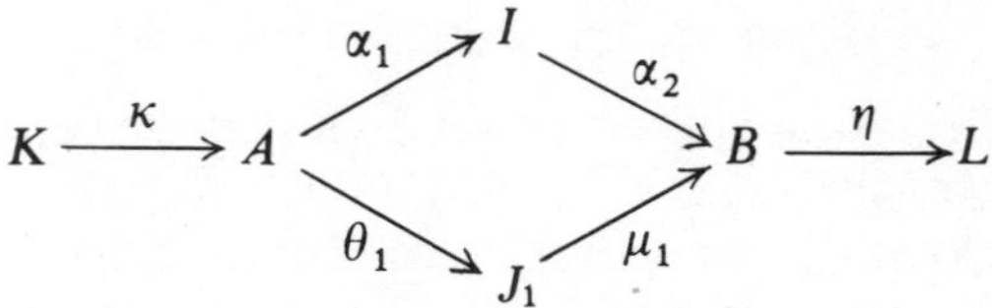

Since $\mu_1\theta_1\kappa = \alpha\kappa = \zeta$ and μ_1 is monic it follows that $\theta_1\kappa = \zeta$. Now α_1 is a cokernel of κ; so there exists a unique morphism φ_1 from I to J_1 such that $\theta_1 = \varphi_1\alpha_1$. By a dual argument, there exists a unique morphism ψ_1 from J_1 to I such that $\mu_1 = \alpha_2\psi_1$. We claim that φ_1 and ψ_1 are mutually inverse isomorphisms.

To see this we remark that $\mu_1\varphi_1\alpha_1 = \mu_1\theta_1 = \alpha = \alpha_2\alpha_1$; since α_1 is epic it follows that $\mu_1\varphi_1 = \alpha_2$. But $\mu_1 = \alpha_2\psi_1$; so $\alpha_2\psi_1\varphi_1 = \alpha_2$ and since α_2 is monic we deduce that $\psi_1\varphi_1$ is the identity morphism of I. The dual of this argument shows that $\varphi_1\psi_1$ is the identity morphism of J_1.

Carrying out a similar procedure for the second factorisation we obtain mutually inverse isomorphisms φ_2 and ψ_2 such that $\theta_2 = \varphi_2\alpha_1$ and $\mu_2 = \alpha_2\psi_2$. Setting $\varphi = \varphi_2\psi_1$ we have $\varphi\theta_1 = \varphi_2\psi_1\varphi_1\alpha_1 = \varphi_2\alpha_1 = \theta_2$ and $\mu_2\varphi = \alpha_2\psi_2\varphi_2\psi_1 = \alpha_2\psi_1 = \mu_1$, as required.

Example 5. If R is any ring with identity the category ${}_R\mathcal{M}$ of left R-modules is an exact category. To see this, let A and B be left R-modules, α an R-homomorphism from A to B. Let K be the kernel of α, in the original sense of §7; if κ is the inclusion monomorphism from K to A then κ is a kernel of α in the sense of the present section. Let I be the image of α, again in the sense of §7; let α_1 be the homomorphism from A to I defined by setting $\alpha_1(a) = \alpha(a)$ for every element a of A and let α_2 be the inclusion monomorphism from I to B. Then $\alpha = \alpha_2\alpha_1$. Let L be the factor module B/I, and let η be the canonical epimorphism from B onto L; then η is a cokernel of α. We now check easily that α_1 is a cokernel of κ and α_2 is a kernel of η. Thus

$$K \xrightarrow{\kappa} A \xrightarrow{\alpha_1} I \xrightarrow{\alpha_2} B \xrightarrow{\eta} L$$

is an analysis of α.

In an exact category, where all morphisms have kernels and images, we can imitate the definition of an exact sequence which we gave in the module situation. Namely, we say that the sequence

$$A \xrightarrow{\alpha} B \xrightarrow{\beta} C \qquad [16.2]$$

is *exact* if every image of α is a kernel of β. Once having made the basic definition of exactness we can proceed to define general exact sequences in an arbitrary exact category C in the same way as we defined exact sequences of left R-modules in §7. One might be tempted to give a dual definition, saying that the sequence [16.2] is *coexact* if every coimage of β is a cokernel of α. The next theorem shows, however, that the two concepts coincide.

THEOREM 16.3 *Let* [16.2] *be a sequence of objects and morphisms in an exact category. Then* [16.2] *is exact if and only if it is coexact.*

Proof. Let $\alpha = \alpha_2\alpha_1$ and $\beta = \beta_2\beta_1$ be factorisations of α and β in the form (image) (coimage).

(1) Suppose the sequence is exact. Then α_2 is a kernel of β, and hence β_1 is a cokernel of α_2. We claim that in fact β_1 is a cokernel of α.

Certainly $\beta_1\alpha = \beta_1(\alpha_2\alpha_1) = (\beta_1\alpha_2)\alpha_1 = \zeta\alpha_1 = \zeta$. Now suppose γ is a morphism such that $\gamma\alpha = \zeta$. Then $(\gamma\alpha_2)\alpha_1 = \gamma(\alpha_2\alpha_1) = \zeta$ and hence, since α_1 is epic, we have $\gamma\alpha_2 = \zeta$. Hence there is a unique morphism γ' such that $\gamma = \gamma'\beta_1$, and β_1 is indeed a cokernel of α.

Thus the sequence is coexact.

(2) The converse follows by the dual argument.

The existence of an analysis for each morphism in an exact category allows us to give various criteria for a morphism in such a category to be monic.

THEOREM 16.4. *Let C be an exact category, α a morphism of C with domain A and codomain B. Then the following conditions are equivalent :*

(a) *α is monic;*
(b) *all kernels of α are zero morphisms;*
(c) *all coimages of α are isomorphisms;*
(d) *α is a kernel of each of its cokernels;*
(e) *for each zero object Z of C the sequence*

$$Z \xrightarrow{\zeta} A \xrightarrow{\alpha} B \qquad [16.3]$$

is exact.

Proof. (1) Suppose α is monic.

Let κ be any kernel of α, with domain K say. Then we have $\alpha\kappa = \zeta_{KB}$; but $\alpha\zeta_{KA} = \zeta_{KB}$. Hence, since α is monic, we have $\kappa = \zeta_{KA}$; thus all kernels of α are zero morphisms, and we have established that (a) implies (b).

(2) Suppose that all kernels of α are zero morphisms.

Then every coimage of α, being a cokernel of a zero morphism, is an isomorphism, by Example 3.

So (b) implies (c).

(3) Suppose that every coimage of α is an isomorphism.

Let η be any cokernel of α, and let $\alpha = \alpha_2\alpha_1$ be a factorisation of α in the form (image) (coimage). Then α_2 is a kernel of η, by definition of an analysis, and hence, since α_1 is an isomorphism, $\alpha = \alpha_2\alpha_1$ is also a kernel of η.

Hence (c) implies (d).

(4) Suppose α is a kernel of each of its cokernels.

Let $\alpha = \alpha_2\alpha_1$ be a factorisation of α as in (3). Since α and α_1 are both kernels of each cokernel of α it follows that α_2 is an isomorphism and hence a cokernel of ζ (by Example 3). So the sequence [16.3] is coexact and hence exact.

So (d) implies (e).

(5) Finally suppose [16.3] is exact, and hence coexact.

Then, in the usual notation, α_2 is a cokernel of ζ and hence an isomorphism. Consequently α_2 is monic; so is α_1; hence so is α.

Thus (e) implies (a).

Dual to Theorem 16.4 we have the following criteria for a morphism in an exact category to be epic.

THEOREM 16.4*. *Let* **C** *be an exact category,* α *a morphism of* **C** *with domain A and codomain B. Then the following conditions are equivalent:*

(a) α *is epic;*

(b) *all cokernels of* α *are zero morphisms;*

(c) *all images of* α *are isomorphisms;*

(d) α *is a cokernel of each of its kernels;*

(e) *for each zero object Z of* **C** *the sequence* $A \xrightarrow{\alpha} B \xrightarrow{\zeta} Z$ *is exact.*

In §15 when we defined monic and epic morphisms and isomorphisms, we mentioned that an isomorphism (in any category) is both monic and epic. We now show that in an exact category the converse holds.

COROLLARY. *A morphism of an exact category is an isomorphism if and only if it is both monic and epic.*

Proof. As we have just said, the 'only if' part holds in every category.

So suppose α is a morphism (in an exact category) which is both monic and epic. Consider a factorisation $\alpha = \alpha_2\alpha_1$ of the usual kind. Since α is monic it follows from Theorem 16.4 that α_1 is an isomorphism, and since α is epic Theorem 16.4* shows that α_2 is an isomorphism. Hence $\alpha = \alpha_2\alpha_1$ is an isomorphism, as required.

We turn now to another important special type of category. A category C is said to be *additive* if it satisfies the following conditions:

AC1. C has at least one zero object.

AC2. For every pair A, B of objects of C the set of morphisms $C(A, B)$ has the structure of an additive abelian group.

AC3. For every triple A, B, C of objects of C the composition mapping $\kappa_{A,B,C}$ from $C(A, B) \times C(B, C)$ to $C(A, C)$ is bilinear, i.e. for all morphisms α, α_1, α_2 in $C(A, B)$ and all morphisms β, β_1, β_2 in $C(B, C)$ we have

$$\beta(\alpha_1 + \alpha_2) = \beta\alpha_1 + \beta\alpha_2 \text{ and } (\beta_1 + \beta_2)\alpha = \beta_1\alpha + \beta_2\alpha.$$

It is clear that for every pair of objects A, B of C the zero element of the abelian group $C(A, B)$ is the zero morphism ζ_{AB}.

In §15, when we discussed products and coproducts of families of objects in a category, we pointed out that there is no guarantee that every family of objects will have either a product or a coproduct. If every finite family of objects in a category C has a product then we say that C *has finite products*; it is clear what we would mean by saying that a category C *has finite coproducts.* An easy inductive argument shows that if every pair of objects of a category C has a product then C has finite products, and similarly for coproducts.

THEOREM 16.5. *An additive category has finite products if and only if it has finite coproducts.*

Proof. (1) Suppose the category has finite coproducts.

If A_1 and A_2 are objects of the category there is a coproduct for the pair (A_1, A_2) consisting of an object S and morphisms ι_1, ι_2 from A_1, A_2 respectively to S. Since this coproduct is universal for morphisms from (A_1, A_2), there exists a unique morphism π_1 from S to A_1 such that $\pi_1\iota_1 = I_{A_1}$ and $\pi_1\iota_2 = \zeta_{A_2A_1}$; similarly there is a unique morphism π_2 from S to A_2 *such that* $\pi_2\iota_1 = \zeta_{A_1A_2}$ and $\pi_2\iota_2 = I_{A_2}$. Then $(\iota_1\pi_1 + \iota_2\pi_2)\iota_1 = \iota_1 = I_S\iota_1$ and $(\iota_1\pi_1 + \iota_2\pi_2)\iota_2 = \iota_2 = I_S\iota_2$, whence (by the universal property of $(S;\iota_1,\iota_2)$) we have $\iota_1\pi_1 + \iota_2\pi_2 = I_S$.

We claim that $(S;\pi_1,\pi_2)$ is a product for (A_1,A_2), i.e. that it is couniversal for morphisms to (A_1,A_2). So let α_1,α_2 be morphisms from an object X to A_1,A_2 respectively. Then $\beta = \iota_1\alpha_1 + \iota_2\alpha_2$ is a morphism from X to S, and we have $\pi_1\beta = \pi_1\iota_1\alpha_1 + \pi_1\iota_2\alpha_2 = \alpha_1$ and similarly $\pi_2\beta = \alpha_2$. This morphism β is plainly unique; for, if β' is another morphism from X to S such that $\pi_1\beta' = \alpha_1$ and $\pi_2\beta' = \alpha_2$, we have $\beta = \iota_1\alpha_1 + \iota_2\alpha_2 = \iota_1\pi_1\beta' + \iota_2\pi_2\beta' = (\iota_1\pi_1 + \iota_2\pi_2)\beta' = I_S\beta' = \beta'$.

Thus $(S;\pi_1,\pi_2)$ is indeed a product.

(2) We leave it to the reader to construct a coproduct from a product.

Example 6. If R is any ring with identity the category ${}_R\mathcal{M}$ of left R-modules is an additive category with finite products. The only condition which we have not already explicitly verified is *AC3*. So let A, B, C be left R-modules, and let α,α_1,α_2 be elements of $\mathrm{Hom}(A,B)$ while β, β_1,β_2 are elements of $\mathrm{Hom}(B,C)$. If a is any element of A we have

$$\begin{aligned}[\beta(\alpha_1+\alpha_2)](a) = \beta((\alpha_1+\alpha_2)(a)) &= \beta(\alpha_1(a)+\alpha_2(a)) \\ &= \beta(\alpha_1(a)) + \beta(\alpha_2(a)) \\ &= \beta\alpha_1(a) + \beta\alpha_2(a) \\ &= (\beta\alpha_1 + \beta\alpha_2)(a).\end{aligned}$$

Thus $\beta(\alpha_1+\alpha_2) = \beta\alpha_1 + \beta\alpha_2$ and similarly we have $(\beta_1+\beta_2)\alpha = \beta_1\alpha + \beta_2\alpha$.

Example 7. Let C be the category of short exact sequences of left R-modules (cf. §15, Example 4). We claim that C is an additive category with finite products. If E_i $(i = 1, 2)$ is the exact sequence

$$0 \to A_i' \xrightarrow{\alpha_i} A_i \xrightarrow{\beta_i} A_i'' \to 0$$

then we recall that the morphisms from E_1 to E_2 are the triples $(\varphi',\varphi,\varphi'')$ of homomorphisms from A_1' to A_2', A_1 to A_2, A_1'' to A_2'' respectively such that $\alpha_2\varphi' = \varphi\alpha_1$ and $\beta_2\varphi = \varphi''\beta_1$. If $(\varphi_1',\varphi_1,\varphi_1'')$ and $(\varphi_2',\varphi_2,\varphi_2'')$ are two such morphisms it is easy to verify that $(\varphi_1'+\varphi_2',\varphi_1+\varphi_2,\varphi_1''+\varphi_2'')$ is also a morphism from E_1 to E_2 and that under the addition operation so defined $C(E_1, E_2)$ is an abelian group. The condition *AC3* is also readily established.

To see that *C* has finite coproducts (and hence also, by Theorem 16.5, finite products) let E_1 and E_2 be the exact sequences of the preceding paragraph. Let $A' = A'_1 \oplus A'_2$, $A = A_1 \oplus A_2$, $A'' = A''_1 \oplus A''_2$, and for $k = 1, 2$ let $\iota'_k, \iota_k, \iota''_k$ be the canonical injection monomorphisms from A'_k to A', A_k to A, A''_k to A'' respectively; let α be the homomorphism from A' to A induced by α_1 and α_2, and β the homomorphism from A to A'' induced by β_1 and β_2. It is easy to show that the sequence $E_1 \oplus E_2$:

$$0 \to A' \xrightarrow{\alpha} A \xrightarrow{\beta} A'' \to 0$$

is exact. It is quickly verified that $E_1 \oplus E_2$ together with the morphisms $(\iota'_1, \iota_1, \iota''_1)$ and $(\iota'_2, \iota_2, \iota''_2)$ is a coproduct for (E_1, E_2).

A category *A* is said to be *abelian* if (1) it is exact, (2) it is additive and (3) it has finite products and coproducts.

Example 8. Referring to Examples 5 and 6 we see that the category of left R-modules ${}_R M$ is abelian.

Example 9. Let I be a directed set; then the category of direct systems of left R-modules indexed by I is an abelian category. If I is any ordered set the category of inverse systems of left R-modules indexed by I is an abelian category.

Example 10. The category of short exact sequences of left R-modules is not an abelian category since it fails to be exact. To see this, we remark first that every morphism in an exact category has a kernel. Now let $(\varphi', \varphi, \varphi'')$ be a morphism from E_1 to E_2 (in the notation of Example 7); it is not hard to see that if this morphism has a kernel then the domain of that kernel must be the sequence

$$0 \to \operatorname{Ker} \varphi' \to \operatorname{Ker} \varphi \to \operatorname{Ker} \varphi'' \to 0$$

with appropriately defined homomorphisms. But we cannot assert in general that this sequence is exact.

Let *C* be an exact category with a zero object Z. If the sequence

$$Z \to A' \xrightarrow{\alpha} A \xrightarrow{\beta} A'' \to Z \qquad [16.4]$$

of objects and morphisms is exact we call it a *short exact sequence.* If [16.4] is a short exact sequence it is said to be *split* if there exists a morphism β'' from A'' to A such that $\beta\beta'' = I_{A''}$; we say that the sequence is *split by* β''. Again, [16.4] is said to be *cosplit* if there exists a morphism α' from A to A' such that $\alpha'\alpha = I_{A'}$; we say that the sequence is *cosplit by* α'. We now show that in an additive exact category the notions of splitting and cosplitting coincide.

THEOREM 16.6. *A short exact sequence of objects and morphisms in an additive exact category is split if and only if it is cosplit.*

Proof. (1) Suppose the short exact sequence [16.4] is split by β''.

Since $Z \to A' \xrightarrow{\alpha} A$ is exact it follows from Theorem 16.4 that every coimage of α is an isomorphism. Since every image of α is a kernel of β it follows that in fact α itself is a kernel of β.

Now $\beta(I_A - \beta''\beta) = \beta I_A - \beta(\beta''\beta) = \beta - (\beta\beta'')\beta = \beta - I_{A''}\beta = \zeta$. Hence, since α is a kernel of β there exists a morphism α' from A to A' such that $I_A - \beta''\beta = \alpha\alpha'$. Then $\alpha(\alpha'\alpha) = (\alpha\alpha')\alpha = (I_A - \beta''\beta)\alpha = I_A\alpha - \beta''(\beta\alpha) = \alpha$; and so, since α is monic, we have $\alpha'\alpha = I_{A'}$. In other words, [16.4] is cosplit by α'.

(2) The converse follows by means of the dual argument.

We now tie up the notions of split and cosplit exact sequences with products and coproducts.

THEOREM 16.7. *Let* **C** *be an additive exact category and let*

$$Z \to A' \xrightarrow{\alpha} A \xrightarrow{\beta} A'' \to Z \qquad [16.5]$$

be a short exact sequence of objects and morphisms of **C**, *split by* β'' *and cosplit by* α' *where* $\alpha\alpha' + \beta''\beta = I_A$. *Then* $(A; \alpha', \beta)$ *is a product for* (A', A'') *and* $(A; \alpha, \beta'')$ *is a coproduct for* (A', A''). *Conversely, if* $(P; \pi', \pi'')$ *is a product for* (A', A'') *and* $(P; \iota', \iota'')$ *is the corresponding coproduct then the sequences*

$$Z \to A' \xrightarrow{\iota'} P \xrightarrow{\pi''} A'' \to Z \qquad [16.6]$$

and

$$Z \to A'' \xrightarrow{\iota''} P \xrightarrow{\pi'} A' \to Z \qquad [16.7]$$

are split exact sequences.

Proof. (1) Suppose [16.5] is split by β'' and cosplit by α', where $\alpha\alpha' + \beta''\beta = I_A$. Then we have

$$\beta'' = I_A\beta'' = (\alpha\alpha' + \beta''\beta)\beta'' = (\alpha\alpha')\beta'' + (\beta''\beta)\beta'' = \alpha(\alpha'\beta'') + \beta''(\beta\beta'')$$
$$= \alpha(\alpha'\beta'') + \beta'' I_{A''}$$

from which it follows that $\alpha(\alpha'\beta'') = \zeta$ and hence $\alpha'\beta'' = \zeta$, since α is monic (by Theorem 16.4).

Now let E be any object of $\boldsymbol{C}$, φ' and φ'' morphisms from E to A' and A'' respectively. Let $\varphi = \alpha\varphi' + \beta''\varphi''$. Then we have

$$\alpha'\varphi = (\alpha'\alpha)\varphi' + (\alpha'\beta'')\varphi'' = I_{A'}\varphi' + \zeta\varphi'' = \varphi'$$

and

$$\beta\varphi = (\beta\alpha)\varphi' + (\beta\beta'')\varphi'' = \zeta\varphi' + I_{A''}\varphi'' = \varphi''.$$

Further, if φ_1 is another morphism from E to A such that $\alpha'\varphi_1 = \varphi'$ and $\beta\varphi_1 = \varphi''$, we have

$$\varphi_1 = I_A\varphi_1 = (\alpha\alpha' + \beta''\beta)\varphi_1 = \alpha(\alpha'\varphi_1) + \beta''(\beta\varphi_1) = \alpha\varphi' + \beta''\varphi'' = \varphi.$$

Thus $(A; \alpha', \beta)$ is a product for (A', A'') and hence $(A; \alpha, \beta'')$ is a coproduct.

(2) Suppose $(P; \pi', \pi'')$ is a product for (A', A'') and $(P; \iota', \iota'')$ the corresponding coproduct. Then π' and π'' are epic, while ι' and ι'' are monic. It follows from Theorems 16.4 and 16.4* that the sequences

$$Z \to A' \xrightarrow{\iota'} P, \qquad P \xrightarrow{\pi''} A'' \to Z,$$

$$Z \to A'' \xrightarrow{\iota''} P, \qquad P \xrightarrow{\pi'} A' \to Z$$

are all exact.

Since ι' is monic, it is itself an image of ι'; we shall show that it is also a kernel of π''. We certainly have $\pi''\iota' = \zeta$. So suppose θ is a morphism such that $\pi''\theta = \zeta$; then we have

$$\theta = I_P\theta = (\iota'\pi' + \iota''\pi'')\,\theta = \iota'(\pi'\theta) + \iota''(\pi''\theta) = \iota'(\pi'\theta),$$

and this representation is plainly unique. Thus every image of ι' is a kernel of π'', i.e. the sequence

$$A' \xrightarrow{\iota'} P \xrightarrow{\pi''} A''$$

is exact. Similarly the sequence

$$A'' \xrightarrow{\iota''} P \xrightarrow{\pi'} A'$$

is exact. Thus [16.6] and [16.7] are exact sequences. Since $\pi''\iota'' = I_{A''}$ and $\pi'\iota' = I_{A'}$, these sequences are split.

This completes the proof.

Finally we show that in an abelian category pullbacks exist (see page 130).

THEOREM 16.8. *Let A, B, C be objects of an abelian category $\mathbf{A}$; let α and β be morphisms from A and B respectively to C. Then there exists a pullback $(X; \xi, \eta)$ for $(A, B, C; \alpha, \beta)$.*

Proof. Let $(P; \pi_A, \pi_B)$ be a product for (A, B) and let κ be a kernel of the morphism $\alpha\pi_A - \beta\pi_B$, with domain X say; define morphisms ξ and η from X to A and B by setting $\xi = \pi_A\kappa$, $\eta = \pi_B\kappa$. Then

$$\alpha\xi - \beta\eta = \alpha\pi_A\kappa - \beta\pi_B\kappa = (\alpha\pi_A - \beta\pi_B)\kappa = \zeta;$$

so the square

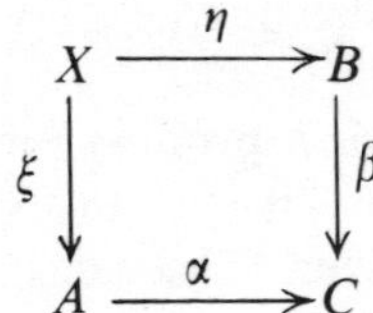

is commutative.

Consider now any commutative square

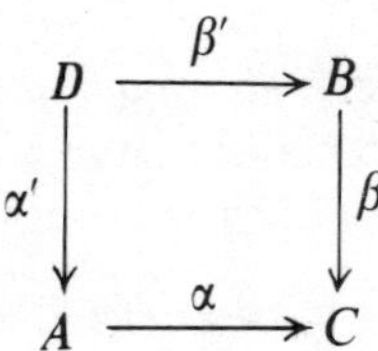

Since $(P; \pi_A, \pi_B)$ is a product for (A, B) there exists a unique morphism γ' from D to P such that $\pi_A\gamma' = \alpha'$ and $\pi_B\gamma' = \beta'$. Then we have

$$(\alpha\pi_A - \beta\pi_B)\gamma' = \alpha(\pi_A\gamma') - \beta(\pi_B\gamma') = \alpha\alpha' - \beta\beta' = \zeta,$$

i.e. γ' annihilates $\alpha\pi_A - \beta\pi_B$ on the right, and hence there exists a unique morphism γ from D to X such that $\gamma' = \kappa\gamma$. The diagram

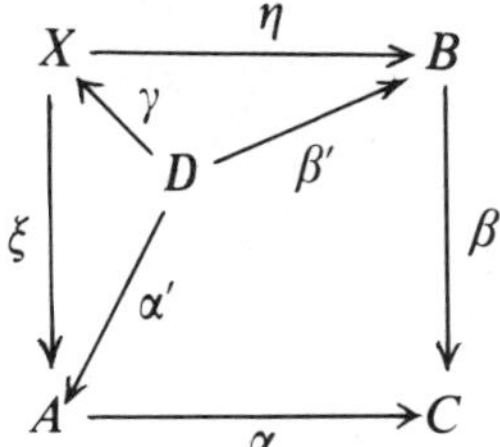

is now commutative, since

$$\xi\gamma = \pi_A\kappa\gamma = \pi_A\gamma' = \alpha'$$

and similarly

$$\eta\gamma = \beta'.$$

If we have another morphism γ_1 from D to X such that $\xi\gamma_1 = \alpha'$ and $\eta\gamma_1 = \beta'$ then $\pi_A(\kappa\gamma_1) = \alpha' = \pi_A(\kappa\gamma)$ and $\pi_B(\kappa\gamma_1) = \beta' = \pi_B(\kappa\gamma)$; so, by the couniversal property of $(P; \pi_A, \pi_B)$, we have $\kappa\gamma_1 = \kappa\gamma$ and hence $\gamma_1 = \gamma$ since κ is monic.

Thus $(X; \xi, \eta)$ is a pullback for $(A, B, C; \alpha, \beta)$ as required.

We can deduce from the discussion in this section that if $\boldsymbol{C}$ is an abelian category then it enjoys the following properties:

(1) $\boldsymbol{C}$ has a zero object;

(2) Every pair of objects in $\boldsymbol{C}$ has a product and a coproduct in $\boldsymbol{C}$;

(3) Every morphism in $\boldsymbol{C}$ has a kernel and a cokernel;

(4) Every monic morphism in $\boldsymbol{C}$ is the kernel of a morphism in $\boldsymbol{C}$.

It can be shown that, conversely, every category which enjoys these properties is an abelian category according to our definition. For the details we refer the reader to Freyd, *Abelian Categories*.

§17 Functors

Let $\boldsymbol{C}_1$ and $\boldsymbol{C}_2$ be two categories. By a *covariant functor* T from $\boldsymbol{C}_1$ to $\boldsymbol{C}_2$ we mean a procedure which assigns to each object A of $\boldsymbol{C}_1$ an object of $\boldsymbol{C}_2$, which we denote by $T(A)$, and to each morphism α of $\boldsymbol{C}_1$ with domain A and codomain B a morphism $T(\alpha)$ of $\boldsymbol{C}_2$ with domain $T(A)$ and codomain $T(B)$ such that the following two conditions are satisfied:

F1. For each object A of $\boldsymbol{C}_1$, $T(I_A)$ is the identity morphism of $T(A)$;

F2. If the composition $\beta\alpha$ of the morphisms α and β of $\mathbf{C}_1$ is defined then the composition $T(\beta)T(\alpha)$ of the morphisms $T(\alpha)$ and $T(\beta)$ of $\mathbf{C}_2$ is defined and we have $T(\beta\alpha) = T(\beta)T(\alpha)$.

A *contravariant functor* from $\mathbf{C}_1$ to $\mathbf{C}_2$ is defined to be a covariant functor from $\mathbf{C}_1$ to the dual category $\mathbf{C}_2^*$. Suppose T' is such a functor. Then to each object A of $\mathbf{C}_1$ the functor T' assigns an object $T'(A)$ of the category $\mathbf{C}_2^*$; but this is, by the definition of the dual category, simply an object of the category $\mathbf{C}_2$. Next let α be any morphism of $\mathbf{C}_1$ with domain A and codomain B; the functor T' assigns to α a morphism $T'(\alpha)$ in the set $\mathbf{C}_2^*(T'(A), T'(B)) = \mathbf{C}_2(T'(B), T'(A))$, i.e. a morphism of $\mathbf{C}_2$ with domain $T'(B)$ and codomain $T'(A)$. Condition *F1* implies easily that for each object A of $\mathbf{C}_1$ the morphism $T'(I_A)$ is the identity morphism of $T'(A)$. According to condition *F2* if the composition $\beta\alpha$ of the morphisms α and β of $\mathbf{C}_1$ is defined then the composition $T'(\beta)\,\mathbf{C}_2^*T'(\alpha)$ is defined and $T'(\beta\alpha) = T'(\beta)\,\mathbf{C}_2^*T'(\alpha) = T'(\alpha)\,\mathbf{C}_2T'(\beta)$.

Example 1. Let ${}_R\mathbf{M}$ and $\mathbf{A}$ be the categories of left R-modules and abelian groups respectively. Let A be a right R-module. For each left R-module V set $T(V) = A \otimes_R V$ and for each R-homomorphism α from V to another left R-module V' set $T(\alpha) = I_A \otimes \alpha$. Then it follows from Theorem 12.3 that T is a covariant functor from ${}_R\mathbf{M}$ to $\mathbf{A}$. We shall denote T by $\mathrm{Ten}_{A}.$; it is sometimes also referred to as 'the functor $A \otimes$—'.

Example 2. In the same way if $\mathbf{M}_R$ and $\mathbf{A}$ are the categories of right R-modules and abelian groups respectively and B is a left R-module, we may define a covariant functor $\mathrm{Ten}._B$ from $\mathbf{M}_R$ to $\mathbf{A}$ by setting $\mathrm{Ten}._B(V) = V \otimes {}_R B$ for each right R-module V and $\mathrm{Ten}._B(\alpha) = \alpha \otimes I_B$ for each R-homomorphism α from V to another right R-module V'. This functor is sometimes called 'the functor $- \otimes B$'.

Example 3. Let ${}_R\mathbf{M}$ and $\mathbf{A}$ be as in Example 1. Let A be a left R-module. For each left R-module V set $T(V) = \mathrm{Hom}_R(A, V)$ and for each R-homomorphism α from V to another left R-module V' set $T(\alpha) = \mathrm{Hom}(I_A, \alpha)$. Then it follows from Theorem 8.2 that T is a covariant functor from ${}_R\mathbf{M}$ to $\mathbf{A}$. We shall denote it by $\mathrm{Hom}_{A}.$; another commonly used notation is $\mathrm{Hom}(A,\ )$.

Example 4. Still keeping the same notation, let B be a left R-module. For each left R-module V set $T(V) = \mathrm{Hom}_R(V, B)$ and for each

R-homomorphism α from V to another left R-module V' set $T(\alpha) = \text{Hom}(\alpha, I_B)$. Using Theorem 8.2, we can show that T is a contravariant functor from ${}_R\mathsf{M}$ to A. It is denoted by $\text{Hom}_{\cdot B}$ or $\text{Hom}(\ , B)$.

Example 5. Let ${}_R\mathsf{M}$ and M_R be the categories of left and right R-modules respectively. For each left R-module V set $D(V) = V^* = \text{Hom}_R(V, R_l)$, the dual of V with the right R-module structure defined in §8; for each R-homomorphism α from V to another left R-module V' set $D(\alpha) = \alpha^* = \text{Hom}(\alpha, I_R)$. It can be shown that α^* is not only an abelian group homomorphism, but actually a right R-module homomorphism from $D(V')$ to $D(V)$. Then D is a contravariant functor from ${}_R\mathsf{M}$ to M_R, which we naturally call the *dual functor*.

Example 6. Let ${}_R\mathsf{M}$ be the category of left R-modules. For each left R-module V set $T(V) = V^{**}$, the bidual of V; for each R-homomorphism α from V to another left R-module V' set $T(\alpha) = \alpha^{**} = \text{Hom}(\alpha^*, I_R)$, which is an R-homomorphism from V^{**} to $(V')^{**}$. Then T is a covariant functor from ${}_R\mathsf{M}$ to itself, which we denote by *Bd* and call the *bidual functor*.

Let $\mathbf{C}_1$ and $\mathbf{C}_2$ be categories, T_1 and T_2 covariant functors from $\mathbf{C}_1$ to $\mathbf{C}_2$. A *natural transformation* from the functor T_1 to the functor T_2 is a procedure η which assigns to each object X of $\mathbf{C}_1$ a morphism $\eta(X)$ of $\mathbf{C}_2$ with domain $T_1(X)$ and codomain $T_2(X)$ such that for every pair of objects X and Y of $\mathbf{C}_1$ and every morphism α of $\mathbf{C}_1$ from X to Y the diagram

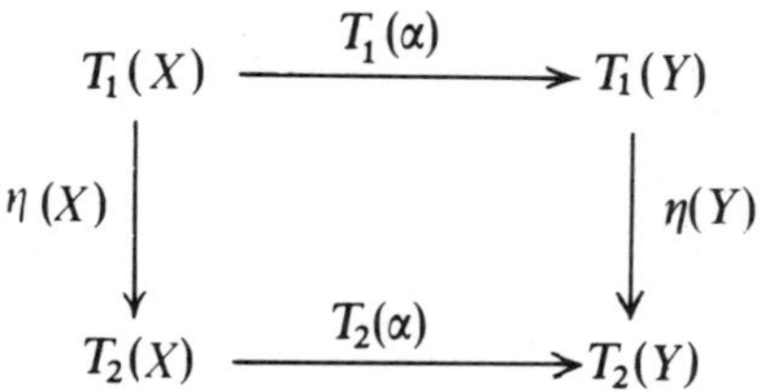

is commutative. If for every object X of $\mathbf{C}_1$ the morphism $\eta(X)$ is an isomorphism then the natural transformation η is called a *natural equivalence* between the functors T_1 and T_2.

Example 7. Let $\mathbf{C}_1 = \mathbf{C}_2 = {}_R\mathsf{M}$. Let T_1 be the identity functor I defined by setting $I(V) = V$ and $I(\alpha) = \alpha$ for each left R-module V

and each R-homomorphism α; let T_2 be the bidual functor Bd of Example 6. We shall define a natural transformation from I to Bd. Let V be any left R-module; we define the mapping η_V from $I(V) = V$ to $Bd(V) = V^{**}$ as follows. For each element x of V, $\eta_V(x)$ is the mapping from V^* to R given by

$$[\eta_V(x)](t^*) = t^*(x)$$

for all elements t^* of V^*. We verify that $\eta_V(x)$, so defined, is an R-homomorphism from V^* to R_r and hence belongs to V^{**}; then we check that η_V is actually an R-homomorphism from V to V^{**}. Now, let V and V' be left R-modules, α an R-homomorphism from V to V'. We claim that the diagram

$$\begin{array}{ccc} V & \xrightarrow{\alpha} & V' \\ \eta_V \downarrow & & \downarrow \eta_{V'} \\ V^{**} & \xrightarrow{\alpha^{**}} & (V')^{**} \end{array}$$

is commutative. To see this, let x be any element of V, t^* any element of $(V')^*$; then

$$\begin{aligned}[\alpha^{**}\eta_V(x)](t^*) = [\eta_V(x)](\alpha^*(t^*)) &= [\alpha^*(t^*)](x) \\ &= t^*(\alpha(x)) = [\eta_{V'}(\alpha(x))](t^*).\end{aligned}$$

Hence, if we define η by setting $\eta(V) = \eta_V$, we see that η is a natural transformation.

It can be shown that if $C_1 = C_2 =$ the category of finite-dimensional vector spaces over a division ring and η is the natural transformation from the identity functor to the bidual functor which we have just described then η is a natural equivalence.

Let C_1 and C_2 be additive categories, T a functor from C_1 to C_2 (covariant or contravariant); T is said to be an *additive functor* if for every pair of morphisms α, β of C_1 with the same domain and the same codomain we have $T(\alpha + \beta) = T(\alpha) + T(\beta)$. If T is covariant this is equivalent to saying that for every pair of objects A, B of C_1 the functor T induces a homomorphism from the abelian group $C_1(A, B)$ to the abelian group $C_2(T(A), T(B))$; if T is contravariant, this induced homomorphism maps $C_1(A,B)$ to $C_2(T(B), T(A))$.

Example 8. It follows from Theorem 12.4 that the covariant functors $\text{Ten}_{A\cdot}$ and $\text{Ten}_{\cdot B}$ described in Examples 1 and 2 are additive functors.

Example 9. Theorem 8.3 shows that the functors $\text{Hom}_{A\cdot}$ and $\text{Hom}_{\cdot B}$ of Examples 3 and 4 are additive functors.

It is easy to see that if Z is a zero object of C_1 and T is an additive functor from C_1 to C_2 then $T(Z)$ is a zero object of C_2. For if Z is a zero object then $C_1(Z, Z)$ consists of a single element, namely the zero element ζ_{ZZ}. Thus $I_Z = \zeta_{ZZ}$ and so we have $I_{T(Z)} = T(I_Z) = T(\zeta_{ZZ}) = \zeta_{T(Z), T(Z)} = \zeta'$ say; it follows that if φ is any element of $C_2(T(Z), T(Z))$ we have $\varphi = \varphi I_{T(Z)} = \varphi\zeta' = \zeta'$. So $C_2(T(Z), T(Z))$ consists of the zero morphism alone and we easily deduce that $T(Z)$ is a zero object of C_2.

Let T be a covariant functor from C_1 to C_2 where C_1 and C_2 are now exact categories. Then T is said to be an *exact functor* if for every short exact sequence

$$0 \to A \xrightarrow{\alpha} B \xrightarrow{\beta} C \to 0 \qquad [17.1]$$

of objects and morphisms of C_1 the sequence

$$0 \to T(A) \xrightarrow{T(\alpha)} T(B) \xrightarrow{T(\beta)} T(C) \to 0 \qquad [17.2]$$

is an exact sequence in C_2. (Here we fall easily into the harmless and useful habit of denoting all zero objects in all exact categories by the symbol 0.) We say that T is *right exact* if for every short exact sequence [17.1] the sequence

$$T(A) \xrightarrow{T(\alpha)} T(B) \xrightarrow{T(\beta)} T(C) \to 0$$

is exact; T is said to be *left exact* if for every short exact sequence [17.1] the sequence

$$0 \to T(A) \xrightarrow{T(\alpha)} T(B) \xrightarrow{T(\beta)} T(C)$$

is exact.

If T is a contravariant functor it is said to be exact, right exact or left exact if for every short exact sequence [17.1] the sequences

$$0 \to T(C) \to T(B) \to T(A) \to 0$$

$$T(C) \to T(B) \to T(A) \to 0$$

$$0 \to T(C) \to T(B) \to T(A)$$

respectively are exact.

Example 10. Let $\mathcal{M}_R$ and $\mathcal{A}$ be the abelian categories of right R-modules and abelian groups respectively. Let B be a left R-module and $\text{Ten}_{\cdot B}$ the covariant additive functor of Example 2. Then Theorem 12.5 shows that $\text{Ten}_{\cdot B}$ is a right exact functor from $\mathcal{M}_R$ to $\mathcal{A}$. According to the definition of flatness (see page 101). $\text{Ten}_{\cdot B}$ is exact if and only if B is flat.

Example 11. Let ${}_R\mathcal{M}$ and $\mathcal{A}$ be the abelian categories of left R-modules and abelian groups respectively. Let A be a right R-module and $\text{Ten}_{A\cdot}$ the covariant additive functor of Example 1. According to Theorem 12.6, this is a right exact functor from ${}_R\mathcal{M}$ to $\mathcal{A}$. Again we see that it is exact if and only if A is flat.

Example 12. Let ${}_R\mathcal{M}$ and $\mathcal{A}$ be as before and let A be a left R-module. In Example 3 we defined a covariant functor $\text{Hom}_{A\cdot}$ from ${}_R\mathcal{M}$ to $\mathcal{A}$, which is in fact an additive functor. According to Theorem 8.4, this functor is left exact, and Theorem 11.6 shows that it is exact if and only if A is projective.

Example 13. Let ${}_R\mathcal{M}$ and $\mathcal{A}$ be as before, B a left R-module, and consider the contravariant functor $\text{Hom}_{\cdot B}$ from ${}_R\mathcal{M}$ to $\mathcal{A}$ which we defined in Example 4; this is also an additive functor. Theorem 8.4 shows that this functor is left exact, and it follows from Theorem 11.8 that it is exact if and only if B is injective.

Example 14. Let $\mathcal{D}$ be the abelian category of direct systems of left R-modules indexed by a given directed set. Then we may define a covariant functor T from $\mathcal{D}$ to ${}_R\mathcal{M}$ by setting $T(V_i, \alpha_{ij}) = \varinjlim V_i$ and $T(\varphi) = \varinjlim \varphi_i$ for all direct systems (V_i, α_{ij}) and all morphisms $\varphi = (\varphi_i)$. This functor T is additive and it is an easy exercise to show that T is exact.

Example 15. Let $\mathcal{I}$ be the abelian category of inverse systems of left R-modules indexed by a given ordered set. We define a functor T from $\mathcal{I}$ to ${}_R\mathcal{M}$ by setting $T(V_i, \alpha_{ji}) = \varprojlim V_i$ and $T(\varphi) = \varprojlim \varphi_i$ for all inverse systems (V_i, α_{ji}) and all morphisms $\varphi = (\varphi_i)$. Then T is a covariant additive functor and it is not hard to show that T is left exact.

Although we have seen from these examples that an additive functor does not necessarily transform short exact sequences into short exact sequences, nevertheless there is a special situation in which this occurs.

THEOREM 17.1. *Let $\mathbf{C}_1$ and $\mathbf{C}_2$ be additive exact categories, T a covariant functor from $\mathbf{C}_1$ to $\mathbf{C}_2$. If*

$$0 \to A' \xrightarrow{\alpha} A \xrightarrow{\beta} A'' \to 0 \qquad [17.3]$$

is a split short exact sequence in $\mathbf{C}_1$ then

$$0 \to T(A') \xrightarrow{T(\alpha)} T(A) \xrightarrow{T(\beta)} T(A'') \to 0 \qquad [17.4]$$

is a split short exact sequence in $\mathbf{C}_2$.

Proof. Suppose [17.3] is split (by β'' say) and hence cosplit (by α' say). Then, as in the proof of Theorem 16.7 we have

$$\beta\alpha = \zeta, \quad \alpha'\beta'' = \zeta \quad \text{and} \quad \alpha\alpha' + \beta''\beta = I_A.$$

Since T is an additive functor, it follows that

$$T(\beta)\,T(\alpha) = \zeta,\ T(\alpha')\,T(\beta'') = \zeta,\ T(\alpha)\,T(\alpha') + T(\beta'')\,T(\beta) = I_{T(A)}.$$

We deduce easily from these relations that $(T(A); T(\alpha'), T(\beta))$ is a product for $(T(A'), T(A''))$ and $(T(A); T(\alpha), T(\beta''))$ is the corresponding coproduct. It follows from Theorem 16.7 that the sequence [17.4] is a short exact sequence.

§18 Relations in an abelian category

A relation between two sets A and B is usually defined to be a subset of the product set $A \times B$; among these relations are the graphs of the mappings from A to B (the functional relations between A and B). There is a natural generalisation of this notion to the case of left R-modules. Namely, if A and B are left R-modules we define an *additive relation* between A and B to be a submodule of the product module $A \times B$; among these are the graphs of the R-homomorphisms from A to B. We should now like to generalise the notion of additive relation still further—to an arbitrary abelian category. Here, however, the difficulty arises that although we have a product for each pair of objects in such a category we have not yet defined 'subobjects'.

To remedy this deficiency we notice first that if A_0 is a submodule of a left R-module A then of course it gives rise to a monomorphism from A_0 to A, namely the inclusion monomorphism ι, and A_0 is the image of this monomorphism. Further, if θ is an isomorphism from any module X onto A_0, then $\iota' = \iota\theta$ is also a monomorphism into A, and A_0 is the image of this monomorphism also.

Guided by these remarks we turn now to an arbitrary category $\mathbf{C}$. Let A be an object of $\mathbf{C}$; if ι_1 and ι_2 are monic morphisms from objects A_1 and A_2 respectively to A, we shall say that ι_1 is *right-equivalent* to ι_2 if there exists an isomorphism θ from A_1 to A_2 such that $\iota_1 = \iota_2\theta$. We now make the hypothesis that for every monic morphism ι with codomain A there exists a monic morphism $Sb(\iota)$ right-equivalent to it, such that if ι_1 is right-equivalent to ι_2 then $Sb(\iota_1) = Sb(\iota_2)$. We call $Sb(\iota)$ the *subobject* of A associated with the monic morphism ι.

Now, let $\mathbf{A}$ be an abelian category. Then every pair (A, B) of objects of $\mathbf{A}$ has a product in $\mathbf{A}$; but this product is not uniquely determined—Example 19 of §15 gives a description of all the products of (A, B). We propose now to choose one of the products for (A, B), which we shall call *the* product of (A, B); we shall usually denote this specially chosen product by $(A \times B; \pi_A, \pi_B)$ and the associated coproduct by $(A \times B; \iota_A, \iota_B)$. If several products involving A and B as factors occur in the same discussion, more elaborate notation may be required in order to avoid confusion. It is easy to check that if A is any object and Z any zero object of $\mathbf{A}$ then $(A; \zeta_{AZ}, I_A)$ is a product for (Z, A) and $(A; \zeta_{ZA}, I_A)$ the corresponding coproduct; we shall agree to take $Z \times A = A$ and similarly $A \times Z = A$.

With these preliminaries established we can now define relations in an arbitrary abelian category. Namely, if A and B are objects of an abelian category $\mathbf{A}$, a *relation between A and B* is defined to be a subobject of the product $A \times B$. Let ρ be such a relation; then ρ is a monic morphism from an object R to $A \times B$. Clearly ρ gives rise to a pair of morphisms $(\pi_A\rho, \pi_B\rho)$ from R to A and B respectively, and of course $\rho = \iota_A(\pi_A\rho) + \iota_B(\pi_B\rho)$. Further, if η is any epic morphism with codomain R, then ρ is easily seen to be an image of $\iota_A(\pi_A\rho\eta) + \iota_B(\pi_B\rho\eta)$. The pairs $(\pi_A\rho\eta, \pi_B\rho\eta)$ are said to be *representative pairs* for the relation ρ. If, conversely, we start with a pair (α, β) of morphisms from an object S to A and B respectively, then any image ι of $\iota_A\alpha + \iota_B\beta$ is a monic morphism with codomain $A \times B$ and so $\sigma = Sb(\iota)$ is a relation between A and B. We call σ the relation *determined by* the pair (α, β); clearly (α, β) is a representative pair for this relation.

From now on we restrict our attention to abelian categories with the property that for each object A the class of subobjects of A is actually a set. Such a category is said to be *locally small.*

If A is a locally small abelian category, we may define a new category $R = R(A)$, the *category of relations* in A as follows. The objects of R are defined to be the same as the objects of A, and for each pair of objects A, B of R the set of R-morphisms $R(A, B)$ is defined to be the set of relations between A and B, i.e. the set of subobjects of $A \times B$. Composition of R-morphisms is defined as follows. If ρ, σ are relations between A and B, B and C respectively, let (α, β), (β', γ) be representative pairs for ρ, σ with domains R, S respectively. According to Theorem 16.8 there exists a pullback $(L; \xi, \eta)$ for $(R, S, B; \beta, \beta')$. We then define $\sigma \circ \rho$ to be the subobject of $A \times C$ determined by the pair $(\alpha\xi, \gamma\eta)$. The situation is illustrated in the diagram

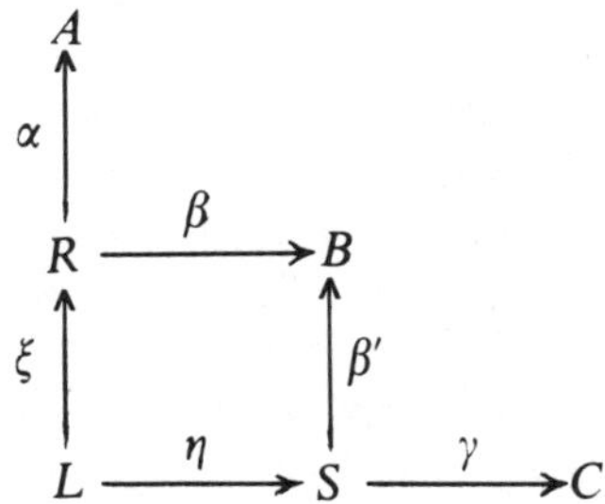

(Notice that we are using the symbol $\circ$ to denote composition of morphisms in R while continuing to denote composition in A by juxtaposition.)

It is clear from our remarks in §15 when we defined pullbacks that the subobject $\sigma \circ \rho$ does not depend on the particular pullback $(L; \xi, \eta)$ used for its construction; we must also show that it does not depend on the choice of representative pairs (α, β), (β', γ). To this end, we consider first the commutative diagram

$$\begin{array}{ccccc} L' & \xrightarrow{\lambda'} & P' & \xrightarrow{\varphi'} & Q' \\ & & \downarrow{\scriptstyle\theta} & & \downarrow{\scriptstyle\kappa} \\ L & \xrightarrow{\lambda} & P & \xrightarrow{\varphi} & Q \end{array} \qquad [18.1]$$

of objects and morphisms of A in which λ is a kernel of φ, λ' is a kernel of φ', θ is epic and κ is monic. Since $\varphi(\theta\lambda') = (\varphi\theta)\lambda' =$

$(\kappa\varphi')\lambda' = \kappa(\varphi'\lambda') = \zeta$ and λ is a kernel of φ, there exists a unique morphism ω from L' to L such that $\theta\lambda' = \lambda\omega$. We shall show that ω is epic.

To this end let $\varphi = \varphi_2\varphi_1$ and $\varphi' = \varphi_2'\varphi_1'$ where φ_1, φ_1' are epic and φ_2, φ_2' are monic; then we have $\varphi_2(\varphi_1\theta) = (\varphi_2\varphi_1)\theta = \varphi\theta = \kappa\varphi' = \kappa(\varphi_2'\varphi_1') = (\kappa\varphi_2')\varphi_1'$. Since φ_2, $\kappa\varphi_2'$ are monic and $\varphi_1\theta$, φ_1' are epic, it follows from Theorem 16.2 that there is an isomorphism ε such that $\varphi_1\theta = \varepsilon\varphi_1'$. Now φ_1' is a coimage for φ'; hence so is $\varphi_1\theta$. It follows that $\varphi_1\theta$ is a cokernel of λ'. We shall now show that φ_1 is a cokernel of $\theta\lambda'$. Certainly we have $\varphi_1(\theta\lambda') = \zeta$; next, if $\nu(\theta\lambda') = \zeta$ we have $(\nu\theta)\lambda' = \zeta$ whence there is a unique morphism ν_1 such that $\nu\theta = \nu_1(\varphi_1\theta) = (\nu_1\varphi_1)\theta$ and so $\nu = \nu_1\varphi_1$ (since θ is epic). Thus φ_1 is a cokernel of $\theta\lambda' = \lambda\omega$. Let now $\lambda\omega = \psi_2\psi_1$ where ψ_1 is epic and ψ_2 is monic; then ψ_2 is an image of $\lambda\omega$ and hence a kernel of φ_1. But λ is a kernel of φ_1; so there is an isomorphism ι such that $\lambda\iota = \psi_2$. Thus $\lambda\omega = \psi_2\psi_1 = (\lambda\iota)\psi_1 = \lambda(\iota\psi_1)$; since λ is monic it follows that $\omega = \iota\psi_1$, which is epic.

We return now to the problem of showing that the relation $\sigma \circ \rho$ which we defined above does not depend on the choice of representative pairs for ρ and σ. So let θ_1 and θ_2 be epic morphisms from R' to R and S' to S respectively. If $P = R \times S$, $P' = R' \times S'$ and ι_R, ι_S, π_R, π_S, $\iota_{R'}$, $\iota_{S'}$, $\pi_{R'}$, $\pi_{S'}$, have the obvious significance then $\theta = \iota_R\theta_1\pi_{R'} + \iota_S\theta_2\pi_{S'}$ is easily seen to be an epic morphism from P' to P. Let λ, with domain L, be a kernel of the morphism $\varphi = \beta\pi_R - \beta'\pi_S$ from P to B; then, as we saw in Theorem 16.8, $(L; \pi_R\lambda, \pi_S\lambda)$ is a pullback for $(R, S, B; \beta, \beta')$. Similarly, if λ', with domain L', is a kernel of $\varphi' = \beta\theta_1\pi_{R'} - \beta'\theta_2\pi_{S'}$, then $(L'; \pi_{R'}\lambda', \pi_{S'}\lambda')$ is a pullback for $(R', S', B; \beta\theta_1, \beta'\theta_2)$. Now in [18.1] take $Q = Q' = B$ and $\kappa = I_B$. It follows from our discussion of [18.1] that there exists an epic morphism ω from L' to L such that $\lambda\omega = \theta\lambda'$.

Now we have $\pi_R\lambda\omega = \pi_R\theta\lambda' = \theta_1\pi_{R'}\lambda'$ and similarly $\pi_S\lambda\omega = \theta_2\pi_{S'}\lambda'$. So $(\alpha\theta_1\pi_{R'}\lambda', \gamma\theta_2\pi_{S'}\lambda') = (\alpha\pi_R\lambda\omega, \gamma\pi_S\lambda\omega)$, which represents the same subobject as $(\alpha\pi_R\lambda, \gamma\pi_S\lambda)$. If we take $(\alpha, \beta) = (\pi_A\rho, \pi_B\rho)$ and a similar pair for (β', γ), it now follows from our discussion that all choices of representative pairs for ρ and σ lead to the same subobject $\sigma \circ \rho$.

We have now to check that $\mathsf{R} = \mathsf{R}(\mathsf{A})$, consisting of the objects, morphisms and composition defined above, is in fact a category. The condition *C1* of §15 is clearly satisfied.

To verify *C2* let A, B, C, D be objects of R (i.e. objects of A) and let ρ, σ, τ be relations between A and B, B and C, C and D respectively. Choose representative pairs (α, β), (β', γ), (γ', δ) with domains R, S, T for ρ, σ, τ respectively. Then we construct the diagram

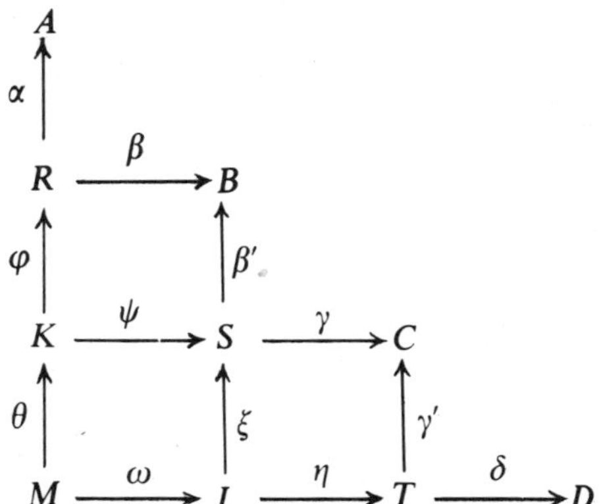

in which the three squares are pullbacks. Then $(\alpha\varphi, \gamma\psi)$ and $(\beta'\xi, \delta\eta)$ are representative pairs for $\sigma \circ \rho$ and $\tau \circ \sigma$ respectively.

According to Theorem 15.3, $(M; \theta, \eta\omega)$ is a pullback for $(K, T, C; \gamma\psi, \gamma')$. Hence $((\alpha\varphi)\theta, \delta(\eta\omega))$ is a representative for $\tau \circ (\sigma \circ \rho)$. Again by Theorem 15.3, $(M; \varphi\theta, \omega)$ is a pullback for $(R, L, B; \beta, \beta'\xi)$ and hence $(\alpha(\varphi\theta), (\delta\eta)\omega)$ is a representative pair for $(\tau \circ \sigma) \circ \rho$. Thus $\tau \circ (\sigma \circ \rho) = (\tau \circ \sigma) \circ \rho$ and $C2$ is established.

Finally, to verify that $C3$ holds, let A be any object of R and consider the relation J_A between A and A represented by the pair (I_A, I_A). Let ρ be a relation between A and an object B of R, represented by the pair (α, β) where α and β are morphisms from an object R to A and B respectively. We check easily that $(R; \alpha, I_R)$ is a pullback for $(A, R, A; I_A, \alpha)$. Thus to form $\rho \circ J_A$ we look at the diagram

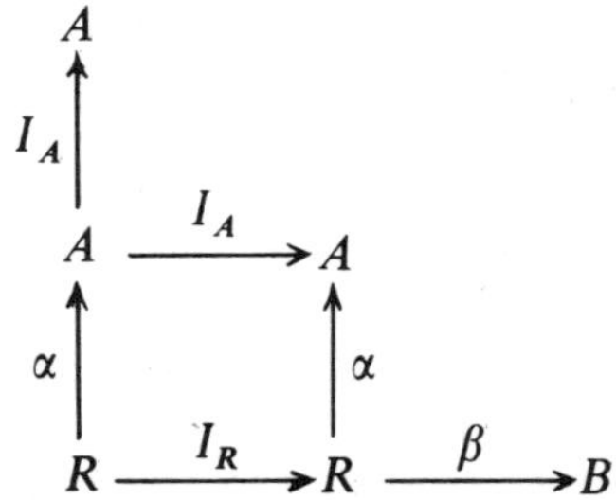

and read off that $\rho \circ J_A$ is the relation represented by $(I_A\alpha, \beta I_R) = (\alpha, \beta)$. So $\rho \circ J_A = \rho$ and similarly $J_A \circ \sigma = \sigma$ for every relation σ between B and A. Thus J_A is an identity and *C3* is satisfied.

Now let ρ be a relation between A and B. If (α, β) is a representative pair for ρ, then (β, α) determines a relation between B and A, which clearly depends only on ρ, not on the choice of the pair (α, β); we denote this relation by ρ^* and call it the relation *opposite* to ρ. Clearly $(\rho^*)^* = \rho$, and it follows easily from the definition of composition in R that if σ is a relation between B and another object C then $(\sigma \circ \rho)^* = \rho^* \circ \sigma^*$.

We now define a covariant functor G from the original category A to its category of relations R. Namely, for each object A of A we set $G(A) = A$ and for each morphism α in $\mathsf{A}(A, B)$ we define $G(\alpha)$ to be the relation between A and B (i.e. R-morphism from A to B) determined by the pair (I_A, α). We call $G(\alpha)$ the *graph* of the morphism α.

To verify that G is indeed a covariant functor from A to R, we must show that the conditions *F1* and *F2* of §17 are satisfied. The first condition clearly holds, since the identity R-morphism J_A of each object A is by definition the graph $G(I_A)$ of the identity A-morphism I_A of A.

Next let α and β be A-morphisms from A to B and B to C respectively. We shall show that $G(\beta\alpha) = G(\beta) \circ G(\alpha)$. To this end, let κ, with domain K, be a kernel for the morphism $\alpha\pi_A - \pi_B$ from $A \times B$ to B. Then $(K; \pi_A\kappa, \pi_B\kappa)$ is a pullback for $(A, B, B; \alpha, I_B)$ and so the diagram

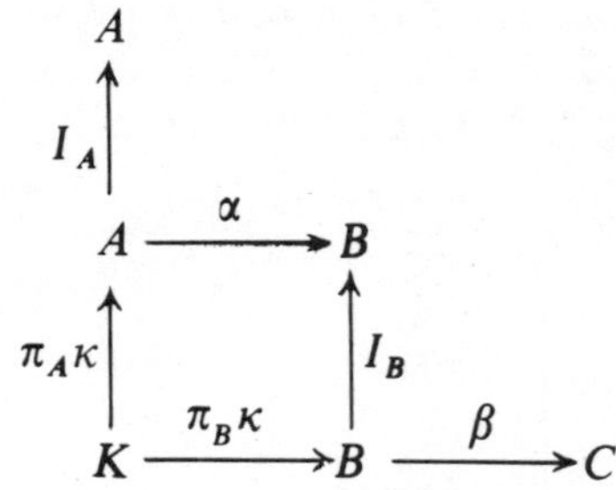

shows that $(\pi_A\kappa, \beta\pi_B\kappa)$ is a representative pair for $G(\beta) \circ G(\alpha)$. But $\pi_B\kappa = \alpha\pi_A\kappa$; so $(\pi_A\kappa, \beta\alpha\pi_A\kappa)$ is a representative pair for $G(\beta) \circ G(\alpha)$.

Now we show that $\pi_A\kappa$ is epic. To this end we notice first that

$(\alpha\pi_A - \pi_B)(\iota_A + \iota_B\alpha) = \zeta$; hence there is a unique morphism φ such that $\iota_A + \iota_B\alpha = \kappa\varphi$. Suppose now that $\theta\pi_A\kappa = \zeta$; then $\theta\pi_A\kappa\varphi = \zeta$ and so $\theta\pi_A(\iota_A + \iota_B\alpha) = \zeta$, whence $\theta = \zeta$. Thus $\pi_A\kappa$ is epic.

Since $(I_A, \beta\alpha)$ determines $G(\beta\alpha)$ and $\pi_A\kappa$ is epic, it follows that $(\pi_A\kappa, \beta\alpha\pi_A\kappa)$ is also a representative pair for $G(\beta\alpha)$. Hence $G(\beta) \circ G(\alpha) = G(\beta\alpha)$ and $F2$ is established.

For each pair of objects A, B of $\boldsymbol{A}$ the functor G gives a mapping g_{AB} from the set $\boldsymbol{A}(A, B)$ to the set $\boldsymbol{R}(A, B)$ defined by setting $g_{AB}(\alpha) = G(\alpha)$ for each morphism α in $\boldsymbol{A}(A, B)$. We claim that these mappings g_{AB} are all injective.

First we remark that if α is any $\boldsymbol{A}$-morphism from A to B then $\iota_A + \iota_B\alpha$ is a monic morphism from A to $A \times B$; for if $(\iota_A + \iota_B\alpha)\,\varphi = \zeta$ then on composing with π_A we deduce at once that $\varphi = \zeta$. It follows that $G(\alpha) = Sb(\iota_A + \iota_B\alpha)$. Thus if α and α' are $\boldsymbol{A}$-morphisms from A to B such that $g_{AB}(\alpha) = g_{AB}(\alpha')$ we have $Sb(\iota_A + \iota_B\alpha) = Sb(\iota_A + \iota_B\alpha')$; hence there is an isomorphism θ such that $\iota_A + \iota_B\alpha = (\iota_A + \iota_B\alpha')\,\theta$. Composing with π_A we deduce that $\theta = I_A$ and then composing with π_B we obtain $\alpha = \alpha'$, as required.

Since the mappings g_{AB} are injective and 'preserve composition' it will be safe to identify the $\boldsymbol{A}$-morphisms α from A to B with their graphs in $\boldsymbol{R}(A, B)$, and we shall do this where it seems convenient.

We should now like to characterize the graphs of $\boldsymbol{A}$-morphisms from A to B among the $\boldsymbol{R}$-morphisms from A to B. For this it is necessary to introduce a little more machinery. Namely, let Z be any zero object of $\boldsymbol{A}$; then for every object X of $\boldsymbol{A}$ we agreed to take X itself as the specially selected product of Z and X. So the relations between Z and X are simply subobjects of X. Two of these relations are of particular importance: $\omega_X = Sb(\zeta_{ZX})$, represented by (ζ_{ZZ}, ζ_{ZX}) and $\Omega_X = Sb(I_X)$, represented by (ζ_{XZ}, I_X).

THEOREM 18.1. *Let $\boldsymbol{A}$ be a locally small abelian category, $\boldsymbol{R}$ its category of relations. If A and B are objects of $\boldsymbol{A}$ then an $\boldsymbol{R}$-morphism ρ from A to B is the graph of an $\boldsymbol{A}$-morphism from A to B if and only if $\rho \circ \omega_A = \omega_B$* and *$\rho^* \circ \Omega_B = \Omega_A$.*

Proof. (1) Suppose there exists an $\boldsymbol{A}$-morphism φ from A to B such that $\rho = G(\varphi)$.

We easily verify that $(Z;\ \zeta_{ZZ}, \zeta_{ZA})$ is a pullback for $(Z, A, A;\ \zeta_{ZA}, I_A)$ and that $(A;\ \varphi, I_A)$ is a pullback for $(B, A, B;\ I_B, \varphi)$. So we read off from the diagrams

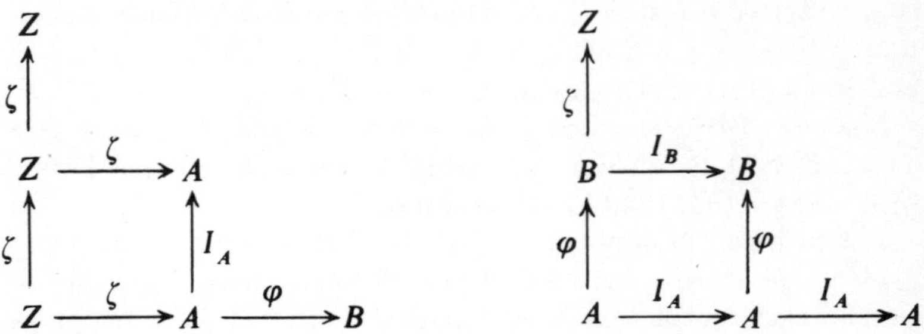

that $\rho \circ \omega_A$ and $\rho^* \circ \Omega_B$ are the relations determined by $(\zeta_{ZZ}\zeta_{ZZ}, \varphi\zeta_{ZA}) = (\zeta_{ZZ}, \zeta_{ZB})$ and $(\zeta_{BZ}\varphi, I_A I_A) = (\zeta_{AZ}, I_A)$ respectively, i.e. $\rho \circ \omega_A = \omega_B$ and $\rho^* \circ \Omega_B = \Omega_A$ as required.

(2) Conversely, suppose $\rho \circ \omega_A = \omega_B$ and $\rho^* \circ \Omega_B = \Omega_A$.

Let ρ have domain R and set $\alpha = \pi_A\rho$, $\beta = \pi_B\rho$.

We verify first that if κ, with domain K, is a kernel of α, then $(K; \zeta_{KZ}, \kappa)$ is a pullback for $(Z, R, A; \zeta_{ZA}, \alpha)$. Then the diagram

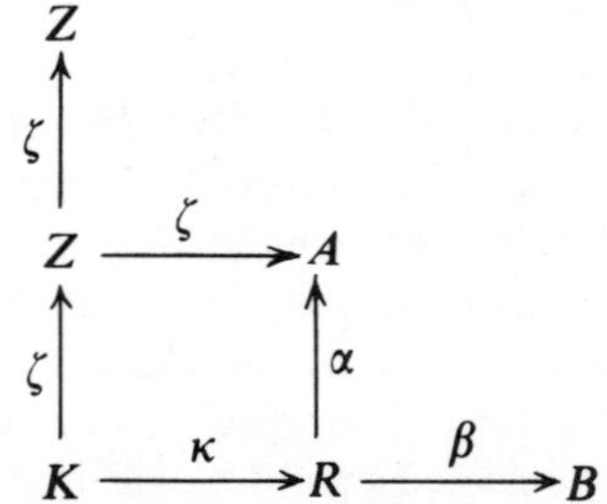

shows that $(\zeta_{ZZ}\zeta_{KZ}, \beta\kappa)$ is a representative pair for $\rho \circ \omega_A$. But since $\rho \circ \omega_A = \omega_B$, it follows that $\beta\kappa = \zeta$.

Next let $(B \times R; \pi'_B, \pi'_R)$ be the product for (B, R) and let λ, with domain L, be a kernel of $\pi'_B - \beta\pi'_R$. Then, according to the proof of Theorem 16.8, $(L; \pi'_B\lambda, \pi'_R\lambda)$ is a pullback for $(B, R, B; I_B, \beta)$. We thus read off from the diagram

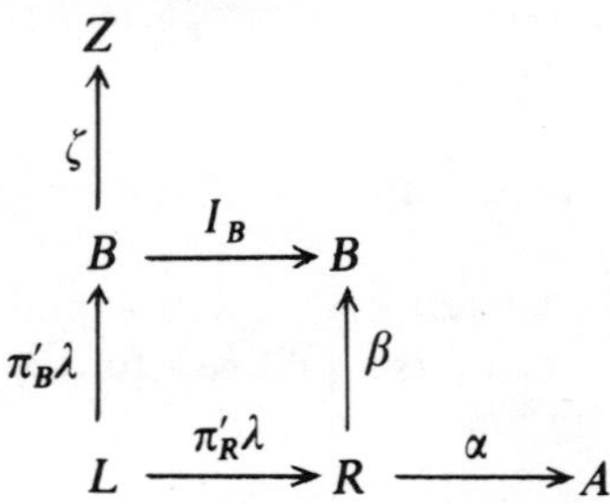

that $(\zeta\pi'_B\lambda, \alpha\pi'_R\lambda)$ is a representative pair for $\rho^* \circ \Omega_B$. Since $\rho^* \circ \Omega_B = \Omega_A$ it follows that $\alpha\pi'_R\lambda$ is epic.

We use these results to prove that α is an isomorphism.

First we show that α is monic. So suppose there is a morphism γ such that $\alpha\gamma = \zeta$; since κ is a kernel for α there exists a unique morphism γ' such that $\gamma = \kappa\gamma'$. Then we have

$$\rho\gamma = (\iota_A\pi_A\rho + \iota_B\pi_B\rho)\gamma = \iota_A\alpha\gamma + \iota_B\beta\gamma = \zeta + \iota_B\beta\kappa\gamma' = \zeta,$$

and since ρ is monic it follows that $\gamma = \zeta$.

Next we show that α is epic. So let θ be a morphism such that $\theta\alpha = \zeta$. Then $\theta(\alpha\pi'_R\lambda) = \zeta$ and so $\theta = \zeta$; thus α is epic.

Since α is monic and epic, it follows from the Corollary to Theorems 16.4 and 16.4* that α is an isomorphism and hence has an inverse α^{-1}. Then, since (α, β) is a representative pair for ρ, so also is $(\alpha\alpha^{-1}, \beta\alpha^{-1}) = (I_A, \beta\alpha^{-1})$. Consequently $\rho = g(\beta\alpha^{-1})$, as required.

This completes the proof.

In the category of sets $\boldsymbol{E}$ all epic morphisms (i.e. surjective mappings) are right-invertible and all monic morphisms (injective mappings) are left-invertible. The epic and monic morphisms in an arbitrary category $\boldsymbol{C}$ do not enjoy these properties—not even when $\boldsymbol{C}$ is abelian. But the next theorem shows that if α is an epic morphism of a locally small abelian category $\boldsymbol{A}$, then its graph $G(\alpha)$ is right-invertible in the category of relations $\boldsymbol{R}(\boldsymbol{A})$, and similarly if α is monic then $G(\alpha)$ is left-invertible.

THEOREM 18.2. *Let α be a morphism from A to B in a locally small abelian category $\boldsymbol{A}$, $G(\alpha)$ its graph in $\boldsymbol{R}(\boldsymbol{A})$. If α is epic then $G(\alpha) \circ G(\alpha)^* = J_B$; if α is monic, then $G(\alpha)^* \circ G(\alpha) = J_A$.*

Proof. (1) Suppose α is epic.

Since $(A; I_A, I_A)$ is a pullback for $(A, A, A; I_A, I_A)$ it follows in the usual way that (α, α) is a representative pair for $G(\alpha) \circ G(\alpha)^*$. But (I_B, I_B) is a representative pair for J_B, and since α is epic, so also is $(I_B\alpha, I_B\alpha) = (\alpha, \alpha)$.

Thus $G(\alpha) \circ G(\alpha)^* = J_B$.

(2) Suppose α is monic.

Then we check that $(A; I_A, I_A)$ is a pullback for $(A, A, B; \alpha, \alpha)$, and deduce that (I_A, I_A) is a representative pair for $G(\alpha)^* \circ G(\alpha)$. So $G(\alpha)^* \circ G(\alpha) = J_A$, as required.

The next theorem will allow us to give a necessary and sufficient condition in R that a given sequence of objects and morphisms of A should be exact.

THEOREM 18.3. *Let φ be morphism from A to B in a locally small abelian category A. If κ is a kernel of φ and ι is an image of φ then $G(\varphi)^* \circ \omega_B = Sb(\kappa)$ and $G(\varphi) \circ \Omega_A = Sb(\iota)$.*

Proof. If κ has domain K then $(K; \zeta_{KZ}, \kappa)$ is a pullback for $(Z, A, B; \zeta_{ZB}, \varphi)$. It follows in the usual way that $(\zeta_{ZZ}\zeta_{KZ}, I_A\kappa) = (\zeta_{KZ}, \kappa)$ is a representative pair for $G(\varphi)^* \circ \omega_B$.

Thus $G(\varphi)^* \circ \omega_B = Sb(\kappa)$.

Clearly $(A; I_A, I_A)$ is a pullback for $(A, A, A; I_A, I_A)$; so it follows that $(\zeta_{AZ}I_A, \varphi I_A) = (\zeta_{AZ}, \varphi)$ is a representative pair for $G(\varphi) \circ \Omega_A$.

Thus $G(\varphi) \circ \Omega_A = Sb(\iota)$.

COROLLARY. *Let A be a locally small abelian category. Then a sequence*

$$A \xrightarrow{\varphi} B \xrightarrow{\psi} C$$

of objects and morphisms of A is exact if and only if $G(\varphi) \circ \Omega_A = G(\psi)^* \circ \omega_C$.

If we were to identify the morphisms φ and ψ with their graphs then this condition would of course appear as $\varphi \circ \Omega_A = \psi^* \circ \omega_C$ and it is in this form that we shall make use of it in §28.

We conclude this section with two results of a technical nature which we shall require in our study of the connecting morphism associated with an exact sequence of chain complexes (see page 249).

THEOREM 18.4. *Let A be a locally small abelian category, R its category of relations. For every relation ρ in the set $\mathsf{R}(A, B)$ we have $\rho \circ \rho^* \circ \omega_B = \rho \circ \omega_A$ and $\rho \circ \rho^* \circ \Omega_B = \rho \circ \Omega_A$.*

Proof. Let (α, β) be a representative pair for ρ, with domain R.

If κ and λ, with domains K and L, are kernels for α and β respectively, then $(K; \zeta, \kappa)$ and $(L; \zeta, \lambda)$ are pullbacks for $(Z, R, A; \zeta, \alpha)$ and $(Z, R, B; \zeta, \beta)$ respectively. So it follows as usual that $(\zeta, \beta\kappa)$ and $(\zeta, \alpha\lambda)$ are representative pairs for $\rho \circ \omega_A$ and $\rho^* \circ \omega_B$ respectively.

Let now μ, with domain M, be a kernel for the morphism $\alpha\pi_R - \alpha\lambda\pi_L$ from $L \times R$ to A; then $(M; \pi_L\mu, \pi_R\mu)$ is a pullback for $(L, R, A; \alpha\lambda, \alpha)$ and we deduce that $\rho \circ \rho^* \circ \omega_B$ is determined by the pair $(\zeta, \beta\pi_R\mu)$.

To establish that $\rho \circ \rho^* \circ \omega_B = \rho \circ \omega_A$ it will be sufficient to show that there exists an epic morphism η from M to K such that $(\beta\kappa)\eta = \beta\pi_R\mu$. To this end, consider the diagram

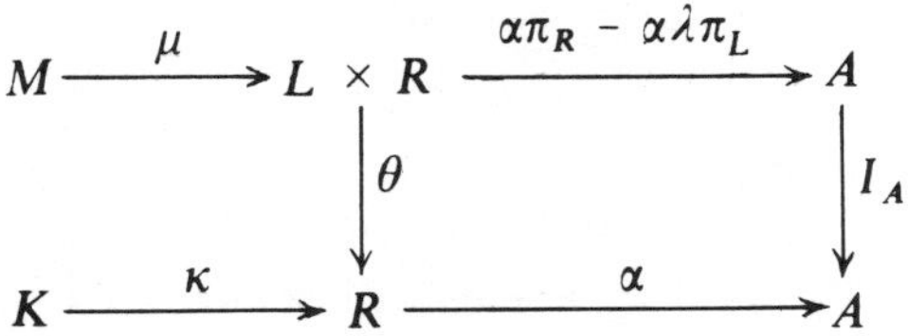

where $\theta = \pi_R - \lambda\pi_L$. This morphism θ is epic, for if $\gamma\theta = \zeta$ then we have $\gamma\pi_R = \gamma\lambda\pi_L$, whence $\gamma = \gamma\pi_R\iota_R = \gamma\lambda\pi_L\iota_R = \zeta$. We are now in the situation of [18.1]; so, by the argument after [18.1] it follows that there exists an epic morphism η from M to K such that $\kappa\eta = \theta\mu$. Then $(\beta\kappa)\eta = \beta(\kappa\eta) = \beta(\theta\mu) = \beta\pi_R\mu - \beta\lambda\pi_L\mu = \beta\pi_R\mu$, as required, since $\beta\lambda = \zeta$.

This completes the proof of the first assertion; we leave the proof of the second part as an exercise for the reader.

THEOREM 18.5. *Let A and B be objects of a locally small abelian category; let ι_1 and ι_2 be subobjects of A (relations between a zero object Z and A) and ρ a relation between A and B. If there is a monic morphism ι such that $\iota_2 = \iota_1\iota$ then there exists a monic morphism ι' such that $\rho \circ \iota_2 = (\rho \circ \iota_1)\iota'$.*

Proof. Let X_1, X_2 be the domains of ι_1, ι_2 respectively; let (α, β) be a representative pair for ρ, with domain R.

For $k = 1, 2$ let $(X_k \times R; \pi_k, \pi'_k)$ be a product for (X_k, R); if λ_k is a kernel for $\iota_k\pi_k - \alpha\pi'_k$, with domain L_k, then $(L_k; \pi_k\lambda_k, \pi'_k\lambda_k)$ is a pullback for $(X_k, R, A; \iota_k, \alpha)$ and hence in the usual way $\rho \circ \iota_k$ is the relation determined by the pair $(\zeta, \beta\pi'_k\lambda_k)$.

Now there is a morphism μ from $X_2 \times R$ to $X_1 \times R$ such that $\pi_1\mu = \iota\pi_2$ and $\pi'_1\mu = \pi'_2$ and it is easy to check that μ is monic. Then since $(\iota_1\pi_1 - \alpha\pi'_1)(\mu\lambda_2) = (\iota_2\pi_2 - \alpha\pi'_2)\lambda_2 = \zeta$ it follows that there exists a unique morphism ι' such that $\mu\lambda_2 = \lambda_1\iota'$. Clearly ι' is monic and we have $\beta\pi'_2\lambda_2 = \beta\pi'_1\mu\lambda_2 = \beta\pi'_1\lambda_1\iota'$.

The required result follows at once.

Chapter 4

STRUCTURE OF RINGS

§19. Simple Modules

Let R be a ring with an identity element e. We recall from §6 that a left R-module V is said to be *simple* or *irreducible* if (1) it does not consist of the zero element alone and (2) it has no submodules other than V itself and the zero submodule. We begin this section with a simple but useful criterion for a left R-module to be simple.

THEOREM 19.1. *A non-zero left R-module V is simple if and only if for every non-zero element x of V we have* $Rx = V$.

Proof. As in §6, Rx is the submodule of V generated by x.

(1) Suppose V is a simple left R-module.

If x is any non-zero element of V then Rx is not the zero submodule, since it contains $ex = x$. It follows that $Rx = V$, since this is the only other possibility.

(2) Conversely, suppose that for every non-zero element x of V we have $Rx = V$.

Let W be any non-zero submodule of V. If x is any non-zero element of W then of course W includes the submodule Rx generated by x. By hypothesis, $Rx = V$. Hence $W = V$ and so V is simple, as we asserted.

Example 1. Let $V = R_l$, the ring R itself considered as a left R-module. Since, as we have seen in Example 8 of §6, the submodules of R_l are simply the left ideals of R, it follows that the simple submodules of R_l are precisely the minimal left ideals of R. (See §3).

Example 2. It follows from Example 1 that R_l is itself a simple left R-module if and only if R has no left ideals except R_l itself and the zero ideal. It is easy to see that this condition is satisfied if and only if R is a division ring. We have already remarked in §3 that a division ring has no ideals, left, right or two-sided except R and the zero ideal. Conversely, suppose R has no other left ideals than these two. Let x be any non-zero element of R. Then the left ideal generated by x, which consists of all elements of the form rx with r in R, is the whole ring R and hence contains the identity e. Thus there is an element x' of R such that $x'x = e$, i.e. x has a left inverse x'. Similarly x' has a left inverse x''. Then, if we set $y = x''x'xx'$ we have on the one hand $y = (x''x')(xx') = e(xx') = xx'$ and on the other $y = x''(x'x)x' = x''ex' = x''x' = e$; so $xx' = e$ and hence x' is a two-sided inverse of x. Thus R is a division ring.

Example 3. Let M be a left ideal of R; then we may form the factor module $V = R_l/M$. According to Theorem 7.5 there is a one-to-one correspondence between the set of submodules of V and the set of submodules of R_l (i.e. left ideals of R) which include M. It follows at once that V is simple if and only if M is a maximal left ideal of R.

In fact all simple left R-modules are isomorphic to modules of this type. To see this, let V_1 be any simple left R-module and let x be any non-zero element of V_1. Consider the mapping α from R to V_1 defined by setting $\alpha(a) = ax$ for every element a of R; it is easy to check that α is an R-module homomorphism from R_l to V_1, and Theorem 19.1 shows that it is an epimorphism. If $M = \text{Ker } \alpha$ it follows from Theorem 7.3 that there exists an isomorphism α_* from R_l/M onto V_1. Since V_1 is simple, so also is R_l/M and hence M is a maximal left ideal.

Example 4. It is easily shown that an ideal of the ring of integers $\mathbf{Z}$ is maximal if and only if it is the principal ideal generated by a prime number p. It follows that a left $\mathbf{Z}$-module V is simple if and only if there is a prime number p such that V is isomorphic to the left $\mathbf{Z}$-module $\mathbf{Z}/(p)$, which of course has exactly p elements.

Example 5. Let R be a division ring. Then a left R-module (i.e. a left vector space over R) is simple if and only if its R-dimension is 1.

In §8 we saw how to impose a ring structure on the set $\text{Hom}_R(V, V)$ of all R-homomorphisms from a left R-module V to itself. When V is simple this ring has an interesting property.

THEOREM 19.2. (*Schur's Lemma*). *If V is a simple left R-module, then the ring* $\text{Hom}_R(V, V)$ *is a division ring.*

Proof. Suppose V is a simple left R-module.

Let α be a non-zero element of $\text{Hom}_R(V, V)$. Then $\text{Im } \alpha$ is not the zero submodule of V; hence $\text{Im } \alpha = V$, i.e. α is an epimorphism. Further, $\text{Ker } \alpha$ is not the whole module V itself; hence $\text{Ker } \alpha$ is the zero submodule and so α is a monomorphism. Thus α is an automorphism of V. Its inverse α^{-1} is also an automorphism of V; so $\mathrm{Hom}_R(V, V)$ is a division ring.

We now make the assumption that for every ring R we can associate with each simple left R-module V a simple left R-module isomorphic to it, which we call the *isomorphism type* of V and denote by Type V, such that if V is isomorphic to V' then Type V = Type V'. It can be shown that the property expressed by 'X is an isomorphism type of simple left R-modules' is set-forming; so it is legitimate to talk of the set of isomorphism types of simple left R-modules. We shall denote this set by $\mathscr{T}(R)$ or simply by $\mathscr{T}$.

§20. Semisimple Modules

Let R be a ring with identity, V a left R-module. Then V is said to be *semisimple* if it is the sum (not necessarily direct) of a family of simple submodules. Thus V is a semisimple module if and only if there exists a family $(S_i)_{i \in I}$ of simple submodules of V such that every element x of V can be expressed (not necessarily uniquely) in the form $x = \sum_{i \in I} s_i$, where $s_i \in S_i$ for each index i and $s_i = 0$ for all but a finite number of indices i. We propose now to give alternative characterizations of semisimple modules; the following result will be useful for this purpose.

THEOREM 20.1. *Let V be a semisimple left R-module which is the sum of the family $(S_i)_{i \in I}$ of simple submodules. If W is any submodule of V, there exists a subset J of I such that the sum $S_J = \sum_{i \in J} S_i$ is direct and V is the internal direct sum of S_J and W.*

Proof. Let E be the set of subsets K of I such that (1) the sum $S_K = \sum_{i \in K} S_i$ is direct and (2) $S_K \cap W = 0$. We shall show that E is inductively ordered by the inclusion relation.

So let E_0 be a totally ordered subset of E; we must prove that E_0 has a least upper bound in E. Let K_0 be the union of all the subsets K of I which belong to E_0; we claim that $K_0 \in E$, from which it easily follows that K_0 is the required least upper bound of E_0.

If x is any element of $S_{K_0} = \sum_{i \in K_0} S_i$, it follows that x can be expressed in the form $x = \sum_{i \in K_0} s_i$ where $s_i \in S_i$ for each index i in K_0 and only finitely many of the elements s_i are non-zero. Let $s_{i_1}, \ldots, s_{i_n}$ be the non-zero elements occurring in this representation. Since each of the indices i_j ($j = 1, \ldots, n$) belongs to K_0, it must belong to one of the subsets of I in E_0; say $i_j \in I_j$ ($j = 1, \ldots, n$). We recall that E_0 is a totally ordered set, from which it follows that the subsets $I_1, \ldots, I_n$ are all included in one of their number, say I_1. Then $x \in S_{I_1}$.

From this discussion we deduce first that $S_{K_0} \cap W = 0$. For if we have $x \in S_{K_0} \cap W$ it follows that $x \in S_{I_1} \cap W$; but $S_{I_1} \cap W = 0$, since I_1 belongs to E. Thus $x = 0$.

Next we show that the sum S_{K_0} is direct. Suppose first that the element $x = \sum_{i \in K_0} s_i$ is zero. Since $x \in S_{I_1}$ and the sum S_{I_1} is direct, it follows from Theorem 9.3 that $s_i = 0$ for every index i. Next, if x is now any element of S_{K_0} which has two expressions of the form $x = \sum_{i \in K_0} s_i = \sum_{i \in K_0} s_i'$ (with s_i and s_i' in S_i for each index i and only finitely many of the elements s_i, s_i' non-zero) we have $\sum_{i \in K_0} (s_i - s_i') = 0$, whence $s_i - s_i' = 0$, i.e. $s_i = s_i'$ for all indices i. Hence the expression for x is unique, and Theorem 9.3 shows at once that the sum S_{K_0} is direct.

This long argument now allows us to say that K_0 is the least upper bound of E_0. Thus E is inductively ordered and hence (by Zorn's Lemma) has a maximal element, J say.

Certainly the sum $S_J = \sum_{i \in J} S_i$ is direct and $S_J \cap W = 0$. We claim that the sum $S_J + W$ (which is direct, by Theorem 9.3) is in fact the whole module V. To prove this we shall show that $S_J + W$ includes every simple submodule of the family $(S_i)_{i \in I}$.

Suppose this is not the case; then there is an index i_0 in I such that S_{i_0} is not included in $S_J + W$. The intersection $S_{i_0} \cap (S_J + W)$ is of course a submodule of S_{i_0}, but it cannot be S_{i_0} itself; since S_{i_0} is simple it follows that the intersection is the zero submodule. Set $J' = J \cup \{i_0\}$. Then we see that the sum $S_{J'} = S_J + S_{i_0}$ is direct and $S_{J'} \cap W = 0$. Hence J' belongs to the set E. But $J' \supset J$, and J is a maximal element of E, so this is a contradiction.

Hence V is the internal direct sum of S_J and W.

THEOREM 20.2. *Let V be a left R-module. Then the following conditions are equivalent*:

(a) *V is a semisimple left R-module*;

(b) *V is the internal direct sum of a family of simple submodules*;

(c) *every submodule of V is a direct summand.*

Proof. (1) Suppose V is a semisimple left R-module.

Then, by definition, V is the sum of a family $(S_i)_{i \in I}$ of simple submodules. We now apply the preceding theorem in the special case where W is the zero submodule. It follows that there exists a subset J of I such that the sum $S_J = \sum_{i \in J} S_i$ is direct and $V = S_J \oplus 0$, whence $V = S_J$.

Thus (a) implies (b).

(2) Suppose V is the internal direct sum of a family of simple submodules, say $(S_i)_{i \in I}$. Then of course V is semisimple. Hence, if W is any submodule of V, we may deduce from Theorem 20.1 that for some subset J of I we have $W \oplus S_J = V$. That is to say, W is a direct summand of V.

So (b) implies (c).

(3) Finally, suppose that every submodule of V is a direct summand.

We show first that this property is hereditary, i.e. if V_0 is any submodule of V, then every submodule of V_0 is a direct summand of V_0.

So let W_0 be any submodule of V_0. Since W_0 is *a fortiori* a submodule of V, there exists a submodule W_1 of V such that V is the internal direct sum of W_0 and W_1. If $W_1' = V_0 \cap W_1$, it is easily verified that V_0 is the internal direct sum of W_0 and W_1'; in other words, W_0 is a direct summand of V_0.

Our next move is to show that under condition (c) every non-zero submodule of V includes a simple submodule.

If W is a non-zero submodule of V, let x be any non-zero element of W and form the submodule $W_0 = Rx$ generated by x. Clearly if we can prove that W_0 includes a simple submodule it will follow that W includes a simple submodule.

To do this we consider the set E of submodules of W_0 which do not contain x. We claim that E is inductively ordered by the inclusion relation; so let E_0 be any totally ordered subset of E. The union of all the submodules of W_0 in E_0 is easily seen to be a submodule of W_0, and it clearly does not contain x; hence this union is a least upper bound for E_0 in E. Thus E is inductively ordered and so, by Zorn's Lemma, has a maximal element, W_1 say. According to our earlier remarks, it follows that there is a submodule W_2 of W_0 such that W_0 is the internal direct sum of W_1 and W_2.

We claim that W_2 is simple. It is clearly non-zero, since $x \in W_0$

but x is not in W_1. Suppose W'_2 is a non-zero submodule of W_2. Then $W_1 + W'_2$ includes W_1 as a proper submodule, and hence, since W_1 is a maximal element of E, we see that $W_1 + W'_2$ does not belong to E. So $W_1 + W'_2$ contains x, and consequently it includes $Rx = W_0$; so $W_1 + W'_2 = W_0$. Now we easily check that $W_2 = W'_2 + (W_2 \cap W_1)$, and since $W_2 \cap W_1 = 0$, we deduce that $W_2 = W'_2$. Thus W_2 is simple, as we asserted.

We can now conclude our proof that (c) implies (a). Let S be the sum of the family of all simple submodules of V. By hypothesis, there exists a submodule T of V such that V is the internal direct sum of S and T. Since $S \cap T = 0$, T cannot include any simple submodule and hence must be zero. That is to say, $V = S$ and so is semisimple.

This completes the proof.

We now draw some easy consequences from Theorem 20.2.

THEOREM 20.3. *Every submodule and every factor module of a semisimple module are semisimple.*

Proof. Let the semisimple module V be the sum of the family $(S_i)_{i \in I}$ of simple submodules.

Let V_0 be any submodule of V and let η be the canonical epimorphism from V onto the factor module V/V_0. For each index i in I the module $\eta(S_i)$ is clearly either a zero module or else simple; let I' be the set of indices i for which $\eta(S_i)$ is non-zero. Then $\eta(V) = V/V_0$ is the sum of the family $(\eta(S_i))_{i \in I'}$ of simple modules and hence is semisimple.

Next let V_1 be any submodule of V. According to Theorem 20.2 V_1 is a direct summand of V, i.e. there exists a submodule V_2 of V such that V is the internal direct sum of V_1 and V_2. It follows that V_1 is isomorphic to the factor module V/V_2 and hence (by the first part of the proof) is semisimple.

THEOREM 20.4. *Let V be a semisimple left R-module. If V is the sum of the family $(S_i)_{i \in I}$ of simple submodules, then every simple submodule S of V is isomorphic to one of the modules in the family (S_i).*

Proof. According to Theorem 20.2, there is a subset J of I such that V is the internal direct sum of the family $(S_i)_{i \in J}$.

As in the second part of the preceding theorem we see that S is isomorphic to a factor module V/V_0 of V. If η is the canonical epimorphism from V onto V/V_0 we see that V/V_0 is the internal direct sum of the family $(\eta(S_i))_{i \in J}$. Since each of the modules $\eta(S_i)$ is either zero or simple (and isomorphic to S_i), and V/V_0 (being isomorphic to S) is simple, it follows that there is only one non-zero member of the family $(\eta(S_i))$, say $\eta(S_{i_0})$. Thus S is isomorphic to S_{i_0}.

Let T be an isomorphism type of simple left R-modules. A left R-module V which is the sum of a family of simple submodules all of which are of type T is said to be *isotopic of type T*. If V is any left R-module, the sum of all simple submodules of V which are of type T is of course isotypic of type T; we call it the *isotypic component* of type T or the *T-component* of V, and denote it by V_T.

THEOREM 20.5. *Let V be a left R-module. Then the sum* $\sum_{T \in \mathscr{T}(R)} V_T$ *of its isotypic components is direct.*

Proof. Let T_1 be an R-isomorphism type of simple left R-modules; let $\mathscr{T}'$ be the set of isomorphism types distinct from T_1. Thus we have

$$\sum_{T \in \mathscr{T}(R)} V_T = V_{T_1} + \sum_{T \in \mathscr{T}'} V_T.$$

Consider now the intersection $W = V_{T_1} \cap \sum_{T \in \mathscr{T}'} V_T$. Since W is a submodule of V_{T_1}, it follows from Theorem 20.4 that every simple submodule of W is of type T_1. But since W is also a submodule of $\sum_{T \in \mathscr{T}'} V_T$, every simple submodule of W belongs to one of the types T in $\mathscr{T}'$. But this is a contradiction; so W has no simple submodules. It follows that W must be the zero submodule; for W, being a submodule of the semisimple module $\sum_{T \in \mathscr{T}(R)} V_T$ is semisimple and hence is either zero or else includes simple submodules.

According to Theorem 9.3 we see that the sum $\sum_{T \in \mathscr{T}(R)} V_T$ is direct.

COROLLARY. *A semisimple left R-module is the internal direct sum of its isotypic components.*

Proof. Let V be a semisimple left R-module. Then V is the sum of a family of simple submodules and hence of the family of all simple submodules of V. Hence V is the sum of its isotypic components. According to the theorem this sum is direct; so the result follows.

As we saw in Theorem 20.2, every semisimple module can be

expressed as a direct sum of simple submodules; our next theorem investigates the extent to which this expression is unique.

THEOREM 20.6. *Let V be a semisimple left R-module. If V is the direct sum of each of the families $(S_i)_{i\in I}$ and $(S'_j)_{j\in J}$ of simple submodules, then I and J are equipotent.*

Proof. We consider first the case where I is a finite set. If Card $I = n$ we may suppose that $I = [1, n]$.

For $i = 0, \ldots, n-1$ set $V_i = \sum_{k=1}^{n-i} S_k$, and set $V_n = 0$. Then

$$V = V_0 \supseteq V_1 \supseteq \ldots \supseteq V_n = 0$$

is a normal series of length n for V. It is in fact a composition series, since for $i = 1, \ldots, n$ the factor module V_{i-1}/V_i is isomorphic to the simple module S_{n-i+1}. According to Theorem 13.1, V has no non-repeating normal series of length greater than n.

It follows at once that Card $J \leqslant n$. For if Card $J > n$, then J has a subset of cardinal $n+1$, say $\{j_1, \ldots, j_{n+1}\}$. Setting $V'_{n+1} = 0$ and $V'_i = \sum_{k=1}^{n+1-i} S'_{j_k}$ $(i = 0, \ldots, n)$, we have, as above, a normal series

$$V \supseteq V'_0 \supset V'_1 \supset \ldots \supset V'_{n+1} = 0$$

of length at least $n+1$ for V. But this contradicts the result of the preceding paragraph.

Hence Card $J \leqslant n =$ Card I. Interchanging the roles of the families $(S_i)_{i\in I}$ and $(S'_j)_{j\in J}$, we obtain the reversed inequality Card $I \leqslant$ Card J. Thus Card $I =$ Card J, as required.

Now suppose Card I is infinite.

For each index i of I let x_i be a non-zero element of the simple submodule S_i; according to Theorem 19.1, we have $Rx_i = S_i$. Since $x_i \in \sum_{j\in J} S'_j$, x_i can be expressed in the form $x_i = \sum_{j\in J} x_{ij}$ where $x_{ij} \in S'_j$ for every index j in J and at most finitely many of the elements x_{ij} are non-zero. Let K_i be the subset of J consisting of those indices j for which x_{ij} is non-zero; set $K = \bigcup_{i\in I} K_i$. Since I is infinite and each of the sets K_i is finite, we have

$$\text{Card } K \leqslant \sum_{i\in I} \text{Card } K_i \leqslant (\text{Card } I)^2 = \text{Card } I.$$

(For every infinite cardinal $\mathbf{a}$ we have $\mathbf{a}^2 = \mathbf{a}$.)

We contend that $K = J$. To see this, we remark that for each index i in I we have $x_i \in \sum_{j \in K} S'_j$. Hence $S_i = Rx_i$ is included in $R(\sum_{j \in K} S'_j) = \sum_{j \in K} S'_j$. Thus we have $\sum_{j \in J} S'_j = V = \sum_{i \in I} S_i$ included in $\sum_{j \in K} S'_j$. Consequently, since these sums are direct, we have $K = J$.

Thus Card $J \leqslant$ Card I. Again interchanging the roles of the families (S_i) and (S'_j) we obtain the reversed inequality. Hence in this case also we have Card $I =$ Card J, as asserted.

Let V be any semisimple left R-module, and express it as a direct sum of a family $(S_i)_{i \in I}$ of simple submodules. Then the cardinal of I which, as we have just seen, depends only on V and not on the choice of the family $(S_i)_{i \in I}$ is called the *length* of V and is denoted by $l(V)$. As we have seen in the first part of the proof of Theorem 20.6, this definition coincides in the case that $l(V)$ is finite with the definition of the length of a module which we have in §13; it refines the previous definition in the case of *semisimple* modules which have no composition series.

If T is an isomorphism type of simple left R-modules, let V_T be the T-component of a left R-module V; then V_T is semisimple. The length of V_T as defined in the preceding paragraph, $l(V_T)$, is called the *T-length* of V and is denoted by $l_T(V)$. It follows at once from the Corollary to Theorem 20.5 that if V is a semisimple left R-module then $l(V) = \sum_{T \in \mathscr{T}(R)} l_T(V)$.

In terms of these definitions we have a criterion for semisimple modules to be isomorphic.

THEOREM 20.7. *Let V and V' be semisimple left R-modules. Then V is isomorphic to V' if and only if $l_T(V) = l_T(V')$ for every isomorphism type T of simple left R-modules.*

Proof. Let T be any isomorphism type of simple left R-modules. Then the T-component V_T of V is the internal direct sum of a family $(S_i)_{i \in I_T}$ of simple submodules S_i of V, all isomorphic to T. Similarly the T-component V'_T of V' is the internal direct sum of a family $(S'_j)_{j \in J_T}$ of simple submodules S'_j of V', again all isomorphic to T.

(1) Suppose the condition of the theorem is satisfied. Then for each isomorphism type T we have Card $I_T = l_T(V) = l_T(V') =$ Card J_T; so there exists a bijection π_T from I_T onto J_T.

For each index i in I_T the submodules S_i and $S'_{\pi_T(i)}$ of V and V' respectively are both isomorphic to T; hence there exists an isomorphism α_i from S_i onto $S'_{\pi_T(i)}$. We now piece these isomorphisms together to produce an isomorphism α_T from V_T onto V'_T. Each element x of V_T can be expressed uniquely in the form $x = \sum_{i \in I_T} x_i$ where $x_i \in S_i$ for each index i in I_T and at most finitely many of the elements x_i are non-zero. For each element x of V_T set $\alpha_T(x) = \sum_{i \in I_T} \alpha_i(x_i)$; then it is easily verified that α_T is an isomorphism from V_T onto V'_T.

In the same way we fit together the family of isomorphisms $(\alpha_T)_{T \in \mathscr{T}(R)}$ to form an isomorphism α from V onto V'.

(2) Conversely, suppose there exists an isomorphism α from V onto V'.

For each isomorphism type T we have $\alpha(V_T) = \alpha(\sum_{i \in I_T} S_i) = \sum_{i \in I_T} \alpha(S_i)$. Since α is an isomorphism, this sum is direct and all the submodules $\alpha(S_i)$ are isomorphic to T. So $\alpha(V_T)$ is included in V'_T and hence we have $l_T(V) = l(V_T) = l(\alpha(V_T)) \leqslant l(V'_T) = l_T(V')$. Similarly, since α^{-1} is an isomorphism from V' onto V, we have $l_T(V') \leqslant l_T(V)$.

Thus for every type T we have $l_T(V) = l_T(V')$.

Let V be a semisimple left R-module. We shall find it useful to have a characterization of the submodules W of V such that $\varphi(W) \subseteq W$ for every endomorphism φ of V. The next theorem gives such a characterisation in terms of the isotypic components of V.

THEOREM 20.8. *Let W be a submodule of a semisimple left R-module V. Then $\varphi(W) \subseteq W$ for every endomorphism φ of V if and only if W is a sum of isotypic components of V.*

Proof. (1) Suppose $W = \sum_{T \in \mathscr{T}} V_T$ where $\mathscr{T}$ is a subset of $\mathscr{T}(R)$.

Let φ be any endomorphism of V. For each isomorphism type T in $\mathscr{T}$, let φ_T be the restriction of φ to V_T. If K_T is the kernel of φ_T it follows from Theorem 20.2 that V_T can be expressed as the direct sum of K_T and another submodule L_T, which is also of course a sum of simple submodules of type T. Thus $\varphi(V_T) = \varphi_T(V_T)$, which is isomorphic to V_T/K_T and so to L_T, is a sum of simple submodules of type T and hence is included in V_T.

Hence $\varphi(W) = \sum_{T \in \mathscr{T}} \varphi(V_T) \subseteq \sum_{T \in \mathscr{T}} V_T = W$, as required.

(2) Conversely, suppose $\varphi(W)$ is included in W for every endomorphism φ of V. Let $\mathscr{T}$ be the set of isomorphism types of simple

left R-modules T such that W includes a simple submodule of type T. We shall prove that $W = \sum_{T \in \mathcal{T}} V_T$.

Let T be one of the types in the set $\mathcal{T}$, and let S be a simple submodule of W of type T. We must show that every simple submodule of type T in V is actually included in W; let S_1 be such a submodule. Then there exists an isomorphism α from S onto S_1. Now, since S is a submodule of the semisimple module V, there is a submodule S' of V such that V is the internal direct sum of S and S'. We now define an endomorphism φ of V as follows. Let x be any element of V; then x can be expressed uniquely in the form $x = s + s'$ where $s \in S$ and $s' \in S'$; we set $\varphi(x) = \alpha(s)$. Then we have $S_1 = \alpha(S) = \varphi(S) \subseteq \varphi(W) \subseteq W$. It follows that $V_T \subseteq W$.

The desired result now follows at once.

§21. Centralisers and Bicentralisers of Modules

Let R be a ring with identity, V a left R-module. Then V is of course an additive abelian group, i.e. a left $\mathbf{Z}$-module, and so we may form the ring $L(V) = \operatorname{Hom}_{\mathbf{Z}}(V, V)$ of abelian group endomorphisms of V.

For each element a of R let λ_a be the mapping from V to itself defined by setting $\lambda_a(x) = ax$ for each element x of V. It is clear that λ_a belongs to $L(V)$; we call it the *left translation* of V by a. This now allows us to define a mapping λ_V from R to $L(V)$ by setting $\lambda_V(a) = \lambda_a$ for each element a of R; it is easy to verify that λ_V is a (ring) homomorphism.

Let S be any subset of $L(V)$. Then we define the *centraliser* of S in $L(V)$ to be the subset S^* of $L(V)$ consisting of those $\mathbf{Z}$-endomorphisms of V which commute with every $\mathbf{Z}$-endomorphism in the set S; in other words, the $\mathbf{Z}$-endomorphism φ of V belongs to S^* if and only if $\varphi\sigma = \sigma\varphi$ for every $\mathbf{Z}$-endomorphism σ in S. We can see at once that S^* is a subring of $L(V)$, even though S is merely an arbitrary subset.

Example 1. Let $S = \lambda_V(R)$. Then a $\mathbf{Z}$-endomorphism φ of V belongs to S^* if and only if $\varphi\lambda_a = \lambda_a\varphi$ for every element a of R, i.e. if and only if for every element a of R and every element x of V we have $\varphi(ax) = \varphi\lambda_a(x) = \lambda_a\varphi(x) = a\varphi(x)$. So φ belongs to S^* if and only if it is an R-endomorphism of V; that is to say, the $\mathbf{Z}$-endomorphisms of V which commute with all the left translations by elements of R are just the R-endomorphisms of V: $(\lambda_V(R))^* = \operatorname{Hom}_R(V, V)$.

Example 2. If S_1 is any subset of $L(V)$, let $S = S_1^*$. We claim that S_1 is included in $S^* = S_1^{**}$. To see this, let φ_1 be any element of S_1, σ any element of $S = S_1^*$. Then σ commutes with every element of S_1, in particular with φ_1; so φ_1 commutes with every element σ of S, and hence $\varphi_1 \in S^* = S_1^{**}$, as we said.

Example 3. Let $S = \mathrm{Hom}_R(V, V)$. Then S^* is called the *bicentraliser* of the module V and is denoted by $B(V)$. According to Example 1 we have $\mathrm{Hom}_R(V, V) = (\lambda_V(R))^*$; hence $B(V) = (\lambda_V(R))^{**}$ and so, according to Example 2, $B(V)$ includes $\lambda_V(R)$ as a subring.

We now specialise our discussion to the case in which the module V is R_l, the ring R itself considered as a left R-module.

THEOREM 21.1. *Let R be a ring with identity. Then the bicentraliser of the left R-module R_l is isomorphic to the ring R itself.*

Proof. Let us write λ as an abbreviation for λ_{R_l}. Thus λ is the (ring) homomorphism from R to $L(R_l)$ such that $(\lambda(a))(x) = ax$ for all elements a, x of R. As we have seen in Example 3, λ maps R into $B(R_l)$. We claim that in fact λ is an isomorphism from R onto $B(R_l)$.

First we show that λ is injective. So suppose a is any element of the kernel of λ, so that $\lambda(a)$ is the zero endomorphism of R_l. We have in particular

$$0 = (\lambda(a))(e) = ae = a$$

(where e is the identity of R); thus the kernel of λ consists of the zero element alone and λ is a monomorphism.

Now to show that λ is surjective, let φ be any element of $B(R_l) = \mathrm{Hom}_R(R_l, R_l)^*$. We claim that φ is an R-endomorphism of the right R-module R_r, i.e. that $\varphi(xa) = \varphi(x)a$ for all elements a, x of R. To see this, let a be any element of R and consider the mapping ρ_a from R to itself defined by setting $\rho_a(t) = ta$ for every element t of R; a trivial verification shows that ρ_a belongs to $\mathrm{Hom}_R(R_l, R_l)$. It follows that φ commutes with each of the mappings ρ_a; so for all elements a, x of R we have

$$\varphi(xa) = \varphi\rho_a(x) = \rho_a\varphi(x) = \varphi(x)a$$

as asserted. Now let t be any element of R; then we have

$$\varphi(t) = \varphi(et) = \varphi(e)t = [\lambda(\varphi(e))](t)$$

from which we deduce that $\varphi = \lambda(\varphi(e))$, i.e. that λ is surjective.

This completes the proof.

We now consider a situation which we shall have to face in the next section when we investigate the structure of simple rings.

THEOREM 21.2. *Let V be a left R-module, $(V_k)_{k\in K}$ a family of left R-modules all of which are isomorphic to V. If S is the external direct sum of this family the bicentraliser of S is isomorphic to the bicentraliser of V.*

Proof. For each index k in K let α_k be an isomorphism from V onto V_k. If ι_k and π_k are as usual the canonical injection from V_k to S and the canonical projection from S onto V_k respectively, we set $\eta_k = \iota_k\alpha_k$ and $\varphi_k = \alpha_k^{-1}\pi_k$. Then η_k is a monomorphism from V to S and φ_k is an epimorphism from S onto V; it is easily verified that for each index k we have $\varphi_k\eta_k = I_V$ and for each pair of distinct indices k, l the mapping $\varphi_k\eta_l$ is the zero homomorphism of V. Furthermore, for each element x of S we have

$$\sum_{k\in K}\eta_k\varphi_k(x) = \sum_{k\in K}\iota_k\pi_k(x) = x.$$

We choose one of the indices in K, say k_0, and write η_0 and φ_0 as abbreviations for η_{k_0} and φ_{k_0}.

To prove that the bicentraliser $B(S)$ is isomorphic to the bicentraliser $B(V)$ we proceed by defining a mapping Φ from $B(S)$ to $L(V)$ as follows. For each element β of $B(S)$ we set $\Phi(\beta) = \varphi_0\beta\eta_0$. It is clear that Φ is a ring homomorphism from $B(S)$ to $L(V)$.

We claim that in fact Φ maps $B(S)$ into $B(V)$. So let β be any element of $B(S)$, θ' any element of $\text{Hom}_R(V, V)$. Then we have $\eta_0\theta'\varphi_0 \in \text{Hom}_R(S, S)$ and hence $\beta\eta_0\theta'\varphi_0 = \eta_0\theta'\varphi_0\beta$. It follows that $\varphi_0\beta\eta_0\theta'\varphi_0\eta_0 = \varphi_0\eta_0\theta'\varphi_0\beta\eta_0$, and this equation reduces at once to $\Phi(\beta)\theta' = \theta'\Phi(\beta)$ when we refer to the first paragraph of the proof. Thus $\Phi(\beta)$ belongs to $B(V)$.

Next we show that Φ is a monomorphism. To do this, let β be any element of Ker Φ; then $\varphi_0\beta\eta_0$ is the zero homomorphism of V. For each index k in K the mapping $\eta_0\varphi_k$ belongs to $\text{Hom}_R(S, S)$ and hence we have $\eta_0\varphi_k\beta = \beta\eta_0\varphi_k$; then $\varphi_0\eta_0\varphi_k\beta = \varphi_0\beta\eta_0\varphi_k$, whence $\varphi_k\beta$ is the zero homomorphism from S to V for each index k. So β is the zero endomorphism of S and Φ is a monomorphism, as required.

Finally we show that the image of Φ is $B(V)$. So let β' be any element of $B(V)$. Consider the mapping β from S to S defined by setting for each element x of S

$$\beta(x) = \sum_{k \in K} \eta_k \beta' \varphi_k(x);$$

since $\varphi_k(x)$ is non-zero for only finitely many indices k, this sum is well-defined. It is clear that β belongs to $L(S)$; we shall prove that in fact it belongs to $B(S)$. So let γ be any element of $\mathrm{Hom}_R(S, S)$.

If k and l are any two indices in K the mapping $\varphi_k \gamma \eta_l$ belongs to $\mathrm{Hom}_R(S, S)$ and hence commutes with β'. Thus if x' is any element of V we have

$$\beta\gamma\eta_l(x') = \sum_{k \in K} \eta_k \beta' \varphi_k \gamma \eta_l(x') = \sum_{k \in K} \eta_k \varphi_k \gamma \eta_l \beta'(x') = \gamma\eta_l\beta'(x')$$

while

$$\gamma\beta\eta_l(x') = \gamma\Big(\sum_{k \in K} \eta_k \beta' \varphi_k \eta_l(x')\Big) = \gamma\eta_l\beta'(x').$$

If now x is any element of S we have $x = \sum_{l \in K} \eta_l \varphi_l(x)$ and so

$$\beta\gamma(x) = \sum_{l \in K} \beta\gamma\eta_l\varphi_l(x) = \sum_{l \in K} \gamma\beta\eta_l\varphi_l(x) = \gamma\beta(x).$$

Thus $\beta\gamma = \gamma\beta$, i.e. β belongs to $B(S)$.

Clearly $\Phi(\beta) = \varphi_0 \beta \eta_0 = \beta'$; so Φ is surjective, as we claimed. This completes the proof.

As an immediate application we prove the so-called *Density Theorem* for semisimple modules.

THEOREM 21.3. *Let V be a semisimple left R-module, β any element of its bicentraliser $B(V)$. For each finite family $(x_i)_{1 \leqslant i \leqslant n}$ of elements of V there exists an element a of R such that $\beta(x_i) = ax_i\ (=\lambda_a(x_i))$ for $i = 1, \ldots, n$.*

Proof. First let V' be a semisimple left R-module, x' an element of V' and β' an element of the bicentraliser of V'.

Then $W_1 = Rx'$ is a submodule of V' and hence is a direct summand of V', i.e. there is a submodule W_2 of V' such that every element y of V' can be expressed uniquely in the form $y = y_1 + y_2$ where $y_1 \in W_1$ and $y_2 \in W_2$. We may thus define an R-endomorphism φ of V' by setting $\varphi(y) = y_1$ for each element y of V'; clearly y belongs to W_1 if and only if $\varphi(y) = y$. Since $\varphi \in \mathrm{Hom}_R(V', V')$ and $\beta' \in B(V')$, we have $\varphi(\beta'(x')) = \beta'(\varphi(x')) = \beta'(x')$; so $\beta'(x')$ belongs to

$W_1 = Rx'$. That is to say, there exists an element a of R such that $\beta'(x') = ax'$.

Now let V, $(x_i)_{1 \leqslant i \leqslant n}$ and β be as in the statement of the theorem. For $i = 1, \ldots, n$ let $V_i = V$ and let V' be the external direct sum of the family $(V_i)_{1 \leqslant i \leqslant n}$. Then V' is semisimple and according to Theorem 21.2 there exists an isomorphism Φ from $B(V')$ onto $B(V)$. We may thus apply the result of the preceding paragraph to the element $x' = (x_1, \ldots, x_n)$ of V' and the homomorphism $\beta' = \Phi^{-1}(\beta)$ in $B(V')$. It follows that there exists an element a of R such that $\beta'(x') = ax' = (ax_1, \ldots, ax_n)$. We easily verify that $\beta'(x') = (\beta(x_1), \ldots, \beta(x_n))$. Hence we have $\beta(x_i) = ax_i$ $(i = 1, \ldots, n)$.

§22. Semisimple and Simple Rings

Let R be a ring with identity element e, which we assume to be distinct from the zero element 0, so that $R \neq \{0\}$; and let R_l as usual be the ring R itself considered as left R-module. It may happen that this left R-module is semisimple, and in this case we propose to say that R is a *left semisimple ring*. Similarly we shall say that R is a *right semisimple ring* if the right R-module R_r is semisimple.

Although this appears to be a fairly natural terminology, its use may prove confusing to the reader who dips into some other books concerned with the structure of rings, which use the term 'semisimple ring' in a different sense. Namely, one can develop various procedures for associating with each ring R an ideal which is called the radical of R; a ring is then said to be semisimple (relative to the particular radical procedure under consideration) if its radical is the zero ideal. For most (though not all) known ways of defining radicals it turns out that if a ring R is left Artinian and semisimple in the radical sense then it is left semisimple according to our definition and conversely. We could of course avoid all possibility of confusion by saying that R is a 'left module semisimple ring' if R_l is a semisimple left R-module; but this surely would be unbearably cumbersome.

With this warning about the terminology we proceed to obtain some of the elementary properties of left semisimple rings.

THEOREM 22.1. *A ring R with identity is left semisimple if and only if every left R-module is semisimple.*

Proof. Clearly if every left R-module is semisimple then in particular R_l is semisimple and so R is a left semisimple ring.

Conversely, suppose R is a left semisimple ring, and let V be any left R module. Let X be a generating system for V (at the worst we can take $X = V$), and for each element x of X let V_x be the module R_l. If S is the external direct sum of the family $(V_x)_{x \in X}$, it is clear that S is a semisimple left R-module. Consider the mapping η from S to V defined by setting $\eta(s) = \sum_{x \in X} \pi_x(s)x$ for each element s of S (π_x is of course the canonical projection epimorphism from S onto $V_x = R$, and the sum is essentially finite, since $\pi_x(s)$ is non-zero for at most a finite number of 'indices' x). Certainly η is a homomorphism, and since X is a generating system for V, it follows that η is surjective. Thus if K is the kernel of η, we see that V is isomorphic to S/K and hence is semisimple by Theorem 20.3.

COROLLARY 1. *A ring R with identity is left semisimple if and only if every left R-module is projective.*

Proof. (1) Suppose R is left semisimple.

Let M be a left R-module; consider a short exact sequence

$$0 \to V' \xrightarrow{\alpha} V \xrightarrow{\beta} M \to 0 \qquad [22.1]$$

where V' and V are left R-modules. According to the theorem, V is semisimple and so, by Theorem 20.2, every submodule of V is a direct summand; that is to say, [22.1] is split. It follows from Theorem 11.5 that M is projective.

(2) Conversely, suppose every left R-module is projective.

Let V be any left R-module, V' any submodule of V. If ι is the inclusion monomorphism from V' to V and η is the canonical epimorphism from V onto V/V' the sequence

$$0 \to V' \xrightarrow{\iota} V \xrightarrow{\eta} V/V' \to 0$$

is exact. By hypothesis V/V' is projective; so this sequence splits. Thus $\iota(V') = V'$ is a direct summand of V. It follows from Theorem 20.2 that V is semisimple.

Since every left R-module is semisimple, R is a left semisimple ring.

A somewhat similar argument serves to establish an analogous result.

COROLLARY 2. *A ring R with identity is left semisimple if and only if each left R-module is injective.*

The next theorem shows, somewhat surprisingly, that although we have apparently made no 'finiteness' assumptions in defining left semisimple rings, nevertheless such rings have in fact strong finiteness properties.

THEOREM 22.2. *A left semisimple ring is both left Artinian and left Noetherian.*

Proof. Let R be a left semisimple ring; thus the left R-module R_l is the sum of a family $(L_i)_{i \in I}$ of simple submodules (minimal left ideals of R). It follows that every element of R, and in particular the identity element e, can be expressed as the sum of a family of elements $(x_i)_{i \in I}$ in which $x_i \in L_i$ for every index i, and only finitely many of the elements x_i are non-zero. Thus we may write

$$e = x_{i_1} + \ldots + x_{i_n}$$

where $x_{i_1}, \ldots, x_{i_n}$ are non-zero and $x_{i_j} \in L_{i_j}$ $(j = 1, \ldots, n)$. Hence if a is any element of R we have $a = ae = \sum_{j=1}^{n} ax_{i_j} \in \sum_{j=1}^{n} L_{i_j}$. So $R_l = \sum_{j=1}^{n} L_{i_j}$, i.e. R_l is the sum of a finite family of simple submodules. From this we deduce easily that R_l has a composition series. According to Theorem 14.5 it follows that the left module R_l is both Artinian and Noetherian; that is to say, the ring R is both left Artinian and left Noetherian.

We now approach the problem of discovering the structure of left semisimple rings. The first step in this direction is to describe the isotypic components of the semisimple left R-module R_l.

THEOREM 22.3. *Let R be a left semisimple ring. Then the non-zero isotypic components of the left R-module R_l are the minimal two-sided ideals of R.*

Proof. We propose to make use of Theorem 20.8; so we begin by determining the endomorphisms of R_l. It is easily seen that for every element a of R the right translation ρ_a (defined by setting $\rho_a(x) = xa$ for every element x of R) belongs to $\mathrm{Hom}_R(R_l, R_l)$. But conversely, if φ belongs to $\mathrm{Hom}_R(R_l, R_l)$ then for each element x of R we have $\varphi(x) = \varphi(xe) = x\varphi(e) = \rho_a(x)$ where we write $a = \varphi(e)$; thus every element of $\mathrm{Hom}_R(R_l, R_l)$ is a right translation.

Now let L be any submodule of R_l (i.e. a left ideal of R). The preceding paragraph shows that $\varphi(L) \subseteq L$ for every endomorphism φ of R if and only if $xa \in L$ for every element of L and every element a of R, i.e. if and only if L (besides being a left ideal) is also a right ideal and hence two-sided.

It follows from Theorem 20.8 that the left ideals of R which can be expressed as sums of isotypic components of R_l are just the two-sided ideals of R. The minimal elements in the set of non-zero left ideals which can be expressed as sums of isotypic components are plainly the single non-zero isotypic components themselves. Hence these non-zero isotypic components are precisely the minimal two-sided ideals of R.

COROLLARY. *Every two-sided ideal of a left semisimple ring R is a direct sum of minimal two-sided ideals of R.*

Now let R be a ring which does not consist of the zero element alone. Such a ring is said to be a *left simple ring* if (1) R is left semisimple and (2) the only two-sided ideals of R are R itself and the zero ideal. We must enter two words of warning about this definition. First, the reader must note carefully that it is not analogous to the definition of left semisimple rings: we have not defined a ring to be left simple if the left R-module R_l is simple—as we saw in Example 2 of §19, this condition implies that R is a division ring. Again the reader must exercise caution when he consults other books on ring structure, for the term 'simple ring' is used with different meanings by different authors: our condition (2) is common to them all, but some have other conditions in place of (1) while others have no additional condition at all. We shall of course call a ring *right simple* if it is right semisimple and satisfies condition (2).

We now produce some simple criteria for a left semisimple ring to be simple.

THEOREM 22.4. *Let R be a left semisimple ring. Then the following conditions are equivalent:*

(a) *R is a left simple ring;*

(b) *the left R-module R_l is isotypic;*

(c) *there is only one isomorphism type of simple left R-modules.*

Proof. (1) Suppose R is a left simple ring.

Then R has only one minimal two-sided ideal, namely R itself. So R_l has only one isotypic component, namely R_l itself. Thus R_l is isotypic.

(2) Suppose R_l is isotypic of type T say.

Let V be any simple left R-module. As in the proof of Theorem 22.1, we see that V is isomorphic to a quotient module of the direct sum of a certain family of modules all equal to R_l. It follows that V is isotypic of type T, and hence, since it is simple, V is isomorphic to T.

So there is only one isomorphism type of simple left R-modules.

(3) Suppose there is only one isomorphism type of simple left R-modules. Then R_l is isotypic, and hence is itself the only non-zero isotypic component of R. So R is the only minimal two-sided ideal of R, which is therefore a simple ring.

In our next theorem we describe a method for constructing left simple rings; we shall see later that all left simple rings may be obtained by this method. Once this has been established, we shall be able to use this theorem to prove that a ring is left simple if and only if it is right simple: we may then talk simply of 'simple rings'.

THEOREM 22.5. *Let V be a left vector space of finite dimension n over a division ring D. Then the ring R of D-endomorphisms of V is both left and right simple, and the isotypic R-modules R_l and R_r have length n.*

Proof. We may consider V as a left R-module by setting $\alpha x = \alpha(x)$ for every element x of V and every D-homomorphism α from V to V (cf. Example 5 of §6). We shall show that when V is considered as a left R-module in this way it is simple.

So let x be any non-zero element of V. According to Theorem 6.5 there exists a basis $\{x_1, \ldots, x_n\}$ of V over D with $x_1 = x$. If y is any element of V the mapping α_y of V into V defined by setting

$$\alpha_y(d_1x_1 + \ldots + d_nx_n) = d_1y$$

for all elements $d_1, \ldots, d_n$ of D is clearly a D-homomorphism from V to V. Then we have $y = \alpha_y(x_1) = \alpha_y x \in Rx$. Consequently $V = Rx$ and hence, by Theorem 19.1, is a simple left R-module.

For $i = 1, \ldots, n$ let $V_i = V$ and let V^n be the direct product module of the family $(V_i)_{1 \leqslant i \leqslant n}$. Let $\{y_1, \ldots, y_n\}$ be a basis for V over D. Then we define a mapping φ from R_l to V^n by setting

$$\varphi(\alpha) = (\alpha(y_1), \ldots, \alpha(y_n))$$

for all elements α of $R = \mathrm{Hom}_D(V, V)$. This mapping φ is clearly a (left R-module) homomorphism; we claim that it is in fact an isomorphism.

Suppose first that $\varphi(\alpha) = (0, \ldots, 0)$, i.e. $\alpha(y_i) = 0$ for $i = 1, \ldots, n$. Then if y is any element of V and we write y in the form $y = d_1y_1 + \ldots + d_ny_n$ with $d_1, \ldots, d_n$ in D we have

$$\alpha(y) = d_1\alpha(y_1) + \ldots + d_n\alpha(y_n) = 0.$$

So α is the zero homomorphism from V to V and hence φ is a monomorphism. Next let $(z_1, \ldots, z_n)$ be any element of V^n. The mapping β from V to V defined by setting

$$\beta(d_1y_1 + \ldots + d_ny_n) = d_1z_1 + \ldots + d_nz_n$$

is a D-homomorphism, i.e. $\beta \in R$ and obviously

$$\varphi(\beta) = (\beta(y_1), \ldots, \beta(y_n)) = (z_1, \ldots, z_n).$$

Hence φ is an isomorphism from R_l onto V^n.

Since V^n is an isotypic semisimple left R-module, so is R_l and hence we may conclude from Theorem 22.4 that R is a left simple ring. Finally, since V^n has length n, so also does R_l.

Now let V^* be the dual space of V; according to its construction, V^* is a right vector space over D, and we proved in §8 that its dimension is n. If R^* is the ring of D-endomorphisms of V^* the preceding argument can be modified in the obvious way to show that R^* is a right simple ring and that $(R^*)_r$ has length n. (Note that in order to give V^* the structure of a right R^*-module we must define the product $\rho^* \circ \sigma^*$ of each pair of elements ρ^*, σ^* of R^* to be the composed mapping $\sigma^*\rho^*$.)

To complete the proof of the theorem we show that the rings R and R^* are isomorphic. For this purpose we define a mapping θ from R to R^* as follows: for each element ρ of $R = \mathrm{Hom}_D(V, V)$ let $\theta(\rho)$ be the mapping from V^* to V^* given by setting

$$[\theta(\rho)](\xi^*) = \xi^*\rho$$

for each linear functional ξ^* in V^*. Then clearly $\theta(\rho) \in R^*$ and further θ is a ring homomorphism from R to R^*: if ρ_1, ρ_2 are elements of R, ξ^* any element of V^* we have

$$[\theta(\rho_1 + \rho_2)](\xi^*) = \xi^*(\rho_1 + \rho_2) = \xi^*\rho_1 + \xi^*\rho_2$$

and

$$[\theta(\rho_1) + \theta(\rho_2)](\xi^*) = [\theta(\rho_1)](\xi^*) + [\theta(\rho_2)](\xi^*) = \xi^*\rho_1 + \xi^*\rho_2,$$

whence

$$\theta(\rho_1 + \rho_2) = \theta(\rho_1) + \theta(\rho_2);$$

similarly

$$[\theta(\rho_1\rho_2)](\xi^*) = \xi^*(\rho_1\rho_2)$$

and

$$\begin{aligned}[\theta(\rho_1) \circ \theta(\rho_2)](\xi^*) &= [\theta(\rho_2)\,\theta(\rho_1)](\xi^*)\\ &= [\theta(\rho_2)](\xi^*\rho_1) = (\xi^*\rho_1)\rho_2\end{aligned}$$

whence

$$\theta(\rho_1\rho_2) = \theta(\rho_1) \circ \theta(\rho_2).$$

To prove that θ is an isomorphism, let $\{\eta_1^*, \ldots, \eta_n^*\}$ be the basis of V^* which is dual to the basis $\{y_1, \ldots, y_n\}$ of V. Suppose first that ρ is an element of $R = \operatorname{Hom}_D(V, V)$ such that $\theta(\rho) = \zeta$, the zero homomorphism from V^* to V^*; then for every element ξ^* of V^* the mapping $[\theta(\rho)](\xi^*) = \xi^*\rho$ is the zero linear functional of V and so $\xi^*\rho(x) = 0$ for every element x of V. In particular, for $i = 1, \ldots, n$ we have

$$0 = \eta_i^*\rho(x) = i\text{th coordinate of } \rho(x) \text{ relative to } \{y_1, \ldots, y_n\}$$

and so $\rho(x) = 0$. It follows that ρ is the zero homomorphism from V to V; thus θ is a monomorphism.

Finally let ρ^* be any element of R^*. If we have $\rho^*(\eta_j^*) = \sum_{i=1}^{n} a_{ij}\eta_i^*$ $(j = 1, \ldots, n)$ with coefficients a_{ij} in D, let ρ be the D-homomorphism from V to V defined by setting $\rho(y_k) = \sum_{l=1}^{n} a_{kl}y_l (k = 1, \ldots, n)$. Then it is easy to check that $\theta(\rho) = \rho^*$. So θ is an epimorphism, and hence an isomorphism as required.

In order to make some progress towards our aim of describing the structure of left semisimple rings we must introduce the notions of direct products of families of rings and internal direct sums of families of subrings.

First let $(R_k)_{k \in K}$ be a family of rings. We form the product set $P = \prod_{k \in K} R_k$ and define operations of addition and multiplication in P in the natural way: if $x = (x_k)$, $y = (y_k)$ are elements of P we set

$x + y = (s_k)$ and $xy = (p_k)$ where

$$s_k = x_k + y_k \text{ and } p_k = x_k y_k$$

for every index k in K. It is the work of a moment to check that with these operations the set P becomes a ring, which we call the *direct product ring* of the family $(R_k)_{k \in K}$ and which we denote by $\prod_{k \in K} R_k$. For each index j we have a *canonical injection monomorphism* ι_j from R_j to P defined by setting for each element x_j of R_j

$$\iota_j(x_j) = y = (y_k)$$

where $y_j = x_j$ and $y_k = 0$ for all indices $k \neq j$. It is clear that for each index j the image $\iota_j(R_j)$ is a subring of P; but we can in fact say more than this, namely that $\iota_j(R_j)$ is a two-sided ideal of P. To see this, let $a = (a_k)$ be any element of P, $y = \iota_j(x_j)$ any element of $\iota_j(R_j)$. Then we have $ay = (z_k)$ where $z_k = a_k y_k$ for all indices k in K, so that $z_j = a_j x_j$ and $z_k = 0$ for all indices $k \neq j$; it follows that $ay = \iota_j(a_j x_j) \in \iota_j(R_j)$ and similarly $ya \in i_j(R_j)$.

In the notation of the preceding paragraph let S be the subset of P consisting of elements $x = (x_k)$ such that x_k is non-zero for at most finitely many indices k in K. Then S is easily seen to be a subring of P, which we call the *external direct sum* of the family $(R_k)_{k \in K}$ and denote by $\oplus_{k \in K} R_k$. Of course if K is a finite set we have $S = P$; if $K = [1, n]$ we usually write

$$P = S = R_1 \oplus R_2 \oplus \ldots \oplus R_n.$$

If A is any ring, we may disregard its multiplication operation and consider only the addition operation, under which A is an abelian group. We shall denote this group by A^+ and call it the *additive group of the ring A*.

Now let R be any ring, $(R_k)_{k \in K}$ a finite family of subrings of R. Suppose first that the additive group R^+ of R is the internal direct sum of the family $(R_k^+)_{k \in K}$ of the additive groups of the subrings R_k; thus every element x of R can be expressed uniquely in the form $x = \sum_{k \in K} x_k$ where $x_k \in R_k$ for every index k. If this condition is satisfied and if, further, for every pair of elements $x = \sum x_k$ and $y = \sum y_k$ of R we have $xy = \sum x_k y_k$, we say that R is the *internal direct sum* of the family of subrings (R_k). In this situation it is clear that the mapping α from the external direct sum $S = \oplus_{k \in K} R_k$ to R

defined by setting $\alpha(s) = \sum_{k \in K} s_k$ for every element $s = (s_k)$ of S is an isomorphism from S onto R.

The next theorem gives alternative criteria for a ring to be the internal direct sum of a family of subrings.

THEOREM 22.6. *Let $(R_k)_{k \in K}$ be a finite family of subrings of a ring R such that R^+ is the internal direct sum of the family (R_k^+). Then the following conditions are equivalent.*

(a) *R is the internal direct sum of the family (R_k);*

(b) *each of the subrings R_k is a two-sided ideal of R;*

(c) *if i, j are distinct indices in K, then for all elements x_i, x_j in R_i, R_j respectively we have $x_i x_j = 0$.*

Proof. (1) Suppose R is the internal direct sum of the family (R_k).

Let k be any index in K, x any element of R_k, a any element of R. Then we may write x and a uniquely in the form $x = \sum_{j \in K} x_j$, $a = \sum_{j \in K} a_j$ with $x_j, a_j \in R_j$ for every index j in K. Since x belongs to R_k, it follows that $x_k = x$ and $x_j = 0$ for all indices $j \neq k$. Hence we have

$$ax = (\sum a_j)(\sum x_j) = \sum a_j x_j = a_k x \in R_k$$

and similarly $xa \in R_k$.

Thus R_k is a two-sided ideal of R.

(2) Suppose the subrings R_k are all two-sided ideals of R.

Let i, j be distinct indices in K, x_i and x_j any elements of R_i and R_j respectively. Then since R_i is a right ideal we have $x_i x_j \in R_i$ and since R_j is a left ideal we have $x_i x_j \in R_j$. So $x_i x_j \in R_i \cap R_j$. But since R^+ is the internal direct sum of the family (R_k^+) we have $R_i \cap R_j = R_i^+ \cap R_j^+ = \{0\}$.

So $x_i x_j = 0$, as required.

(3) Suppose that if i, j are distinct indices in K and x_i, x_j are any elements of R_i, R_j respectively then $x_i x_j = 0$.

Let a and b be any two elements of R. Since R^+ is the internal direct sum of the family (R_k^+) we can express a and b uniquely in the form $a = \sum_{k \in K} a_k$, $b = \sum_{k \in K} b_k$ with $a_k, b_k \in R_k$ for every index k in K. Then we have

$$ab = (\sum_{i \in K} a_i)(\sum_{j \in K} b_j) = \sum_{k \in K} a_k b_k + \sum_{i \neq j} a_i b_j = \sum_{k \in K} a_k b_k.$$

So R is the internal direct sum of the family (R_k), as required.

This completes the proof.

With these preliminaries we can take a major step towards determining the structure of left semisimple rings.

THEOREM 22.7. *Let R be a left semisimple ring. Then R_l has only finitely many minimal two-sided ideals, $S_1, \ldots, S_n$ say. Each of these, considered as a ring, has an identity element and is left simple. Further, R is the internal direct sum of the family $(S_i)_{1 \leqslant i \leqslant n}$.*

Proof. According to Theorem 22.3 the minimal two-sided ideals of R are precisely the non-zero isotypic components of the left R-module R_l. The Corollary to Theorem 20.5 shows that R_l is the internal direct sum of its isotypic components, which must therefore be finite in number, since R_l is Artinian (Theorem 22.2). Hence we have established that R has only finitely many minimal two-sided ideals, $S_1, \ldots, S_n$ say.

Since R_l is the (module) internal direct sum of the family $(S_i)_{1 \leqslant i \leqslant n}$, it follows that R^+ is the (abelian group) internal direct sum of the family (S_i^+). All the subrings S_i $(i = 1, \ldots, n)$ are two-sided ideals of R; so it follows from Theorem 22.6 that R is the (ring) internal direct sum of the family of subrings $(S_i)_{1 \leqslant i \leqslant n}$.

If e is the identity element of R, we may express e uniquely in the form $e = \sum_{i=1}^{n} e_i$ where $e_i \in S_i$ $(i = 1, \ldots, n)$. Then for every element s_j of S_j we have

$$s_j = es_j = (\sum_{i=1}^{n} e_i)\, s_j = \sum_{i=1}^{n} e_i s_j = e_j s_j.$$

Thus e_j is an identity element for S_j $(j = 1, \ldots, n)$.

It remains to show that each of the rings S_i is left simple. To do this we prove first that a subgroup V of S_i^+ is an S_i-submodule of $(S_i)_l$ if and only if it is an R-submodule of S_i considered as left R-module. Suppose that V is an S_i-submodule of $(S_i)_l$; let x_i be any element of V, a any element of R. Then a can be expressed uniquely in the form $a = \sum_{k=1}^{n} a_k$ where $a_k \in S_k$ $(k = 1, \ldots, n)$, and we have

$$ax_i = (\sum_{k=1}^{n} a_k)\, x_i = \sum_{k=1}^{n} a_k x_i = a_i x_i,$$

which belongs to S_i. So S_i is an R-submodule of S_i. The converse is of course trivial. Since R is left semisimple, S_i is semisimple as a left

R-module and hence, by what we have just proved, $(S_i)_l$ is a semisimple left S_i-module. So S_i is a semisimple ring.

A similar argument to that in the preceding paragraph shows that every two-sided ideal of S_i is also a two-sided ideal of R. But S_i is a minimal two-sided ideal of R; so the only two-sided ideals of R included in S_i are S_i itself and the zero ideal. These are therefore the only two-sided ideals of S_i, which is therefore a left simple ring.

This completes the proof.

The proof of the following result, which is essentially a converse of Theorem 22.7, is left to the reader as an easy exercise.

THEOREM 22.8. *Let $S_1, \ldots, S_n$ be left simple rings. Then the external direct sum R of the family $(S_i)_{1 \leqslant i \leqslant n}$ is a left semisimple ring. The minimal two-sided ideals of R are the images of the rings S_i under the canonical inclusion monomorphisms.*

We have just shown that the minimal two-sided ideals of a left semisimple ring R, considered as rings, are left simple. We call these left simple rings the *simple components* of the ring R. In terms of these we shall be able to give a description of the isomorphism types of simple left R-modules.

Before doing this we break off to introduce a small piece of terminology which will be useful here and in the next section. Let R be any ring, V a left R-module; the *annihilator* of V is defined to be the subset $A(V)$ of R consisting of those elements n of R such that $nx = 0$ for every element x of V. It is easy to show that $A(V)$ is a two-sided ideal of R. Suppose now that α is an isomorphism from V onto another left R-module V'. If $n \in A(V)$ and x' is any element of V' there exists an element x of V such that $x' = \alpha(x)$; hence we have $nx' = n\alpha(x) = \alpha(nx) = \alpha(0) = 0$, so $n \in A(V')$. Thus $A(V) \subseteq A(V')$. Similarly $A(V') \subseteq A(V)$. So we have established that isomorphic left R-modules have the *same* annihilator (not just, as one might expect, isomorphic annihilators).

THEOREM 22.9. *Let R be a left semisimple ring. If R has n simple components there are exactly n isomorphism types of simple left R-modules.*

Proof. Let $S_1, \ldots, S_n$ be the simple components of R.

Since each of the rings S_i is left simple it follows from Theorem 22.4 that there is exactly one isomorphism type of simple left

S_i-modules. Let V_i be a simple left S_i-module. We now give V_i the structure of a left R-module, as follows. For every element r of R we write as usual $r = \sum_{i=1}^{n} r_i$ where $r_i \in S_i$ $(i = 1, \ldots, n)$; then for each element x_i of V_i we set $rx_i = r_i x_i$. It is easy to see that under this left multiplication by R the abelian group V_i becomes a left R-module, which is clearly a simple left R-module. We denote this left R-module by $(V_i)_R$.

It is plain that the annihilator of $(V_i)_R$ is the sum (actually direct) of all the simple components S_k except S_i. Hence if $i \neq j$ the simple left R-modules $(V_i)_R$ and $(V_j)_R$ have different annihilators and consequently are not isomorphic. Thus we have shown that there are at least n isomorphism types of simple left R-modules.

To show that there are no more, let V be any simple left R-module. As we saw in Example 3 of §19, there exists a maximal left ideal of R such that V is isomorphic to the factor module R_l/M. Since R_l is a semisimple left R-module, M is a direct summand of R_l, i.e. there exists a submodule L of R_l (left ideal of R) such that R_l is the internal direct sum of L and M. Then V is isomorphic to L. Now, since M is a maximal left ideal of R it follows easily that L is a minimal left ideal of R (for if L' were a left ideal such that $0 \subset L' \subset L$ we should have $M \subset M + L' \subset R$, contradicting the maximal property of M). But the minimal left ideals of R are just the minimal left ideals of the simple components of R. If L is a minimal left ideal of S_i say, then L is S_i-isomorphic to V_i and hence R-isomorphic to $(V_i)_R$.

This completes the proof.

Having shown that every left semisimple ring is expressible as the internal direct sum of a finite family of simple subrings, we now complete our description of the structure of left semisimple rings by proving the *Artin-Wedderburn Theorem* which establishes the structure of left simple rings.

THEOREM 22.10. *Let R be a left simple ring such that the isotypic left R-module R_l has length r. Let V be a simple left R-module and let D be the division ring of R-endomorphisms of V. Then V, considered as left vector space over D, has dimension r and R is isomorphic to the ring of D-endomorphisms of V.*

Proof. We recall that Schur's Lemma (Theorem 19.2) is our warrant for the assertion that $D = \operatorname{Hom}_R(V, V)$ is a division ring.

Since R is a left simple ring, the left R-module R_l is a direct sum of submodules all isomorphic to V (Theorem 22.4). It follows from Theorem 21.2 that the bicentraliser $B(R_l)$ of R_l is isomorphic to the bicentraliser $B(V)$ of V. According to Theorem 21.1, $B(R_l)$ is isomorphic to the ring R itself.

In order to identify the bicentraliser $B(V) = \mathrm{Hom}_R(V, V)^* = D^*$, we first give V the structure of a left vector space over D by setting $\tau . x = \tau(x)$ for every element x of V and every homomorphism τ in D. Then an abelian group endomorphism α of V belongs to D^* if and only if $\alpha\tau = \tau\alpha$ for every element τ of D, that is to say if and only if

$$\alpha(\tau . x) = \alpha(\tau(x)) = \tau(\alpha(x)) = \tau . \alpha(x),$$

i.e. if and only if α is in $\mathrm{Hom}_D(V, V)$.

It follows that R is isomorphic to $\mathrm{Hom}_D(V, V)$, as we asserted.

To show that V is a finite-dimensional vector space over D, let $(x_i)_{i \in I}$ be a basis for V over D. For each index i in I let L_i be the subset of $\mathrm{Hom}_D(V, V)$ consisting of those D-endomorphisms α of V such that $\alpha(x_j) = 0$ for all indices $j \neq i$. These subsets are actually left ideals of $\mathrm{Hom}_D(V, V)$ and it is easy to see that the sum $\sum_{i \in I} L_i$ is direct. If I were an infinite set, it would follow that $\mathrm{Hom}_D(V, V)$ is not left Noetherian; but R, which is isomorphic to $\mathrm{Hom}_D(V, V)$, is left Noetherian, according to Theorem 22.2. Hence I is a finite set.

If V has dimension r over D it follows from Theorem 22.5 that the left module $(\mathrm{Hom}_D(V, V))_l$ has length r. So R_l has length r. This completes the proof.

According to Theorem 22.5, if V is a finite-dimensional vector space over a division ring D then $R = \mathrm{Hom}_D(V, V)$ is both left simple and right simple. Linking this result with the Artin-Wedderburn Theorem, we have the following result.

COROLLARY 1. *Every left simple ring is right simple and conversely. Every left semisimple ring is right semisimple and conversely.*

It follows that we may drop the words 'left' and 'right' and talk only of 'simple rings' and 'semisimple rings'.

Applying the result of Theorem 8.5 we have the following 'concrete' characterization of simple rings.

COROLLARY 2. *A ring R is simple if and only if there is a division ring D and a natural number n such that R is isomorphic to the ring $M_n(D)$ of $n \times n$ matrices with elements in D.*

In this situation, for $k = 1, \ldots, n$ let L_k be the set of $n \times n$ matrices $A = [a_{ij}]$ with elements in D such that $a_{ij} = 0$ for all pairs (i, j) such that $j \neq k$. It is easily verified that $L_1, \ldots, L_n$ are minimal left ideals of $M_n(D)$ and that the left module $(M_n(D))_l$ is the direct sum of the family $(L_k)_{1 \leqslant k \leqslant n}$. Again for $k = 1, \ldots, n$ let R_k be the set of matrices $A = [a_{ij}]$ such that $a_{ij} = 0$ for all pairs (i, j) such that $i \neq k$. Then $R_1, \ldots, R_n$ are minimal right ideals of $M_n(D)$ and the right module $(M_n(D))_r$ is the direct sum of the family $(R_k)_{1 \leqslant k \leqslant n}$.

Let R be any ring. The *centre* of R is the subset Z of R consisting of those elements which commute with every element of R; so an element z belongs to Z if and only if $zx = xz$ for every element x of R. It is an easy matter to verify that the centre of any ring R is a subring of R, that the centre of a division ring is a field and that the centre of the direct sum (internal or external) of a family of rings $(R_i)_{i \in I}$ is the direct sum of the family $(Z_i)_{i \in I}$ where, for each index i in I, Z_i is the centre of R_i.

THEOREM 22.11. *Let R be a simple ring, V a simple left R-module and D the ring of R-endomorphisms of V. Then the centre of R is isomorphic to the centre of D and hence is a field.*

Proof. According to the Artin-Wedderburn Theorem, R is isomorphic to $\mathrm{Hom}_D(V, V)$. So it will clearly be sufficient to show that the centre of $\mathrm{Hom}_D(V, V)$ is isomorphic to the centre of D.

Let κ be any element of the centre of $\mathrm{Hom}_D(V, V)$.

We show that if x is any element of V there exists an element a_x of D such that $\kappa(x) = a_x x$. It is easy to show that if this were not the case the set $\{x, \kappa(x)\}$ would be linearly independent and (by Theorem 6.5) there would be a basis $\{x_1, \ldots, x_r\}$ for V over D with $x_1 = x$, $x_2 = \kappa(x)$. We could then define a D-endomorphism τ of V by setting

$$\tau(d_1 x_1 + d_2 x_2 + \ldots + d_r x_r) = d_2 t$$

where t is a non-zero element of V; thus $\tau(x) = 0$ and $\tau(\kappa(x)) = t$. But since κ is in the centre of $\mathrm{Hom}_D(V, V)$ we have

$$0 = \kappa(\tau(x)) = \kappa\tau(x) = \tau\kappa(x) = \tau(\kappa(x)) = t \neq 0,$$

which is a contradiction. So there is indeed an element a_x of D such that $\kappa(x) = a_x x$.

Next we show that the element a_x is independent of x. So let x and y be any two non-zero elements of V and suppose $\kappa(x) = a_x x$,

$\kappa(y) = a_y y$ where a_x, a_y are elements of V. Since x is non-zero, there exists a basis $\{x_1, \ldots, x_r\}$ of V with $x_1 = x$, and we may define a D-endomorphism γ of V by setting

$$\gamma(d_1 x_1 + \ldots + d_r x_r) = d_1 y.$$

In particular $\gamma(x) = y$. Then, since κ is in the centre of $\mathrm{Hom}_D(V, V)$, we have $\kappa\gamma = \gamma\kappa$ and so

$$a_y y = \kappa(y) = \kappa\gamma(x) = \gamma\kappa(x) = \gamma(a_x x) = a_x \gamma(x) = a_x y.$$

We deduce easily that $a_y = a_x = a$ say.

So there is an element a of D such that $\kappa = \lambda_a$ (left translation by a). It is clear that this element a lies in the centre of D. For if d is any element of D we have

$$(ad)\,x = a(dx) = \lambda_a(dx) = \kappa(dx) = d\kappa(x) = d\lambda_a(x) = d(ax) = (da)\,x$$

for every element x of V; hence $ad = da$ for every element d of D.

It is now an easy exercise to establish that the mapping φ from the centre of D to $\mathrm{Hom}_D(V, V)$ defined by setting $\varphi(a) = \lambda_a$ for every element a of the centre of D is an isomorphism onto the centre of $\mathrm{Hom}_D(V, V)$.

This completes the proof.

COROLLARY. *The centre of a semisimple ring is a direct sum of fields, the centres of its simple components.*

§23. The Radical of a Ring

Let R be a ring with identity element e. The *left radical* of R is defined to be the intersection $J_l(R)$ of the family of all maximal left ideals of R. If R does not consist of the zero alone, then since the identity element of R is not contained in any maximal left ideal of R it is clear that $J_l(R) \neq R$. If the left radical of R is the zero ideal, we say then that 'R has zero left radical'. The *right radical* of R is similarly defined to be the intersection $J_r(R)$ of the family of all maximal right ideals of R. If R does not consist of the zero element alone, then $J_r(R) \neq R$. We shall soon see that the left and right radicals of R coincide, so we shall be able to talk simply of 'the radical' of R. As we mentioned at the beginning of §22, there are various procedures by which we can associate with each ring an ideal called its radical; these procedures do not all lead to the same result. Thus it may sometimes be necessary to distinguish the radicals we have just defined from those formed by alternative procedures. This situation will not arise in this book, for we shall not be discussing any of the other radicals; but in any situation where a distinction must be made the radicals defined above are called the *Jacobson* (left and right) *radicals.*

THEOREM 23.1. *Let R be a ring. Then the left radical $J_l(R)$ is the intersection of the family of annihilators of all simple left R-modules.*

Proof. (1) Let r be any element of $J_l(R)$.

Let V be any simple left R-module, x any non-zero element of V, so that $V = Rx$ (Theorem 19.1). As we saw in Example 3 of §19, the set of elements a of R such that $ax = 0$ is a maximal left ideal M of R; since r belongs to every maximal left ideal of R it follows that r belongs to M and hence $rx = 0$. Thus r annihilates every element of V and so belongs to $A(V)$.

Hence r belongs to the annihilator of every simple left R-module.

(2) Conversely, suppose that r belongs to the annihilator of every simple left R module.

Let M be any maximal left ideal of R. According to Example 3 of §19, the left R-module R_l/M is simple, and hence is annihilated by r. Thus if C is any coset of R_l modulo M the coset rC is the zero coset modulo M, i.e. for every element a of R we have $ra \in$ M. In particular $r = re \in M$.

Thus r belongs to $J_l(R)$.

This completes the proof.

COROLLARY. *Let R be a ring. Then the left radical of R is a two-sided ideal of R.*

Proof. As we remarked in §22, the annihilator of any left R-module is a two-sided ideal of R. Since $J_l(R)$ is the intersection of a family of annihilators, it follows that it is a two-sided ideal.

Our next theorem will enable us to give another characterization of the radical.

THEOREM 23.2. *Let R_1 and R_2 be rings, α an epimorphism from R_1 onto R_2. Then $\alpha(J_l(R_1))$ is included in $J_l(R_2)$.*

Proof. Let n_1 be any element of $J_l(R_1)$.

If V is any simple left R_2-module we may give V the structure of a left R_1-module by defining r_1x to be $\alpha(r_1)\,x$ for all elements r_1 of R_1

and all elements x of V; when we regard V as a left R_1-module in this way we shall denote it by V_1. Since α is an epimorphism every R_1-submodule of V_1 is an R_2-submodule of V; hence V_1 is a simple left R_1-module.

Now since $n_1 \in J_l(R_1)$ it follows that n_1 belongs to the annihilator of V_1 and so $\alpha(n_1)x = n_1 x = 0$ for every element x of V; thus $\alpha(n_1)$ belongs to the annihilator of V.

Hence $\alpha(n_1) \in J_l(R_2)$, as required.

COROLLARY 1. *Let R be a ring, I any two-sided ideal of R. Then $J_l(R/I)$ includes $(J_l(R) + I)/I$.*

Proof. Let η be the canonical epimorphism from R onto the residue class ring R/I. Then, according to the theorem,

$$(J_l(R) + I)/I = \eta(J_l(R) + I) = \eta(J_l(R)) \subseteq J_l(R/I)$$

as required.

COROLLARY 2. *Let R be a ring. Then the left radical of R is the least element of the set of two-sided ideals I of R such that R/I has zero left radical.*

Proof. Let us write $J = J_l(R)$. According to Theorem 4.5 the canonical epimorphism η from R onto R/J sets up a one-to-one correspondence between the set of ideals of R which include J and the set of all ideals of R/J. It is clear that in this correspondence the maximal left ideals of R/J correspond precisely to the maximal left ideals of R which include J. This discussion shows that the left radical $J_l(R/J)$, which is the intersection of the family of all maximal left ideals of R/J, corresponds to the intersection K of the family of all maximal left ideals of R which include J. But every maximal left ideal of R includes J; hence $K = J$ and so $J_l(R/J) = \eta(J) =$ the zero ideal of R/J.

Hence R/J has zero left radical.

Now suppose N is a two-sided ideal of R such that R/N has zero left radical. We claim that $J \subseteq N$. If this is not the case, then $J + N \supset N$ and hence $(J + N)/N$ is not the zero ideal of R/N. But according to Corollary 1 the left radical of R/N includes $(J + N)/N$; so we have a contradiction to the hypothesis that R/N has zero radical.

Hence $J \subseteq N$ and the Corollary is established.

Yet another characterization of the radical can be deduced from our next theorem.

THEOREM 23.3. *Let R be a ring with identity element e. The left radical $J_l(R)$ consists precisely of those elements x of R such that for every element a of R the element $e - ax$ is left invertible in R.*

Proof. As before let us write $J = J_l(R)$.

(1) Let x be any element of J, a any element of R. We shall show that $e - ax$ is left invertible in R.

To this end we consider the left ideal A of R generated by $e - ax$. Since $e - ax \in A$ and $ax \in J$ it follows that the identity $e = (e - ax) + ax \in A + J$, whence $A + J = R$. According to Corollary 1 of Theorem 23.2 we have

$$J_l(R/A) \supseteq (J + A)/A = R/A.$$

Thus $J_l(R/A) = R/A$; which is impossible unless R/A consists of the zero element alone. Hence $A = R$ and, in particular, e belongs to A; thus there is an element y of R such that $y(e - ax) = e$. This, however, is precisely what is meant by saying that $e - ax$ is left invertible.

(2) Conversely, suppose x is an element of R which does not belong to J; we claim that there is an element a of R such that $e - ax$ is not left invertible.

Since x is not in J there exists a maximal left ideal M of R such that x does not belong to M. It follows from the maximality of M that the left ideal generated by $M \cup \{x\}$ is the whole ring R. Consequently there is an element a of R and an element m of M such that $m + ax = e$. Then $e - ax$ is not left invertible; if it were, with left inverse y say, we would have $e = y(e - ax) = ym \in M$, which is a contradiction.

This completes the proof.

COROLLARY. *Let R be a ring with identity element e. The left radical $J_l(R)$ is the greatest element of the set of two-sided ideals I of R such that $e - t$ is invertible for every element t of I.*

Proof. (1) Let I be any two-sided ideal of R such that $e - t$ is invertible for every element t of I. Then, if x is any element of I, a any element of R, the element ax belongs to I and hence $e - ax$ is invertible. It follows from the theorem that $x \in J$; thus $I \subseteq J$.

(2) Let t be any element of J. According to the theorem, $e - t$ is left invertible, with left inverse y say. Then $y - yt = e$; so, if we

write $x = e - y$, we have $x = -yt \in J$. Hence, using the theorem yet again, we see that $e - x = y$ is left invertible. But y is of course right invertible, with right inverse $e - t$. Thus $e - t$ is the unique two-sided inverse of y (Theorem 1.8); consequently y is a two-sided inverse of $e - t$.

So J itself belongs to the set of two-sided ideals described in the Corollary.

The characterization of the left radical which we have just obtained is completely left-right symmetric: in other words, if we were to carry out the same discussion for the right radical we should obtain precisely the same characterization. We thus have at once the result which we mentioned at the beginning of the section.

THEOREM 23.4. *Let R be a ring with identity. Then the left and right radicals of R coincide.*

We now talk simply of 'the radical' of R and we shall denote it by $J(R)$.

The reader will no doubt recall that when we defined semisimple rings at the beginning of §22 we mentioned that some authors use the term 'semisimple ring' for what we have called a ring with zero radical. We now propose to show that for Artinian rings the two senses coincide.

First we make some definitions, which are applicable to any ring, not necessarily Artinian. An element a of a ring R is said to be *nilpotent* if there exists a positive integer n such that $a^n = 0$. Now let I be an ideal of R, left, right or two-sided. We say that I is a *nil ideal* if every element of I is nilpotent; we say that I is a *nilpotent ideal* if there exists a positive integer n such that I^n is the zero ideal. If a is any element of a nilpotent ideal I such that $I^n = \{0\}$, then since $a^n \in I^n$ it follows that a is nilpotent. So every nilpotent ideal is a nil ideal. In general the converse of this result is false, but we shall show that in an Artinian ring every nil ideal is nilpotent.

THEOREM 23.5. *Let R be a ring with identity element e. Then every nil ideal of R (left, right or two-sided) is included in the radical $J(R)$.*

Proof. Let N be a nil left ideal of R, x any element of N. For every element a of R the element ax belongs to N, and hence is nilpotent; say $(ax)^n = 0$. Then we readily compute that

$$(e + ax + (ax)^2 + \ldots + (ax)^{n-1})(e - ax) = e.$$

So $e - ax$ is left invertible and hence, by Theorem 23.3, x belongs to $J(R)$. Hence $N \subseteq J(R)$.

By using the characterization analogous to Theorem 23.3 for $J(R)$ as right radical we deduce in the same way that every nil right ideal is included in $J(R)$.

We must enter two words of caution concerning this result. First, although we have established that every nil ideal (and so every nilpotent ideal) of R is included in the radical $J(R)$ it does not follow in general that $J(R)$ is itself nilpotent, or even nil. We shall show, however, that in an Artinian ring R the radical $J(R)$ is nilpotent.

Next, we remark that although every ideal consisting entirely of nilpotent elements of R is included in the radical $J(R)$ it does not follow from this that every nilpotent element of R belongs to the radical—for a nilpotent element need not belong to any nil ideal. Of course, if a is a nilpotent element of the centre of R, with $a^n = 0$ say, then for every element r of R we have $(ra)^n = r^n a^n = 0$; so in this case the left ideal generated by a is nil and hence a belongs to $J(R)$.

Now we turn for the rest of this section to the case of Artinian rings.

THEOREM 23.6. *Let R be a left Artinian ring with identity element e. Then the radical of R is the greatest element in the set of nilpotent two-sided ideals of R.*

Proof. Let J be the radical of R.

It follows from Theorem 23.5 that J includes every nilpotent two-sided ideal of R. So all that remains to be shown is that J is itself nilpotent.

The sequence $(J^n)_{n \in \mathbf{N}}$ of left (actually two-sided) ideals of R is clearly an infinite descending chain; since R is left Artinian, this sequence must level off. That is to say, there is an integer p say such that $J^r = J^p$ for all integers $r \geqslant p$; set $K = J^p$. We should like to prove that $K = \{0\}$.

Suppose, to the contrary, that K is not the zero ideal. Let E be the set of left ideals I included in K such that KI is not the zero ideal; E is non-empty since K itself belongs to E. Hence, since R satisfies the minimum condition on left ideals, E has a minimal element I_0 say. Then KI_0 is not the zero ideal and hence there is an element x_0 of I_0 such that Kx_0 is not the zero ideal. Now of course Kx_0 is a left ideal of R such that $Kx_0 \subseteq I_0$ and, since $K(Kx_0) =$

$K^2x_0 = Kx_0$ is not the zero ideal, Kx_0 belongs to the set E. But I_0 is a minimal element of E; hence $Kx_0 = I_0$ and so there is an element k of K such that $kx_0 = x_0$, i.e. $(e - k)x_0 = 0$. It follows that $x_0 = 0$, for k belongs to the radical J and hence, according to the Corollary of Theorem 23.3, $e - k$ is invertible. This contradicts the fact that Kx_0 is not the zero ideal.

Thus $K = J^p = \{0\}$ and so J is nilpotent, as required.

COROLLARY. *Let R be a left Artinian ring. Then every nil ideal of R (left, right or two-sided) is nilpotent.*

Proof. According to Theorem 23.5 every nil ideal N of R is included in the radical J. Since J is nilpotent—say $J^p = \{0\}$—we have $N^p \subseteq J^p = \{0\}$ and hence N is nilpotent.

We come next to the connexion between the radical and semisimplicity.

THEOREM 23.7. *Let R be a semisimple ring. Then R is left and right Artinian and has zero radical.*

Proof. According to Theorem 22.2 every left semisimple ring is left Artinian. Similarly every right semisimple ring is right Artinian. But by Corollary 1 of Theorem 22.10 a ring is left semisimple if and only if it is right semisimple. So a (left and right) semisimple ring is both left and right Artinian.

It follows that the left R-module R_l is the direct sum of a finite family $(L_k)_{k \in K}$ of simple submodules (left ideals of R). Let x be any non-zero element of R; then x may be expressed uniquely in the form $x = \sum_{k \in K} x_k$ where $x_k \in L_k$ for each index k in K. There exists at least one index k_0 such that x_{k_0} is non-zero; then x does not belong to the left ideal $M_{k_0} = \sum_{k \neq k_0} L_k$, which is a maximal left ideal of R since R_l/M_{k_0} is isomorphic to the simple left ideal L_{k_0}. Thus x does not belong to the radical $J(R)$, which is the intersection of the family of all maximal left ideals of R. Hence $J(R) = \{0\}$.

This completes the proof.

THEOREM 23.8. *Let R be a ring with identity. If R is left Artinian and has zero radical then R is semisimple.*

Proof. Let E be the set of left ideals of R which are expressible as the intersections of finite families of maximal left ideals. Clearly E

is non-empty and hence, since R is left Artinian, has a minimal element, I_0; say $I_0 = \bigcap_{k \in K} M_k$ where K is a finite index set. If M is any maximal left ideal of R the intersection $M \cap I_0 = M \cap (\bigcap_{k \in K} M_k)$ belongs to the set E and hence, since $M \cap I_0 \subseteq I_0$ and I_0 is a minimal element of E, we have $M \cap I_0 = I_0$. It follows that I_0 is included in every maximal left ideal M of R; so $I_0 \subseteq J(R) = \{0\}$, whence $I_0 = \{0\}$.

For each index k in K let η_k be the canonical epimorphism from R_l onto the simple left R-module $S_k = R_l/M_k$. Let S be the direct product module of the family $(S_k)_{k \in K}$ with canonical projection epimorphisms π_k $(k \in K)$; S is a semisimple left R-module of finite length. Since $(S, (\pi_k))$ is couniversal for homomorphisms to the family (S_k), there exists a unique homomorphism φ from R_l to S such that $\pi_k\varphi = \eta_k$ for each index k in K. We claim that φ is a monomorphism. For if x belongs to Ker φ we have $\eta_k(x) = \pi_k\varphi(x) = 0$ for each index k; hence $x \in \bigcap_{k \in K} M_k = I_0 = \{0\}$. It follows that R_l is isomorphic to a submodule of S and hence is a semisimple module of finite length.

COROLLARY 1. *Let R be a left Artinian ring with radical J. Then the residue class ring R/J is semisimple.*

Proof. Since R is left Artinian and J is a two-sided ideal of R it follows from Theorem 14.3 that R/J is left Artinian. By Corollary 2 of Theorem 23.2, R/J has zero radical.

The result follows at once from the theorem.

COROLLARY 2. *Let R be a ring with identity. Then R is simple if and only if it is left Artinian and has no two-sided ideals other than R and $\{0\}$.*

Proof. Since a simple ring is, by definition, semisimple, it follows from Theorem 22.2 that it is left Artinian. Again by definition the only two-sided ideals of a simple ring R are R itself and the zero ideal.

Conversely, suppose the conditions of the Corollary are satisfied. Then the radical of R must be the zero ideal and hence R is semisimple by the theorem. Since its only two-sided ideals are R and $\{0\}$ it follows that R is simple.

We conclude this section with two results on left Artinian rings which are not necessarily semisimple. Our arguments depend on two elementary remarks which we might have made much earlier, but which we have not so far had any occasion to use.

Let R be any ring, I any two-sided ideal; let η be the canonical epimorphism from R onto the residue class ring $R/I = \overline{R}$. If $\overline{V}$ is any left $\overline{R}$-module we may define a scalar multiplication on the left of $\overline{V}$ by elements of R; namely, for each element a of R and each element $\bar{x}$ of $\overline{V}$ we set $a\bar{x} = \eta(a)\,\bar{x}$. We verify easily that when $\overline{V}$ is equipped with its original addition and the scalar multiplication we have just defined it has the structure of a left R-module. When we regard $\overline{V}$ as a left R-module in this way we shall denote it by $(\overline{V})_R$. It is easy to check also that if $\overline{V}$ is a simple left $\overline{R}$-module then $(\overline{V})_R$ is a simple left R-module, and that if $\overline{V}_1$ and $\overline{V}_2$ are isomorphic left $\overline{R}$-modules then $(\overline{V_1})_R$ and $(\overline{V_2})_R$ are R-isomorphic left R-modules.

On the other hand let V be a left R-module such that I is included in the annihilator of V; that is to say, $rx = 0$ for every element r of I and every element x of V. In this situation we define a scalar multiplication on the left of V by elements of $\overline{R}$ as follows: if A is any element of $\overline{R} = R/I$, let a be any element of R in A and for each element x of V set $Ax = ax$. It is clear that the element ax depends only on A and x, not on the choice of the element a; for if a' is another element of A, then $a - a' \in I \subseteq A(V)$ and so $(a - a')\,x = 0$, whence $ax = a'x$. When V is equipped with its original addition and the scalar multiplication which we have just described, it is clearly a left $\overline{R}$-module; when we consider V as a left $\overline{R}$-module in this way we shall denote it by $(V)_{\overline{R}}$. We see easily that if V is a simple left R-module such that $I \subseteq A(V)$ then $(V)_{\overline{R}}$ is a simple left $\overline{R}$-module and that if V_1 and V_2 are R-isomorphic left R-modules annihilated by I then $(V_1)_{\overline{R}}$ and $(V_2)_{\overline{R}}$ are $\overline{R}$-isomorphic left $\overline{R}$-modules.

Furthermore, if V is an Artinian left R-module such that I is included in the annihilator of V, the associated left $\overline{R}$-module $(V)_{\overline{R}}$ is also Artinian. To see this, let $(\overline{V_n})_{n \in \mathbf{N}}$ be a decreasing sequence of $\overline{R}$-modules with $\overline{V}_0 = (V)_{\overline{R}}$. Then $((\overline{V_n})_R)$ is a decreasing sequence of R-modules with $(\overline{V_0})_R = ((V)_{\overline{R}})_R = V$; since V is Artinian, this sequence levels off, and hence so does the original sequence.

With these preliminaries we are in a position to prove our first result on arbitrary left Artinian rings.

THEOREM 23.9. *Let R be a left Artinian ring with radical J. If the semisimple ring R/J has n simple components then there are exactly n isomorphism types of simple left R-modules.*

Proof. According to Theorem 22.9, there are exactly n isomorphism types of simple left (R/J)-modules.

Let $\mathscr{T}$ and $\overline{\mathscr{T}}$ be the sets of isomorphism types of simple left R-modules and (R/J)-modules respectively. We define mappings φ and ψ from $\mathscr{T}$ to $\overline{\mathscr{T}}$ and $\overline{\mathscr{T}}$ to $\mathscr{T}$ respectively as follows: for each isomorphism type T of simple left R-modules we set

$$\varphi(T) = \text{isomorphism type of the simple left } (R/J)\text{-module } (T)_{R/J}$$

and for each isomorphism type $\overline{T}$ of simple left (R/J)-modules we set

$$\psi(\overline{T}) = \text{isomorphism type of the simple left } R\text{-module } (\overline{T})_R.$$

(We remark that since, according to Theorem 23.1, J annihilates every simple left R-module we may form the left (R/J)-module $(T)_{R/J}$.)

It is easily verified that $\psi\varphi$ and $\varphi\psi$ are respectively the identity mappings of $\mathscr{T}$ and $\overline{\mathscr{T}}$. So φ is a bijection from $\mathscr{T}$ onto $\overline{\mathscr{T}}$. Since Card $\overline{\mathscr{T}} = n$ it follows that Card $\mathscr{T} = n$ also, as required.

We now conclude this chapter by proving a result which we have already stated in §14.

THEOREM 23.10 (*Hopkins's Theorem*). *Every left Artinian ring with identity is left Noetherian.*

Proof. Let R be a left Artinian ring with radical J. We write $\overline{R} = R/J$.

If $J = \{0\}$, then Theorem 23.8 shows that R is semisimple; hence, by Theorem 22.2, R is left Noetherian.

Suppose now that J is not the zero ideal. According to Theorem 23.6, J is nilpotent and hence there exists a finite strictly descending chain

$$R \supset J \supset J^2 \supset \ldots \supset J^m = \{0\}.$$

We now apply Theorem 14.3: each of the ideals $J^0 = R, J, J^2, \ldots, J^{m-1}$ is a submodule of the Artinian left R-module R_l and so is itself an Artinian left R-module; hence all the factor modules $V_i = J^{i-1}/J^i$ $(i = 1, \ldots, m)$ are Artinian left R-modules. Clearly, each of these modules V_i is annihilated by J; so we may define the left $\overline{R}$-modules $\overline{V}_i = (V_i)_{\overline{R}}$, and these are Artinian by our earlier discussion (just before Theorem 23.9).

According to Corollary 1 of Theorem 23.8 the ring $\overline{R}$ is semisimple. Hence, by Theorem 22.1, each of the left $\overline{R}$-modules $\overline{V}_i$ is semisimple and is thus the internal direct sum of a family of simple left $\overline{R}$-modules—which must be a finite family since $\overline{V}_i$ is Artinian. Suppose $\overline{V}_i$ is the direct sum of the simple left $\overline{R}$-modules $\overline{S}_{i1}, \ldots, \overline{S}_{in_i}$

$(i = 1, \ldots, m)$; for $i = 1, \ldots, m$ set $\overline{W}_{ij} = \overline{S}_{i1} + \overline{S}_{i2} + \ldots + \overline{S}_{i, n_i - j}$ $(j = 0, \ldots, n_i - 1)$ and let $\overline{W}_{in_i}$ be the zero sunmodule of $\overline{V}_i (i = 1, \ldots, m)$. If we denote the left R-module $(\overline{W}_{ij})_R$ by W_{ij}, it is easy to see that the factor module $W_{ij}/W_{i, j+1}$ is isomorphic to the simple left R-module $(\overline{S}_{i, n_i - j})_R$ $(i = 1, \ldots, m; j = 0, \ldots, n_i - 1)$.

Finally we set $V_{ij} = \eta_i^{-1}(W_{ij})$ $(i = 1, \ldots, m;\ j = 0, \ldots, n_i - 1)$ where η_i is the canonical epimorphism from J^{i-1} onto V_i; then $V_{in_i} = V_{i+1, 0}$ $(i = 1, \ldots, m - 1)$. According to Theorem 7.5 the factor module $V_{ij}/V_{i, j+1}$ is isomorphic to $W_{ij}/W_{i, j+1}$ and hence is a simple left R-module $(i = 1, \ldots, m; j = 0, \ldots, n_i - 1)$. Thus the sequence

$$R_l = V_{10} \supset V_{11} \supset \ldots \supset V_{1n_1} (= V_{20}) \supset V_{21} \supset \ldots \supset V_{2n_2} (= V_{30}) \supset \ldots$$
$$V_{m-1, n_{m-1}} (= V_{m0}) \supset V_{m1} \supset \ldots \supset V_{mn_m} = 0$$

is a composition series for R_l. Hence, by Theorem 14.5, R_l is a Noetherian left R-module, i.e. R is a left Noetherian ring.

Chapter 5

ALGEBRAS

§24. Tensor Products of Algebras

Let K be a field. Then an *algebra over* K or a K-*algebra* is a pair (A, ι) consisting of a ring A with identity and a monomorphism ι from K to A such that (1) the image under ι of the identity element of K is the identity element of A and (2) the image of K under ι is included in the centre of A. If (A, ι) is an algebra over K, we may define a scalar multiplication on the left of A by elements of K by setting

$$kx = \iota(k)\,x$$

for all elements k of K, x of A. (In this equation the product on the right is formed according to the multiplication operation in the ring A). It is easy to verify that under this scalar multiplication and its own addition operation A forms a left vector space over the field K; further, for every element k of K and all elements x, y of A we have

$$(kx)\,y = k(xy) = x(ky). \qquad [24.1]$$

Suppose conversely that a ring A with identity element e has the structure of a left vector space over a field K and that the condition [24.1] is satisfied for every element k of K and all elements x, y of A. If in this situation we define a mapping ι from K to A by setting

$$\iota(k) = ke$$

for each element k of K, it is clear that ι is a monomorphism and that the pair (A, ι) is an algebra over K.

Example 1. Let $K = \mathbf{R}$, the field of real numbers, $A = \mathbf{C}$, the field of complex numbers and let j be the mapping from $\mathbf{R}$ to $\mathbf{C}$ defined by setting

$$j(a) = a + 0i$$

for every real number a. Then $(\mathbf{C}, j)$ is an algebra over $\mathbf{R}$.

Example 2. Let K be any field, V a left vector space over K and let $A = \mathrm{Hom}_K(V, V)$ be the ring of endomorphisms of V. We define a mapping λ from K to A as follows; we remark first that for every element k of K the mapping λ_k from V to V defined by setting

$$\lambda_k(x) = kx$$

for every element x of V is a K-endomorphism of V, i.e. $\lambda_k \in A$. If then we set

$$\lambda(k) = \lambda_k$$

for each element k of K, the pair (A, λ) is easily seen to be an algebra over K. We call it the *algebra of K-linear transformations of V.*

Example 3. Let K be any field, n any positive integer and let $A = M_n(K)$ be the ring of $n \times n$ matrices with coefficients in K. Let δ be the monomorphism from K to $M_n(K)$ defined in Example 3 of §5—for each element k of K the matrix $\delta(k)$ is the diagonal matrix with diagonal elements all equal to k. Then (A, δ) is an algebra over K, which we call the *algebra of $n \times n$ matrices* over K.

If (A, ι) is an algebra over a field K, it is of course an extension of K in the sense of §5—though it also has additional properties not required of the general extensions discussed there. But, as with these general extensions, we shall often identify K with its image under ι and so regard K as a subfield of the centre of A. We shall also say that A is an algebra over K.

Let (A, ι) be an algebra over a field K. Let B be a subring of A having the same identity element as A. If $\iota(K)$ is included in the centre of B then (B, ι) is an algebra over K and we call it a *subalgebra* of (A, ι).

Let (A, ι) and (A', ι') be algebras over the same field K. A mapping φ from A to A' is called an *algebra homomorphism* from (A, ι) to (A', ι') if (1) φ is a ring homomorphism from A to A', (2) the image under φ of the identity element of A is the identity element of A' and (3) for every element k of K and every element x of A we have $\varphi(\iota(k)x) = \iota'(k)\varphi(x)$. If we identify K with its image under ι in A and also with its image under ι' in A', this third condition reduces to the requirement that $\varphi(kx) = k\varphi(x)$ for every element k of K and every element x of A. We shall use the terms algebra monomorphism, algebra epimorphism, . . . in the obvious way.

Example 4. Let V be a vector space of finite dimension n over a field K. Let $(\mathrm{Hom}_K(V, V), \lambda)$ be the algebra of K-linear transformations of

V as described in Example 2 and $(M_n(K), \delta)$ the algebra of $n \times n$ matrices with coefficients in K as described in Example 3. Let $X = \{x_1, \ldots, x_n\}$ be a basis for V over K. For each element α of $\mathrm{Hom}_K(V, V)$ we define an $n \times n$ matrix $M(\alpha) = [m_{ij}(\alpha)]$ by setting

$$\alpha(x_j) = \sum_{i=1}^{n} m_{ij}(\alpha)\, x_i \qquad (j = 1, \ldots, n).$$

As we have seen in Theorem 8.5, the mapping φ_X from $\mathrm{Hom}_K(V, V)$ to $M_n(K)$ defined by setting $\varphi_X(\alpha) = M(\alpha)$ for every K-endomorphism α of V is a ring isomorphism (since K is commutative). It is in fact also an algebra isomorphism. To see this, let a be any element of K; then $\varphi_X(\lambda_a)$ is the matrix $L = [l_{ij}]$ given by

$$ax_j = \lambda_a(x_j) = \sum_{i=1}^{n} l_{ij} x_i \qquad (j = 1, \ldots, n).$$

Since the basis $\{x_1, \ldots, x_n\}$ is linearly independent it follows that for $j = 1, \ldots, n$ we have $l_{jj} = a$ and $l_{ij} = 0$ $(i \neq j)$; in other words, L is the diagonal matrix $\delta(a)$. Hence for each element α of $\mathrm{Hom}_K(V, V)$ and each element a of K we have

$$\varphi_X(\lambda(a)\alpha) = \varphi_X(\lambda_a\alpha) = \varphi_X(\lambda_a)\varphi_X(\alpha) = \delta(a)\varphi_X(\alpha)$$

as required.

We now make some elementary remarks about tensor products of modules over commutative rings, supplementing those we have already made in §12.

So let R be a commutative ring with identity. First let us observe that every left R-module V may be regarded as a right R-module. Namely, if we define a scalar multiplication on the right of V by setting

$$xr = rx$$

for all elements r of R, x of V, then V becomes a right R-module. The point at which the commutativity of R enters is in the verification of condition $RM(3)$ of §6: if r, s are elements of R, x an element of V, then

$$x(rs) = (rs)x = (sr)x = s(rx) = (rx)s = (xr)s.$$

Similarly every right R-module may be regarded as a left R-module. We shall allow ourselves to be rather easy-going in our terminology and notation when dealing with modules over commutative rings—

we shall talk simply of 'modules', omitting any mention of 'left' or 'right' and we shall write the elements of R on the left or on the right of those of V as seems best fitted to the context.

Let V, W and M be R-modules. Then a mapping φ from $V \times W$ to M is said to be *bilinear* if for all elements x, x_1, x_2 of V, y, y_1, y_2 of W and r of R we have

$$\varphi(x_1 + x_2, y) = \varphi(x_1, y) + \varphi(x_2, y)$$
$$\varphi(x, y_1 + y_2) = \varphi(x, y_1) + \varphi(x, y_2)$$
$$\varphi(xr, y) = \varphi(x, ry) = r\varphi(x, y).$$

We recall now from §12 that when R is commutative the tensor product $V \otimes_R W$ can be given the structure of an R-module by prescribing that $r(x \otimes y) = xr \otimes y$ for all elements r of R, x of V, y of W. It follows easily that when we consider $V \otimes_R W$ as an R-module in this way the natural mapping ν from $V \times W$ to $V \otimes W$ is bilinear.

Let φ be a bilinear mapping from $V \times W$ to M, where V, W, M are R-modules. Then φ is of course a balanced mapping (in the sense of §12) from $V \times W$ to M considered simply as an abelian group; hence, according to Theorem 12.1, there exists a unique abelian group homomorphism α from $V \otimes_R W$ to M such that $\alpha\nu = \varphi$. We claim that α is in fact an R-module homomorphism. First let x, y, r be any elements of V, W, R respectively; we have

$$\begin{aligned}\alpha(r(x \otimes y)) = \alpha(xr \otimes y) = \alpha\nu(xr, y) = \varphi(xr, y)&\\ = r\varphi(x, y) = r\alpha\nu(x, y) = r\alpha(x \otimes y).&\end{aligned}$$

Then since every element of $V \otimes_R W$ is expressible as a finite sum of elements of the form $x \otimes y$ it follows easily that α is an R-homomorphism. We shall make use of this fact without further comment.

In the general situation discussed in §12, where V is a right R-module and W a left R-module, we may form the tensor product $V \otimes_R W$ but not the product $W \otimes_R V$. When R is commutative, however, then (as we explained above) V and W may be regarded as left or right R-modules at will and we may form both products $V \otimes_R W$ and $W \otimes_R V$.

THEOREM 24.1. *Let R be a commutative ring and let V and W be R-modules. Then the tensor products $V \otimes_R W$ and $W \otimes_R V$ are isomorphic R-modules.*

Proof. Consider the mapping φ from $V \times W$ to $W \otimes_R V$ defined by setting $\varphi(x, y) = y \otimes x$ for all elements x of V, y of W. It is easy to verify that φ is bilinear; hence there is an R-homomorphism α from $V \otimes_R W$ to $W \otimes_R V$ such that $\alpha(x \otimes y) = y \otimes x$ for all elements x of V, y of W.

Similarly there exists an R-homomorphism β from $W \otimes_R V$ to $V \otimes_R W$ such that $\beta(y \otimes x) = x \otimes y$ for all elements y of W, x of V.

We verify at once that $\beta\alpha$ and $\alpha\beta$ are the identity mappings of $V \otimes_R W$ and $W \otimes_R V$ respectively. Hence α and β are isomorphisms, as required.

This theorem shows that the formation of tensor products of modules over a commutative ring is commutative; next we show that it is also associative.

THEOREM 24.2. *Let R be a commutative ring and let U, V and W be R-modules. Then the tensor products $(U \otimes_R V) \otimes_R W$ and $U \otimes_R (V \otimes_R W)$ are isomorphic R-modules.*

Proof. For each element z of W consider the mapping φ_z from $U \times V$ to $U \otimes_R (V \otimes_R W)$ defined by setting

$$\varphi_z(x, y) = x \otimes (y \otimes z)$$

for each pair (x, y) in $U \times V$. This mapping is easily seen to be bilinear, and hence there exists an R-homomorphism α_z from $U \otimes_R V$ to $U \otimes_R (V \otimes_R W)$ such that

$$\alpha_z(x \otimes y) = x \otimes (y \otimes z)$$

for all elements x of U, y of V.

Now consider the mapping φ from $(U \otimes V) \times W$ to $U \otimes_R (V \otimes_R W)$ defined by setting

$$\varphi(t, z) = \alpha_z(t)$$

for each pair (t, z) in $(U \otimes_R V) \times W$. We claim that φ is bilinear. Since α_z is an R-homomorphism we certainly have

$$\varphi(t_1 + t_2, z) = \alpha_z(t_1 + t_2) = \alpha_z(t_1) + \alpha_z(t_2) = \varphi(t_1, z) + \varphi(t_2, z) \qquad [24.2]$$

and

$$\varphi(tr, z) = \alpha_z(tr) = r\alpha_z(t) = r\varphi(t, z) \qquad [24.3]$$

for all elements t, t_1, t_2 of $U \otimes_R V$, z of W and r of R. Next for all elements x of U, y of V, z, z_1, z_2 of W and r of R we have

$$\begin{aligned} \varphi(x \otimes y, z_1 + z_2) = \alpha_{z_1+z_2}(x \otimes y) &= x \otimes (y \otimes (z_1 + z_2)) \\ &= x \otimes ((y \otimes z_1) + (y \otimes z_2)) \\ &= x \otimes (y \otimes z_1) + x \otimes (y \otimes z_2) \\ &= \alpha_{z_1}(x \otimes y) + \alpha_{z_2}(x \otimes y) \\ &= \varphi(x \otimes y, z_1) + \varphi(x \otimes y, z_2) \end{aligned}$$

and

$$\begin{aligned} \varphi(x \otimes y, rz) = \alpha_{rz}(x \otimes y) &= x \otimes (y \otimes rz) \\ &= x \otimes (yr \otimes z) \\ &= \alpha_z(x \otimes yr) = \varphi((x \otimes y)r, z). \end{aligned}$$

Since every element of $U \otimes_R V$ is a sum of elements of the form $x \otimes y$ with x in U and y in V, it follows that for all elements t of $U \otimes_R V$, z, z_1, z_2 of W and r of R we have

$$\varphi(t, z_1 + z_2) = \varphi(t, z_1) + \varphi(t, z_2) \qquad [24.4]$$

and

$$\varphi(t, rz) = \varphi(tr, z). \qquad [24.5]$$

Equations [24.2]–[24.5] show that φ is bilinear. Hence there exists an R-homomorphism α from $(U \otimes_R V) \otimes_R W$ to $U \otimes_R (V \otimes_R W)$ such that

$$\alpha((x \otimes y) \otimes z) = x \otimes (y \otimes z)$$

for all elements x, y, z of U, V, W respectively.

Similarly there is an R-homomorphism β from $U \otimes_R (V \otimes_R W)$ to $(U \otimes_R V) \otimes_R W$ such that

$$\beta(x \otimes (y \otimes z)) = (x \otimes y) \otimes z$$

for all elements x of U, y of V, z of W. It is an easy matter to verify that $\beta\alpha$ and $\alpha\beta$ are the appropriate identity homomorphisms; hence α and β are isomorphisms, as required.

Now let (A, ι_A) and (B, ι_B) be algebras over the same field K. As described at the beginning of this section we may regard A and B as vector spaces over K by setting

$$kx = \iota_A(k) . x \text{ and } ky = \iota_B(k) . y$$

for all elements x of A, y of B, k of K. Hence we may form the tensor product $A \otimes_K B$. At first this product has only the structure of a vector space over K, but we shall define a multiplication operation under which it becomes a ring and a monomorphism ι from K to this ring such that $(A \otimes_K B, \iota)$ is an algebra over K.

To this end let x and y be elements of A and B respectively and consider the mapping $\varphi_{x \otimes y}$ from $A \times B$ to $A \otimes_K B$ defined by setting

$$\varphi_{x \otimes y}(x', y') = xx' \otimes yy'$$

for each element (x', y') of $A \times B$. We check that $\varphi_{x \otimes y}$ is bilinear and deduce that there is a K-homomorphism $\lambda_{x \otimes y}$ from $A \otimes_K B$ to itself such that

$$\lambda_{x \otimes y}(x' \otimes y') = xx' \otimes yy'$$

for all elements x' of A and y' of B. Similarly if x' and y' are elements of A and B respectively there is a K-homomorphism $\rho_{x' \otimes y'}$ from $A \otimes_K B$ to itself such that

$$\rho_{x' \otimes y'}(x \otimes y) = xx' \otimes yy'$$

for all elements x of A and y of B. Next let t be any element of $A \otimes_K B$ and consider the mapping φ_t from $A \times B$ to $A \otimes_K B$ defined by setting

$$\varphi_t(x', y') = \rho_{x' \otimes y'}(t)$$

for each element (x', y') of $A \times B$. If we use the fact that t can be expressed as a sum of a finite family of terms of the form $x \otimes y$ we can verify that φ_t is bilinear and hence deduce that there is a K-homomorphism λ_t from $A \otimes_K B$ to itself such that

$$\lambda_t(x' \otimes y') = \rho_{x' \otimes y'}(t)$$

for all elements x' of A and y' of B. Similarly for each element t' of $A \otimes_K B$ there exists a K-homomorphism $\rho_{t'}$ from $A \otimes_K B$ to itself such that

$$\rho_{t'}(x \otimes y) = \lambda_{x \otimes y}(t')$$

for all elements x of A and y of B.

If $t = \sum_{i \in I} x_i \otimes y_i$ and $t' = \sum_{j \in J} x'_j \otimes y'_j$ are elements of $A \otimes_K B$ we easily compute that

$$\lambda_t(t') = \sum_{(i,j)\in I\times J} x_i x'_j \otimes y_i y'_j = \rho_{t'}(t).$$

We now define a multiplication operation in $A \otimes_K B$ be setting

$$tt' = \lambda_t(t') = \rho_{t'}(t)$$

for every pair of elements t, t' of $A \otimes_K B$. Since λ_t and $\rho_{t'}$ are homomorphisms it follows that this multiplication is distributive over the addition in $A \otimes_K B$. To check that it is associative it is clearly sufficient to observe that

$$\begin{aligned}((x \otimes y)(x' \otimes y'))(x'' \otimes y'') &= (xx' \otimes yy') \otimes (x'', y'') \\ &= ((xx')x'') \otimes ((yy')y'') \\ &= (x(x'x'')) \otimes (y(y'y'')) \\ &= (x \otimes y)(x'x'' \otimes y'y'') \\ &= (x \otimes y)((x' \otimes y')(x'' \otimes y''))\end{aligned}$$

for all elements x, x', x'' of A and y, y', y'' of B. Hence $A \otimes_K B$ forms a ring under its vector space addition and the multiplication we have just defined. It is clear that if e_A and e_B are the identity elements of A and B respectively then $e_A \otimes e_B$ is an identity element for $A \otimes_K B$.

We notice finally that for all elements t, t' of $A \otimes_K B$ and all elements k of K we have

$$(kt)t' = \rho_{t'}(kt) = k\rho_{t'}(t) = k(tt')$$

and

$$t(kt') = \lambda_t(kt') = k\lambda_t(t') = k(tt')$$

(since $\rho_{t'}$ and λ_t are K-homomorphisms). Consequently, if we set $\iota(k) = k(e_A \otimes e_B)$ for every element k of K, if follows from our remarks at the beginning of this section that $(A \otimes_K B, \iota)$ is an algebra over K.

If A, B and C are algebras over K then according to Theorems 24.1 and 24.2 there exist vector space isomorphisms from $A \otimes_K B$ onto $B \otimes_K A$ and from $(A \otimes_K B) \otimes_K C$ onto $A \otimes_K (B \otimes_K C)$. It is easy to check that the mappings in question are also ring homomorphisms, and hence are algebras isomorphisms. (Note that we have resorted to the lazy but convenient practice of omitting the monomorphisms from K to A, B, C and writing simply A instead of (A, ι_A) etc.)

Let (A, ι) be an algebra over a field K, S any subset of A. Let S^* be the *centraliser* of S in A, i.e. the set of elements of A which commute with every element of S. Then it is easy to verify that (S^*, ι) is a subalgebra of (A, ι).

Let A and B be algebras over the same field K and let S and T be subalgebras of A and B respectively. Since vector spaces over a field have bases, it follows from the Corollary to Theorem 12.8 that algebras over K, when considered in the usual way as vector spaces over K, are flat K-modules; from this we deduce that if ι_S and ι_T are the inclusion monomorphisms from S to A and T to B respectively then $\iota_S \otimes \iota_T$ is a monomorphism from $S \otimes_K T$ to $A \otimes_K B$—and in fact an algebra monomorphism. We may thus identify $S \otimes_K T$ with its image under $\iota_S \otimes \iota_T$ and so regard it as a subalgebra of $A \otimes_K B$.

THEOREM 24.3. *Let A and B be algebras over a field K; let S and T be subalgebras of A and B, S^* and T^* the centralisers of S in A and T in B respectively. Then the centraliser of $S \otimes_K T$ in $A \otimes_K B$ is $S^* \otimes_K T^*$.*

Proof. It is clear from the definition of multiplication in $A \otimes_K B$ that $S^* \otimes_K T^*$ is included in the centraliser of $S \otimes_K T$.

To prove the reverse inclusion we remark first that there exist subspaces S_1 and T_1 of A and B respectively (considered as vector spaces over K) such that A is the internal direct sum of S^* and S_1 and B is the internal direct sum of T^* and T_1 respectively. (We use Theorem 11.10 and the fact that vector spaces over a field are injective.) It follows from Theorem 12.7 that $A \otimes T^*$ is the internal direct sum of $S^* \otimes T^*$ and $S_1 \otimes T^*$, while $S^* \otimes B$ is the internal direct sum of $S^* \otimes T^*$ and $S^* \otimes T_1$. Hence $S^* \otimes T^*$ is the intersection of $A \otimes T^*$ and $S^* \otimes B$.

Now let z be any element of the centraliser of $S \otimes T$ in $A \otimes B$. If $\{b_i\}_{i \in I}$ is a basis for B over K we may write z uniquely in the form $z = \sum_{i \in I} a_i \otimes b_i$ where $(a_i)_{i \in I}$ is a quasi-finite family of elements of A. For every element s of S, z commutes with $s \otimes e_B$ (where e_B is the identity of B) and hence

$$\sum_{i \in I} (a_i s - s a_i) \otimes b_i = 0,$$

whence $a_i s - s a_i = 0$ for all indices i in I (since $\{b_i\}$ is a basis for B).

Thus all the elements a_i belong to S^*; hence $z \in S^* \otimes B$. Similarly $z \in A \otimes T^*$ and hence $z \in S^* \otimes T^*$, as required.

COROLLARY. *Let A and B be algebras over a field K with centres Z_A and Z_B respectively. Then the centre of $A \otimes_K B$ is $Z_A \otimes_K Z_B$.*

Proof. We have only to take $S = A$ and $T = B$ in the theorem.

We conclude this section with two examples which we shall find useful in the sequel.

Example 5. Let K be a field, $(A, \imath)$ an algebra over K; let n be any positive integer, $M_n(K)$ and $M_n(A)$ the rings of $n \times n$ matrices with coefficients in K and A respectively. If δ_K and δ_A are the diagonal mappings from K to $M_n(K)$ and A to $M_n(A)$ respectively, we have already seen in Example 3 of §24 that $(M_n(K), \delta_K)$ is an algebra over K, and we can easily check that $(M_n(A), \delta_A \imath)$ is also an algebra over K. We claim that there is an algebra isomorphism from $A \otimes_K M_n(K)$ onto $M_n(A)$.

To establish this result, let φ be the mapping from $A \times M_n(K)$ to $M_n(A)$ defined by setting

$$\varphi(a, X) = \delta_A(a)\, \imath(X)$$

for all elements a of A and all $n \times n$ matrices X in $M_n(K)$, where, in an obvious notation, $\imath(X)$ is the matrix with coefficients in A obtained by applying the monomorphism $\imath$ to all the coefficients of X. Clearly φ is bilinear and hence there exists a K-homomorphism α from $A \otimes_K M_n(K)$ to $M_n(A)$ such that

$$\alpha(a \otimes X) = \delta_A(a)\, \imath(X)$$

for all elements a of A and X of $M_n(K)$. Since the elements of the matrix $\imath(X)$ all belong to the centre of A and $\delta_A(a)$ is a diagonal matrix, it follows by a simple calculation that $\delta_A(a)\, \imath(X) = \imath(X)\, \delta_A(a)$; hence if a, a' are any two elements of A and X, X' are any two elements of $M_n(K)$ we have

$$\begin{aligned}\alpha(a \otimes X)\alpha(a' \otimes X') &= \delta_A(a)\imath(X)\delta_A(a')\imath(X') \\ &= \delta_A(a)\delta_A(a')\imath(X)\imath(X') \\ &= \delta_A(aa')\imath(XX') \\ &= \alpha(aa' \otimes XX') \\ &= \alpha((a \otimes X)(a' \otimes X')).\end{aligned}$$

From this we deduce easily that α is an algebra homomorphism.

We have now to show that α is actually an isomorphism. So for each pair of indices i, j $(i, j = 1, \ldots, n)$ let $E_{ij}(K)$ be the $n \times n$ matrix which has the identity element of K in the (i, j)th position and the zero element of K in all other positions and define the matrices $E_{ij}(A)$ similarly. It is easy to show that the sets $\{E_{ij}(K)\}$ and $\{E_{ij}(A)\}$ are bases for $M_n(K)$ and $M_n(A)$ as vector spaces over K and left (or right) A-module respectively.

It follows first from Corollary 1 of Theorem 12.7 that every element t of $A \otimes_K M_n(K)$ can be expressed uniquely in the form $t = \sum_{i,j=1}^{n} a_{ij} \otimes E_{ij}(K)$, where a_{ij} $(i, j = 1, \ldots, n)$ are elements of A. Next, according to Theorem 6.4 every element Y of $M_n(A)$ can be expressed uniquely in the form $Y = \sum_{i,j=1}^{n} y_{ij}E_{ij}(A)$ where y_{ij} $(i, j = 1, \ldots, n)$ are elements of A. We may thus define a mapping β from $M_n(A)$ to $A \otimes_K M_n(K)$ by setting

$$\beta(Y) = \sum_{i,j=1}^{n} y_{ij} \otimes E_{ij}(K)$$

for each element Y of $M_n(A)$. Remarking that for each element $t = \sum_{i,j} a_{ij} \otimes E_{ij}(K)$ we have

$$\alpha(t) = \sum_{i,j} \delta_A(a_{ij})\iota(E_{ij}(K)) = \sum_{i,j} \delta_A(a_{ij})E_{ij}(A) = \sum_{i,j} a_{ij}E_{ij}(A)$$

we deduce immediately that $\beta\alpha$ and $\alpha\beta$ are the appropriate identity homomorphisms; hence α is an isomorphism, as required.

Example 6. Let K be a field, m and n positive integers. We shall show that there is an algebra isomorphism from the tensor product $M_m(K) \otimes_K M_n(K)$ onto $M_{mn}(K)$. For $i, j = 1, \ldots, m$ we define $E_m(i, j)$ to be the $m \times m$ matrix which has the identity element of K in the (i, j)th position and the zero element of K in all other positions; in a similar way we define the $n \times n$ matrices $E_n(k, l)$ $(k, l = 1, \ldots, n)$ and the $mn \times mn$ matrices $E_{mn}(r, s)$ $(r, s = 1, \ldots, mn)$. As we remarked in Example 5, the sets $\{E_m(i, j)\}$, $\{E_n(k, l)\}$ and $\{E_{mn}(r, s)\}$ are bases for $M_m(K)$, $M_n(K)$ and $M_{mn}(K)$ respectively, considered as vector spaces over K. According to Corollary 2 of Theorem 12.7 the set $\{E_m(i, j) \otimes E_n(k, l)\}$ is a basis for $M_m(K) \otimes_K M_n(K)$ over K. It follows easily that the mapping α from $M_m(K) \otimes_K M_n(K)$ to

$M_{mn}(K)$ defined by setting

$$\alpha\Big(\sum_{i,j=1}^{m}\sum_{k,l=1}^{n} a_{ijkl}E_m(i,j)\otimes E_n(k,l)\Big) = \sum_{i,j=1}^{m}\sum_{k,l=1}^{n} a_{ijkl}E_{mn}(n(i-1)+k, n(j-1)+l)$$

is a vector space isomorphism.

We claim that α is also a ring homomorphism; to see this it is clearly sufficient to verify that for all indices $i, j, i', j' = 1, \ldots, m$ and $k, l, k', l' = 1, \ldots, n$ we have

$$\begin{aligned}\alpha(E_m(i,j)E_m(i',j') &\otimes E_n(k,l)E_n(k',l')) \\ &= \alpha(E_m(i,j)\otimes E_n(k,l))\alpha(E_m(i',j')\otimes E_n(k',l')) \\ &= E_{mn}(n(i-1)+k, n(j-1)+l)E_{mn}(n(i'-1)+k', n(j'-1)+l').\end{aligned}$$

If $j \neq i'$ then $E_m(i,j)E_m(i',j')$ is the zero matrix. So the first member is the zero matrix in $M_{mn}(K)$; but so also is the last member, for if $j \neq i'$ then $n(j-1)+l \neq n(i'-1)+k'$. A similar argument applies if $l \neq k'$. Suppose next that $j = i'$ and $l = k'$. Then we have $E_m(i,j)E_m(i',j') = E_m(i,j')$ and $E_n(k,l)E_n(k',l') = E_n(k,l')$; hence the first member is

$$\alpha(E_m(i,j')\otimes E_n(k,l')) = E_{mn}(n(i-1)+k, n(j'-1)+l')$$

and the last member is

$$E_{mn}(n(i-1)+k, n(j-1)+l)E_{mn}(n(i'-1)+k', n(j'-1)+l') = E_{mn}(n(i-1)+k, n(j'-1)+l')$$

also.

Consequently α is an algebra isomorphism.

§25. Simple Algebras

Let K be a field, (A, ι) an algebra over K. If the ring A is a simple ring in the sense of §22 we shall say that (A, i) is a *simple algebra*; if the image of K under the monomorphism ι coincides with the centre of A we shall say that (A, ι)—or, briefly, A—is a *central* simple algebra over K and in this case we shall always identify K with the centre of A. If (D, ι) is an algebra over K and D is a division ring then we say that (D, ι) is a *division algebra*.

We begin our study of simple algebras by restating, in a slightly altered notation which will be better suited to our present purposes, some of the results on the structure of simple rings which we established in §22. If A is a simple ring, V a simple left A-module, then the ring $S = \mathrm{Hom}_A(V, V)$ is a division ring and A is isomorphic to the ring of S-endomorphisms of V; further, the centre of A is isomorphic to the centre of S. If D is the opposite ring of S, then $\mathrm{Hom}_S(V, V)$—and hence of course A—is isomorphic to the ring $M_r(D)$ of $r \times r$ matrices with coefficients in D, where r is the length of the isotypic left A-module A_l; clearly D has the same centre as S. Now let (A, ι) be a simple algebra over a field K; by composing ι with the isomorphism from the centre of A onto the centre of D we obtain a monomorphism ι' from K to D such that (D, ι') is an algebra over K. Composing ι' with the diagonal mapping δ_D from D to $M_r(D)$, we obtain a monomorphism δ from K to $M_r(D)$ such that $(M_r(D), \delta)$ is an algebra over K. Finally, as we have seen in Example 5 of §24, this algebra is isomorphic to the tensor product of the algebras $(M_r(K), \delta_K)$ and (D, ι'). Hence, to sum up briefly, if A is a simple algebra over a field K there is a positive integer r and a division algebra D over K such that A is isomorphic to $M_r(K) \otimes_K D$; we call D the *division algebra component* of A. Finally it is clear that if A is finite-dimensional when considered as a vector space over K then so is D and we have $\dim_K A = r^2 \dim_K D$.

Now let A be any ring (not necessarily simple) with centre K; we may regard A as a left or right K-module. For each element a of A the *left translation* λ_a and the *right translation* ρ_a of A by a, i.e. the mappings from A to itself defined by setting

$$\lambda_a(x) = ax \qquad \text{and} \qquad \rho_a(x) = xa.$$

for all elements x of A are clearly K-endomorphisms of A. If $(a_i)_{i \in I}$ and $(b_i)_{i \in I}$ are finite families of elements of A with the same index set I, the mapping $\tau = \sum_{i \in I} \lambda_{a_i} \rho_{b_i}$ from A to itself is also a K-endomorphism of A; a K-endomorphism which can be expressed in this form is said to be *analytic*. It is easy to verify that the set of all analytic K-endomorphisms of A is a subring of $\mathrm{Hom}_K(A, A)$, which we shall denote by $T(A)$.

We may consider the ring A as a left $T(A)$-module under the left scalar multiplication defined by setting $\tau \,.\, a = \tau(a)$ for every analytic K-endomorphism τ and every element a of A. We shall now apply the theory of §21 to the left $T(A)$-module A. To do this we must first identify the bicentraliser of A, i.e. the set of abelian group endomorphisms of A which commute with all the $T(A)$-endomorphisms of A.

For this purpose we prove first that the $T(A)$-endomorphisms of A are precisely the left (or right) translations of A by the elements of the centre K. A simple calculation shows at once that if k is any element of K then λ_k $(=\rho_k)$ is a $T(A)$-endomorphism of A. Suppose conversely that φ is a $T(A)$-endomorphism of A; we shall show that $\varphi(e)$ belongs to K (where e is the identity of A) and that φ is just left translation of A by $\varphi(e)$. So let a be any element of A; since λ_a and ρ_a belong to $T(A)$ and φ is a $T(A)$-endomorphism, we have

$$a\varphi(e) = \lambda_a \,.\, \varphi(e) = \varphi(\lambda_a \,.\, e) = \varphi(ae) = \varphi(a)$$

and

$$\varphi(e)a = \rho_a \,.\, \varphi(e) = \varphi(\rho_a \,.\, e) = \varphi(ea) = \varphi(a).$$

Thus $\varphi(e)$ belongs to K, and for each element a of A we have $\varphi(a) = \lambda_{\varphi(e)}(a)$, i.e. $\varphi = \lambda_{\varphi(e)}$.

It follows immediately that the bicentraliser of A consists simply of all K-endomorphisms of A. For by what we have just said, an abelian group endomorphism β of A belongs to the bicentraliser of A if and only if it commutes with every translation of A by an element of K, in other words if and only if for every element k of K we have

$$k\beta(x) = \lambda_k(\beta(x)) = (\lambda_k\beta)(x) = (\beta\lambda_k)(x) = \beta(\lambda_k(x)) = \beta(kx).$$

Suppose next that A is a simple ring; we contend that A is then a simple left $T(A)$-module. To see this, let x be any non-zero element of A; then the submodule $T(A)x$ of A consists of all elements of A expressible in the form $\sum_{i\in I} a_i x b_i$ where $(a_i)_{i\in I}$ and $(b_i)_{i\in I}$ are finite families of elements of A. Clearly this set is a non-zero two-sided ideal of A and hence, since A is simple, we have $T(A)x = A$. It follows from Theorem 19.1 that A is a simple left $T(A)$-module.

The preceding discussion shows that we may apply the Density Theorem (Theorem 21.3) to the left $T(A)$-module A; we deduce that if $(x_i)_{1\leqslant i\leqslant n}$ is any finite family of elements of a simple ring A with centre K and β is any K-endomorphism of A, then there exists an analytic K-endomorphism τ such that for $i = 1, \ldots, n$ we have $\beta(x_i) = \tau \,.\, x_i = \tau(x_i)$.

THEOREM 25.1. *Let A be a simple ring with centre K, considered as a central simple algebra over K; let B be any algebra over K. Then there is a one-to-one correspondence between the set of two-sided ideals of $A \otimes_K B$ and the set of two-sided ideals of B.*

Proof. Let $\mathscr{B}$ be the set of two-sided ideals of B, $\mathscr{T}$ the set of two-sided ideals of $A \otimes_K B$. If J is any two-sided ideal of B it is easy to check that $A \otimes_K J$ (considered as subalgebra of $A \otimes_K B$) is a two-sided ideal of $A \otimes_K B$. Thus we may define a mapping α from $\mathscr{B}$ to $\mathscr{T}$ by setting $\alpha(J) = A \otimes_K J$ for every two-sided ideal J of B.

To see that α is an injection, let J_1 and J_2 be distinct ideals of B; suppose J_1 is not included in J_2, so that there is an element d_1 of J_1 which does not belong to J_2. We claim that $\alpha(J_1) \neq \alpha(J_2)$. Namely, since d_1 does not belong to J_2, the sum $J_2 + Kd_1$ is direct, and hence (by Theorem 12.7) so is the sum $(A \otimes_K J_2) + (A \otimes_K Kd_1)$; thus the non-zero elements of $A \otimes_K Kd_1$, for example $e \otimes d_1$, belong to $A \otimes_K J_1$ but not to $A \otimes_K J_2$. So $\alpha(J_1) \neq \alpha(J_2)$ and α is injective.

To see that α is surjective, let I be any two-sided ideal of $A \otimes_K B$. Let J be the subset of B consisting of those elements d such that $a \otimes d$ belongs to I for every element a of A. It is clear that J is a K-subspace of B and that $A \otimes_K J$ is included in I; we shall show that J is in fact a two-sided ideal of B and that $A \otimes_K J$ is actually equal to I.

We first remark that if τ is an analytic K-endomorphism of A and c is any element of I then $(\tau \otimes I_B)(c)$ is also contained in I. For if $\tau = \sum_{i=1}^{n} \lambda_{a_i}\rho_{a'_i}$, where (a_i), (a'_i) are finite families of elements of A, we compute easily that $(\tau \otimes I_B)(c) = \sum_{i=1}^{n} (a_i \otimes e_B)c(a'_i \otimes e_B)$ which belongs to I since I is a two-sided ideal.

If c is any element of I it follows from Corollary 1 of Theorem 12.7 that we may express c in the form $c = \sum_{i=1}^{r} x_i \otimes b_i$ where $\{x_1, \ldots, x_r\}$ is a linearly independent subset of A and $\{b_1, \ldots, b_r\}$ is a subset of B. By Theorem 6.5 we may extend $\{x_1, \ldots, x_r\}$ to a basis X for A over K. Let a be any element of A. Then for $j = 1, \ldots, r$ there exists a K-endomorphism β_j of A such that $\beta_j(x_j) = a$ and $\beta_j(x) = 0$ for all elements $x \neq x_j$ of X. It follows from the Density Theorem, as described just before the present theorem, that for $j = 1, \ldots, r$ there exists an analytic K-endomorphism τ_j of A such that

$\tau_j(x_i) = \beta_j(x_i)$ $(i = 1, \ldots, r)$. Now we have

$$(\tau_j \otimes I_B)(c) = \sum_{i=1}^{r} \tau_j(x_i) \otimes b_i = a \otimes b_j \, (j = 1, \ldots, r)$$

and according to the preceding paragraph $(\tau_j \otimes I_B)(c)$ belongs to I. So $a \otimes b_j$ belongs to I for every element a of A, i.e. we have $b_j \in J$ $(j = 1, \ldots, r)$. Thus c is contained in $A \otimes_K J$ and hence $I = A \otimes_K J$ as asserted.

To show that J is a two-sided ideal of B, let d be any element of J, b any element of B. Then $e_A \otimes d \in I$ and hence $(e_A \otimes d)(e_A \otimes b) = e_A \otimes db \in I$. The argument of the preceding paragraph shows that db belongs to J. Similarly bd belongs to J. So J is a two-sided ideal of B and $I = \alpha(J)$.

This completes the proof that α is a bijection.

We now proceed to draw some useful consequences from this result.

THEOREM 25.2. *Let A be a central simple algebra over a field K and let B be any simple algebra over K. If B is finite-dimensional when considered as a vector space over K then the tensor product $A \otimes_K B$ is a simple algebra over K.*

Proof. Let $\{b_1, \ldots, b_n\}$ be a basis for B over K. Then according to Corollary 2 of Theorem 12.7, $A \otimes_K B$ (considered as an abelian group) is the internal direct sum of the family of subgroups $\{A \otimes_K Kb_i\}_{1 \leqslant i \leqslant n}$. We may regard $A \otimes_K B$ as a left A-module with left multiplication defined by setting $at = (a \otimes e_B)t$ for all elements a of A and all elements t of $A \otimes_K B$; the subgroups $A \otimes_K Kb_i$ are then A-submodules. It is a routine matter to construct an abelian group homomorphism α_i from $A \otimes_K Kb_i$ to A such that $\alpha_i(a \otimes kb_i) = ak$ for all elements a of A and k of $K (i = 1, \ldots, n)$; and we verify easily that α_i is in fact an A-module isomorphism from $A \otimes_K Kb_i$ onto the left A-module A.

Since A is a simple ring it follows from Theorem 22.2 that A_l is an Artinian left A-module. Hence so are the modules $A \otimes_K Kb_i$ $(i = 1, \ldots, n)$ and consequently, by Corollary 1 of Theorem 14.4, $A \otimes_K B$ is an Artinian left A-module. Every left ideal of $A \otimes_K B$ is clearly an A-submodule; so it follows that $A \otimes_K B$ is a left Artinian ring.

The simple ring B has no two-sided ideals except for B itself and the zero ideal. Consequently, appealing to the preceding theorem, we see that $A \otimes_K B$ has no two-sided ideals except for $A \otimes_K B$ and the zero ideal.

It now follows from Corollary 2 of Theorem 23.8 that $A \otimes_K B$ is simple.

COROLLARY. *If A and B are central simple algebras over a field K, at least one of which is finite-dimensional when considered as a vector space over K, then $A \otimes_K B$ is a central simple algebra over K.*

Proof. According to the theorem, $A \otimes_K B$ is a simple algebra. The Corollary of Theorem 24.3 shows that the centre of $A \otimes_K B$ is $K \otimes_K K$, which is of course isomorphic to K.

For the rest of this section we shall restrict our attention to central simple algebras which are finite-dimensional when considered as vector spaces over their centres.

THEOREM 25.3. *Let A be a central simple algebra over a field K such that A is an n-dimensional vector space over K. Then $A \otimes_K A^{op}$ is isomorphic to the algebra of $n \times n$ matrices with coefficients in K.*

Proof. Consider the mapping φ from $A \times A^{op}$ to $\mathrm{Hom}_K(A, A)$ defined by setting $\varphi(a, b) = \lambda_a \rho_b$ for every pair of elements a in A, b in A^{op}. (The elements of A^{op} are the same as the elements of A, so ρ_b is indeed a K-endomorphism of A.) Remembering that the addition operation in A^{op} is the same as that in A we verify at once that φ is bilinear. Hence there exists a K-homomorphism α from $A \otimes_K A^{op}$ to $\mathrm{Hom}_K(A, A)$ such that $\alpha(a \otimes b) = \lambda_a \rho_b$ for every pair of elements a in A, b in A^{op}.

We claim that α is also a ring homomorphism and hence an algebra homomorphism. So let a, a' be any two elements of A, b and b' any two elements of A^{op} and x any element of A. Then, denoting the multiplication in A^{op} by $*$ we have

$$(a \otimes b)(a' \otimes b') = aa' \otimes (b * b') = aa' \otimes b'b$$

and hence

$$[\alpha((a \otimes b)(a' \otimes b'))](x) = [\alpha(aa' \otimes b'b)](x) = \lambda_{aa'} \rho_{b'b}(x) = aa'xb'b;$$

on the other hand

$$[\alpha(a \otimes b)\alpha(a' \otimes b')](x) = \lambda_a \rho_b \lambda_{a'} \rho_{b'}(x) = aa'xb'b$$

also. Thus we have

$$\alpha((a \otimes b)(a' \otimes b')) = \alpha(a \otimes b)\alpha(a' \otimes b'),$$

from which it follows that α is a ring homomorphism.

Now α is clearly not the zero homomorphism since, for example, $\alpha(e_A \otimes e_B)$ is the identity mapping of A. Hence the kernel of α, which is a two-sided ideal of $A \otimes_K A^{\text{op}}$, is not the whole ring. It follows that Ker α is the zero ideal and hence α is a monomorphism (for since A, and consequently also A^{op}, are finite-dimensional central simple algebras over K, so also is $A \otimes_K A^{\text{op}}$).

Finally we show that α is an epimorphism. We notice first that according to Corollary 2 of Theorem 12.7 we have $\dim_K(A \otimes_K A^{\text{op}}) = (\dim_K A)(\dim_K A^{\text{op}}) = (\dim_K A)^2 = n^2$. Hence, since α is a monomorphism, $\dim_K \alpha(A \otimes_K A^{\text{op}}) = n^2$ also. But $\dim_K \text{Hom}_K(A, A) = n^2$; so $\alpha(A \otimes_K A^{\text{op}}) = \text{Hom}_K(A, A)$, as asserted.

Thus α is an algebra isomorphism from $A \otimes_K A^{\text{op}}$ onto $\text{Hom}_K(A, A)$. Since $\text{Hom}_K(A, A)$ is isomorphic to $M_n(K^{\text{op}}) = M_n(K)$ by Theorem 8.5, the desired result is established.

Now let K be a field, A and B two central simple algebras over K which have finite dimension as vector spaces over K. We say that A and B are *similar* if their division algebra components (see page 213) are isomorphic algebras over K. This is equivalent to saying that A and B are similar if there exist a central division algebra D over K and positive integers m and n such that A and B are isomorphic to the matrix algebras $M_m(D)$ and $M_n(D)$ respectively.

THEOREM 25.4. *Let A, A', B, B' be central simple algebras over a field K which have finite dimension as vector spaces over K. If A and A' are similar and B and B' are similar then $A \otimes_K B$ and $A' \otimes_K B'$ are similar.*

Proof. Since A and A' are similar there exist a central division algebra D over K and positive integers m and m' such that A and A' are isomorphic to the matrix algebras $M_m(D)$ and $M_{m'}(D)$ respectively; in the same way there exists a central division algebra E over K and positive integers n and n' such that B and B' are isomorphic to the matrix algebras $M_n(E)$ and $M_{n'}(E)$ respectively.

Since D and E are central simple algebras of finite dimension over K, so is their tensor product $D \otimes_K E$ (by the Corollary to Theorem 25.2). Hence there is a central division algebra C over K

and a positive integer l such that $D \otimes_K E$ is isomorphic to the matrix algebra $M_l(C)$, and so (by Example 5 of §24) to $C \otimes_K M_l(K)$.

Using Example 5 of §24 again we see that $A \otimes_K B$ is isomorphic to $(D \otimes_K M_m(K)) \otimes_K (E \otimes_K M_n(K))$ and hence, by the commutativity and associativity of tensor products, to the algebra $(D \otimes_K E) \otimes_K (M_m(K) \otimes_K M_n(K))$. This in turn, according to Example 6 of §24, is isomorphic to $(C \otimes_K M_l(K)) \otimes_K M_{mn}(K)$ and hence eventually to $C \otimes_K M_{lmn}(K)$ or simply to $M_{lmn}(C)$. In the same way $A' \otimes_K B'$ is isomorphic to $M_{lm'n'}(C)$.

Hence $A \otimes_K B$ and $A' \otimes_K B'$ are similar, as required.

We now make the assumption that for every central simple algebra A over a field K with finite dimension as vector space over K there exists a central simple algebra similar to it, which we call the *similarity class* of A and denote by $\mathrm{Sim}(A)$, such that if A' is similar to A then $\mathrm{Sim}(A') = \mathrm{Sim}(A)$. It can be shown that the property expressed by 'X is a similarity class of central simple algebras of finite dimension over K' is set-forming. So we may talk of the set of similarity classes of central simple algebras of finite dimension over K; we denote this set by $B(K)$.

It is easy to introduce an internal law of composition into $B(K)$; namely, if C_1 and C_2 are similarity classes of finite-dimensional central simple algebras over K, we define the product C_1C_2 to be the similarity class $\mathrm{Sim}(C_1 \otimes_K C_2)$.

THEOREM 25.5. *Under the multiplication just described the set $B(K)$ forms an abelian group.*

Proof. The multiplication is clearly commutative, for if C_1, C_2 are any two elements of $B(K)$ the tensor products $C_1 \otimes_K C_2$ and $C_2 \otimes_K C_1$ are isomorphic (and hence of course similar). Thus $C_1C_2 = \mathrm{Sim}(C_1 \otimes_K C_2) = \mathrm{Sim}(C_2 \otimes_K C_1) = C_2C_1$.

To show that the multiplication is associative, let C_1, C_2, C_3 be elements of $B(K)$. Since, by definition, $C_1C_2 = \mathrm{Sim}(C_1 \otimes_K C_2)$, it follows that C_1C_2 is similar to $C_1 \otimes_K C_2$; hence, by Theorem 25.4, $(C_1C_2)C_3 = \mathrm{Sim}((C_1C_2) \otimes_K C_3) = \mathrm{Sim}((C_1 \otimes_K C_2) \otimes_K C_3)$. In the same way $C_1(C_2C_3) = \mathrm{Sim}(C_1 \otimes_K (C_2 \otimes_K C_3))$. We deduce at once that $C_1(C_2C_3) = (C_1C_2)C_3$.

Let C_0 be the similarity class of the field K itself, considered as a central simple algebra over K. Then, if C is any element of $B(K)$, we know that $C \otimes_K K$ is isomorphic to C; it follows easily that

$CC_0 = C$. So C_0 is an identity element for the multiplication in $B(K)$.

Finally, let C be any element of $B(K)$. If $C' = \mathrm{Sim}(C^{\mathrm{op}})$, where C^{op} is the opposite algebra of C, then according to Theorem 25.4 CC' is similar to $C \otimes_K C^{\mathrm{op}}$, which (by Theorem 25.3) is isomorphic to $M_n(K)$ where $n = \dim_K C$; since $M_n(K)$ is similar to K it follows that $CC' = \mathrm{Sim}(K) = C_0$. That is to say, C' is an inverse for C.

Hence $B(K)$ is an abelian group.

We call $B(K)$ the *Brauer group* of the field K.

Example 1. Let K be an algebraically closed field (see Adamson, *Introduction to Field Theory*,* §12.) We shall show that the Brauer group $B(K)$ consists of a single element.

So suppose D is a central division algebra of finite dimension over K. If x is any element of D it is easy to verify that the division subring of D generated by $K \cup \{x\}$ is commutative and hence is a field. Since it is clearly finite-dimensional over K it is an algebraic extension of K (*I.F.T.*, Theorem 9.1) and hence coincides with K, since K is algebraically closed. Thus x belongs to K and $D = K$.

It follows that every finite-dimensional central simple algebra over K is similar to the algebra K itself. Hence there is only one similarity class, namely $\mathrm{Sim}(K)$.

Example 2. Let K be a finite field. We shall show here also that $B(K)$ consists of a single element.

If D is any central division algebra of finite dimension over K then D has only finitely many elements and hence by Wedderburn's Theorem (*I.F.T.*, Theorem 24.2) D is commutative and hence $D = K$.

As in Example 1, it follows that $B(K) = \{\mathrm{Sim}(K)\}$.

Example 3. Lest the reader form the idea that the Brauer group of a field is always trivial, we describe without proof the Brauer groups of the real field $\mathbf{R}$ and the rational field $\mathbf{Q}$.

First, $B(\mathbf{R})$ consists of two elements, namely the similarity classes of $\mathbf{R}$ itself and of $\mathbf{H}$, the ring of real quaternions. So $B(\mathbf{R})$ is a cyclic group of order 2.

The group $B(\mathbf{Q})$ is more complicated. Let P be the set of prime numbers, and let $P' = P \cup \{\infty\}$; for each element p' of P' let $G_{p'}$ be the additive abelian group $\mathbf{Q}/\mathbf{Z}$, and let G be the direct sum of the family $(G_{p'})_{p' \in P'}$. Then $B(\mathbf{Q})$ is isomorphic to the subgroup of G

* From now on we shall refer to this book as I.F.T.

consisting of elements $(g_{p'})$ such that g_∞ is the residue class of either 0 or $\frac{1}{2}$ and $\sum_{p' \in P'} g_{p'} = 0$.

We conclude this section with two famous theorems on automorphisms of simple algebras. If R is any ring with identity and u is an invertible element (unit) of R it is easy to verify that the mapping τ_u from R to itself defined by setting $\tau_u(x) = u^{-1}xu$ for all elements x of R is an automorphism of R; we call τ_u the *inner automorphism* of R determined by u.

With this definition we are in a position to give the *Skolem-Noether Theorem.*

THEOREM 25.6. *Let A be a central simple algebra over a field K, and let B be a finite-dimensional simple algebra over K. If α_1 and α_2 are K-algebra monomorphisms from B to A there exists an inner automorphism τ of A such that $\alpha_2 = \tau\alpha_1$.*

Proof. Let $T = A \otimes_K B$; according to Theorem 25.2, T is a simple algebra over K.

We propose now to give A the structure of a left T-module in two different ways. So let x be any element of A and consider the mapping φ_x from $A \times B$ to A defined by setting $\varphi_x(a, b) = ax\alpha_1(b)$ for all elements a of A and b of B. We check at once that φ_x is a bilinear mapping and hence deduce that there exists a K-homomorphism μ_x from T to A such that $\mu_x(a \otimes b) = ax\alpha_1(b)$ for all elements a of A and b of B. This discussion allows us to define a left scalar multiplication of A by elements of T: namely, we set

$$t \wedge_1 x = \mu_x(t)$$

for all elements t of T and all elements x of A. With this scalar multiplication A becomes a left T-module, which we shall denote by A_1. Replacing α_1 by α_2 in the preceding discussion we may define another left scalar multiplication $\wedge_2$ of A by elements of T; when A is considered as a left T-module with this multiplication we shall denote it by A_2.

Since T is a simple ring (and hence semisimple), A_1 and A_2 are semisimple modules (Theorem 22.1); and they are isotypic, since (by Theorem 22.4) a simple ring has only one isomorphism type of simple left modules. Let the lengths of A_1 and A_2 be l_1 and l_2 respectively; if $l_1 \leqslant l_2$, the method used in part (1) of the proof of Theorem 20.7 can be adapted to show that there exists a T-monomorphism σ from A_1 to A_2. Thus for every element t of T and every element x of A we have

$$\sigma(t \wedge_1 x) = t \wedge_2 \sigma(x). \qquad [25.1]$$

Let a be any element of A; taking $t = a \otimes e_B$ and $x = e_A$ in [25.1] we deduce that

$$\sigma(a) = a\sigma(e_A) = au,$$

where we write $u = \sigma(e_A)$. We claim that u is invertible in A. First of all we notice that since σ is a monomorphism u is not a left divisor of zero; for if a is any element of A such that $au = 0$ we have $\sigma(a) = 0$ and so $a = 0$. For each natural number k let I_k be the principal left ideal of A generated by u^k; then $(I_k)_{k \in \mathbf{N}}$ is an infinite descending chain and hence, since A (being a simple ring) is left Artinian by Theorem 22.2, (I_k) levels off—at I_n say. Since $u^n \in I_n = I_{n+1}$ there exists an element u' of A such that $u^n = u'u^{n+1}$; from this equation, using the fact that u is not a left divisor of zero, we see that $u'u = e_A$, i.e. u is left invertible. Similarly, since A is right Artinian, there is a natural number m and an element u'' of A such that $u^m = u^{m+1}u''$. Multiplying this equation on the left by $(u')^m$ we see that $e_A = uu''$; hence u is also right invertible and hence invertible.

We now return to [25.1] and take $t = e_A \otimes b$ where b is any element of B. This yields

$$\sigma(e_A x \alpha_1(b)) = e_A \sigma(x) \alpha_2(b)$$

whence

$$x\alpha_1(b)u = xu\alpha_2(b)$$

for every element x of A. In particular, taking $x = e_A$, we obtain

$$\alpha_1(b)u = u\alpha_2(b)$$

for every element b of B. It follows at once that $\alpha_2 = \tau_u \alpha_1$, where τ_u is the inner automorphism of R determined by u.

This completes the proof.

COROLLARY 1. *Let A be a central simple algebra over a field K, B and B' two simple subalgebras of A of finite dimension over K. If α is an algebra isomorphism from B onto B' there exists an inner automorphism τ of A such that $\tau(b) = \alpha(b)$ for every element b of B.*

Proof. In the theorem let α_1 be the inclusion monomorphism from B to A and let $\alpha_2 = \alpha$.

COROLLARY 2. *Let A be a central simple algebra of finite dimension over a field K. Then every (algebra) automorphism of A is an inner automorphism.*

Proof. In Corollary 1 take $B = B' = A$.

Finally we prove the *Cartan-Brauer-Hua Theorem* on the division subrings of a division ring which are invariant under all inner automorphisms.

THEOREM 25.7. *Let D be a division ring with centre K. If L is a division subring of D such that $\tau(L)$ is included in L for every inner automorphism τ of D then either $L = D$ or else L is included in K.*

Proof. Let x be any non-zero (and hence invertible) element of D, and let a be any element of L. If τ_x is the inner automorphism of D determined by x then, by hypothesis, $\tau_x(a)$ belongs to L; hence there is an element a_1 of L such that $\tau_x(a) = a_1$, whence $ax = xa_1$. Of course this result holds also when $x = 0$: we need only take $a_1 = a$. Similarly there exists an element a_2 of L such that $a(x + e) = (x + e)a_2$, i.e. $ax + a = xa_2 + a_2$. It follows that $x(a_2 - a_1) = a - a_2$.

We deduce from this that if x does not belong to L then $a_1 = a_2$; for otherwise we should have $x = (a - a_2)(a_2 - a_1)^{-1}$, which certainly belongs to L. If $a_1 = a_2$ we also have $a = a_2$; so $a_1 = a$ and hence $ax = xa$. Thus if x does not belong to L it commutes with every element of L.

Suppose now that L is not included in the centre K. Then there exist elements a of L and x of D such that $ax \neq xa$; it follows from the preceding paragraph that x belongs to L. If, further, L is not the whole ring D then there exists an element y of D which does not belong to L and hence, as we have seen, commutes with every element of L; in particular $ay = ya$. Since x belongs to L and y does not it follows that $x + y$ does not belong to L. Hence $a(x + y) = (x + y)a$, whence $ax + ay = xa + ya$ and so $ax = xa$. This is a contradiction; so, as asserted, either L is included in K or else $L = D$.

§26. Splitting Fields of Simple Algebras

Let K be a field, A a central simple algebra of finite dimension over

K. Let L be a field which includes K; then we may think of L as a simple algebra over K (not central, of course) and form the tensor product $A \otimes_K L$, which is again a simple algebra over K, by Theorem 25.2. According to the Corollary of Theorem 24.3 the centre of $A \otimes_K L$ is $K \otimes L$, which is isomorphic to L. Thus if ι is the mapping from L to $A \otimes_K L$ defined by setting $\iota(l) = e_A \otimes l$ for every element l of L we see at once that ι is a ring isomorphism from L onto the centre of $A \otimes_K L$ and that ι maps the identity of L onto the identity of $A \otimes_K L$; so $(A \otimes_K L, \iota)$ is a central simple algebra over L. We can easily verify that if $\{x_1, \ldots, x_n\}$ is a basis for A over K then $\{x_1 \otimes e_L, \ldots, x_n \otimes e_L\}$ is a basis for $A \otimes_K L$ over L; hence $\dim_L (A \otimes_K L) = \dim_K A$. From this quite elementary remark we derive at once an interesting result on central simple algebras.

THEOREM 26.1. *Let A be a central simple algebra over a field K. If the dimension of A over K is finite then it is the square of a natural number.*

Proof. Let L be an algebraically closed extension field of K, for example an algebraic closure of K (see *I.F.T.*, Theorem 12.2). As we saw in Example 1 of §25 the finite-dimensional central simple algebra $A \otimes_K L$ over L is similar to L and hence is isomorphic to a ring of matrices with coefficients in L, say to $M_n(L)$. Then we have

$$\dim_K A = \dim_L(A \otimes_K L) = \dim_L(M_n(L)) = n^2.$$

Now let A and B be central simple algebras of finite dimension over a field K and let L be a field which includes K. Then we may form the central simple algebras $A \otimes_K L$ and $B \otimes_K L$ over L and their tensor product $(A \otimes_K L) \otimes_L (B \otimes_K L)$, which is also a central simple algebra over L. We may also form the tensor product $(A \otimes_K B) \otimes_K L$; this again is a central simple algebra over L. The reader is invited to exercise his ingenuity in proving that $(A \otimes_K L) \otimes_L (B \otimes_K L)$ and $(A \otimes_K B) \otimes_K L$ are isomorphic algebras over L—one way to achieve this would be to use the universal property of tensor products.

From this result we deduce easily that if A and B are similar (as central simple algebras over K) then $A \otimes_K L$ and $B \otimes_K L$ are similar (as central simple algebras over L). For if A and B are similar there exist a central division algebra D over K and natural numbers m and n such that A and B are isomorphic to $M_m(K) \otimes_K D$ and $M_n(K) \otimes_K D$ respectively. Then we see that $A \otimes_K L$ is isomorphic to $(M_m(K) \otimes_K D) \otimes_K L$, hence (by the preceding paragraph) to $(M_m(K) \otimes_K L) \otimes_L (D \otimes_K L)$ and so, according to Example 5 of §24, to $M_m(L) \otimes_L (D \otimes_K L)$; thus $A \otimes_K L$ is similar, as an algebra over L, to $D \otimes_K L$. In the same way $B \otimes_K L$ is similar to $D \otimes_K L$, and consequently $A \otimes_K L$ and $B \otimes_K L$ are similar.

We next define a mapping $\tau_{K,L}$ from the Brauer group of K to the Brauer group of L by setting, for each similarity class C of finite-dimensional central simple algebras over K,

$$\tau_{K,L}(C) = \mathrm{Sim}_L(C \otimes_K L),$$

where by $\mathrm{Sim}_L(C \otimes_K L)$ we mean the similarity class of $C \otimes_K L$ considered as a central simple algebra over L.

THEOREM 26.2. *The mapping $\tau_{K,L}$ just defined is a homomorphism from the Brauer group of K to the Brauer group of L.*

Proof. Let C_1, C_2 be elements of $B(K)$; then (writing τ as an abbreviation for $\tau_{K,L}$) we have

$$\begin{aligned}\tau(C_1)\tau(C_2) &= \mathrm{Sim}_L(\tau(C_1) \otimes_L \tau(C_2)) \\ &= \mathrm{Sim}_L(\mathrm{Sim}_L(C_1 \otimes_K L) \otimes_L \mathrm{Sim}_L(C_2 \otimes_K L)) \\ &= \mathrm{Sim}_L((C_1 \otimes_K L) \otimes_L (C_2 \otimes_K L)) \\ &= \mathrm{Sim}_L((C_1 \otimes_K C_2) \otimes_K L) \\ &= \mathrm{Sim}_L(\mathrm{Sim}_K(C_1 \otimes_K C_2) \otimes_K L) \\ &= \mathrm{Sim}_L(C_1 C_2 \otimes_K L) \\ &= \tau(C_1 C_2).\end{aligned}$$

Thus τ is a homomorphism as required.

If the similarity class C in $B(K)$ belongs to the kernel of $\tau_{K,L}$ we say that the class C is *split* by L and that L is a *splitting field* for C. If A is any central simple algebra such that $\mathrm{Sim}_K(A) = C$, then there is a natural number n such that $A \otimes_K L$ is isomorphic to the ring of matrices $M_n(L)$. We say also that A is split by L and that L is a splitting field for A. Our next task is to determine necessary and sufficient conditions for an extension field of K to be a splitting field for a central simple algebra over K. We begin by establishing the following result on the centraliser of a simple subalgebra of a central simple algebra.

THEOREM 26.3. *Let A be a finite-dimensional central simple algebra over a field K; let B be a simple algebra over K included in A and let B^**

be the centraliser of B in A. Then (1) *B^* is a simple algebra over K,* (2) *the centraliser of B^* in A is the algebra and* (3) $(\dim_K B)(\dim_K B^*) = \dim_K A$.

Proof. If we consider B as a vector space over K we may form the ring $R = \mathrm{Hom}_K(B, B)$; since B is finite-dimensional (being a subspace of the finite-dimensional vector space A) it follows from Theorem 22.5 that R is a simple ring, and by Theorem 22.11 its centre is isomorphic to K. We may thus regard R as a central simple algebra over K.

The mappings λ and ρ from B and B^{op} respectively to R defined by setting

$$\lambda(b) = \lambda_b = \text{left translation by } b$$

and

$$\rho(b) = \rho_b = \text{right translation by } b$$

for each element b of B or B^{op} are easily seen to be algebra monomorphisms. We claim that $\rho(B^{op})$ is the centraliser of $\lambda(B)$ in R. It is clear that every right translation commutes in R with every left translation, so $\rho(B^{op})$ is certainly included in the centraliser of $\lambda(B)$. Conversely, suppose φ is an element of R which commutes with every left translation; then for all elements x of B we have $\varphi\lambda_x = \lambda_x\varphi$ and hence

$$\varphi(x) = \varphi(xe) = \varphi(\lambda_x(e)) = \lambda_x(\varphi(e)) = x\varphi(e),$$

i.e. φ is simply right translation by $\varphi(e)$.

Since A and R are central simple algebras over K so is their tensor product. We have now two K-algebra monomorphisms α_1 and α_2 from B to $A \otimes_K R$ defined by setting

$$\alpha_1(b) = b \otimes e_R \qquad \text{and} \qquad \alpha_2(b) = e_A \otimes \lambda(b)$$

for all elements b of B. According to Theorem 25.6 there exists an inner automorphism τ of $A \otimes_K R$ such that $\alpha_2 = \tau\alpha_1$. We deduce easily from Theorem 24.3 that the centraliser of $\alpha_1(B) = B \otimes_K K$ is $B^* \otimes_K R$ and that the centraliser of $\alpha_2(B) = K \otimes_K \lambda(B)$ is $A \otimes_K \rho(B^{op})$. Since $\tau(\alpha_1(B)) = \alpha_2(B)$, it follows that $\tau(B^* \otimes_K R) = A \otimes_K \rho(B^{op})$, i.e. $B^* \otimes_K R$ is isomorphic to $A \otimes_K \rho(B^{op})$, which is a simple algebra by Theorem 25.2. This implies that B^* is simple; for if not, it would have two-sided ideals other than B^* and the zero ideal, and hence by Theorem 25.1 $B^* \otimes_K R$ would not be simple.

Next, since $B^* \otimes_K R$ and $A \otimes_K \rho(B^{op})$ are isomorphic they have the same dimension over K. Thus

$$(\dim_K B^*)(\dim_K R) = (\dim_K A)(\dim_K \rho(B^{op}))$$

whence

$$(\dim_K B^*)(\dim_K B)^2 = (\dim_K A)(\dim_K B)$$

and so we have the relation asserted in the theorem:

$$\dim_K A = (\dim_K B)(\dim_K B^*).$$

Finally, the centraliser B^{**} of B^* in A certainly includes B and, by the result just established, we have

$$(\dim_K B^*)(\dim_K B^{**}) = \dim_K A = (\dim_K B)(\dim_K B^*).$$

Thus $\dim_K B^{**} = \dim_K B$ and so $B^{**} = B$.

This completes the proof.

COROLLARY. *If, in the situation described in the theorem, B is a central simple algebra over K then so is B^* and A is isomorphic to the tensor product $B \otimes_K B^*$.*

Proof. It is clear that the centre of any subring S of A is the intersection of S with its centraliser. Hence we see that the centre of $B^* = B^* \cap B^{**} = B^* \cap B =$ the centre of $B = K$, as required.

Routine checking shows that the mapping φ from $B \times B^*$ to A defined by setting $\varphi(b, c) = bc$ for all elements b of B, c of B^* is bilinear and hence gives rise to a K-homomorphism α from $B \otimes_K B^*$ to A such that $\alpha(b \otimes c) = bc$ for all elements b of B, c of B^*. This mapping α is actually an algebra homomorphism; so, since it is not the zero homomorphism and $B \otimes_K B^*$ is simple, α is a monomorphism. Consequently we have

$$\dim_K(\alpha(B \otimes_K B^*)) = \dim_K(B \otimes_K B^*) = (\dim_K B)(\dim_K B^*) = \dim_K A.$$

So $\alpha(B \otimes_K B^*) = A$, and the desired result is established.

Our aim now is to obtain a criterion for an extension field L of K to be a splitting field of a given similarity class of central simple algebras over K. The next theorem gives us some information which will be useful for this task.

THEOREM 26.4. *Let A be a finite-dimensional central simple algebra over a field K; let L be a subfield of A which includes the centre K.*

Then the following statements are equivalent:

(a) *L coincides with its centraliser L^* in A;*

(b) $\dim_K A = (\dim_K L)^2$;

(c) *L is a maximal commutative subalgebra of A.*

Proof. (1) Suppose $L = L^*$.

Then, since L is a simple algebra over K, we can apply Theorem 26.3 and obtain $\dim_K A = (\dim_K L)(\dim_K L^*) = (\dim_K L)^2$.

(2) Suppose $\dim_K A = (\dim_K L)^2$.

Let M be a commutative subalgebra of A which includes L; then M is included in the centraliser L^* of L. Hence we have

$$(\dim_K L)(\dim_K M) \leqslant (\dim_K L)(\dim_K L^*) = \dim_K A = (\dim_K L)^2.$$

It follows that $\dim_K M \leqslant \dim_K L$, whence we deduce at once that $M = L$.

Thus L is a maximal commutative subalgebra of A.

(3) Finally suppose L is a maximal commutative subalgebra of A.

Then every element of A which commutes with every element of L belongs to L. That is to say, $L^* \subseteq L$; but of course $L \subseteq L^*$.

Hence $L = L^*$.

COROLLARY. *Let D be a finite-dimensional central division algebra over K. If L is a maximal subfield of D then* $(\dim_K L)^2 = \dim_K D$.

Proof. Let E be any commutative subalgebra of D. Then E is a finite-dimensional vector space over K. If x is any non-zero element of E we verify at once that left translation λ_x by x is a vector space endomorphism of E; since D is a division ring this endomorphism is injective and hence also surjective (because E is finite-dimensional). Thus there exists an element x' of E such that $\lambda_x(x') = xx' = e$; in other words x is invertible in E. So E is a field.

It follows that the maximal subfields of D and the maximal commutative subrings of D coincide; and the Corollary is then an immediate consequence of the theorem.

We are now in a position to give a criterion for a field to be a splitting field.

THEOREM 26.5. *Let C be a similarity class of finite-dimensional central simple algebras over a field K and let L be an extension field of finite degree* (see *I.F.T.* §7) *over K. Then L is a splitting field of C if and only if there exists an algebra A with class C such that L is isomorphic to a maximal commutative subring of A.*

Proof. (1) Suppose L is a splitting field of C.

Thus C belongs to the kernel of the Brauer group homomorphism $\tau_{K,L}$ described above. Hence so does the inverse class C^{-1} of C in $B(K)$; consequently, since $C^{-1} = \mathrm{Sim}_K(C^{\mathrm{op}})$, the tensor product $C^{\mathrm{op}} \otimes_K L$ is isomorphic to an algebra of matrices with coefficients in L. So there exists a finite-dimensional vector space V over L and an isomorphism θ from $C^{\mathrm{op}} \otimes_K L$ onto the ring $R = \mathrm{Hom}_L(V, V)$. We may of course regard V as a vector space over K and form the ring $S = \mathrm{Hom}_K(V, V)$ which is a central simple algebra over K with R as simple subalgebra.

If λ is the monomorphism from L to S defined by setting

$$\lambda(a) = \lambda_a = \text{left translation of } V \text{ by } a$$

then it is immediate that $\lambda(L)$ is included in the centraliser R^* of R in S. But we have

$$\dim_K S = (\dim_K V)^2 = ((\dim_K L)(\dim_L V))^2 = (\dim_K \lambda(L))(\dim_K R)$$

and also, by Theorem 26.3,

$$\dim_K S = (\dim_K R^*)(\dim_K R). \qquad [26.1]$$

Hence $\dim_K \lambda(L) = \dim_K R^*$ and so $R^* = \lambda(L)$.

Now let A be the centraliser of $\theta(C^{\mathrm{op}} \otimes_K K)$ in S. Then A includes $\lambda(L)$ and according to Theorem 26.3 and its Corollary A is a central simple subalgebra of S such that

$$(\dim_K A)(\dim_K C^{\mathrm{op}}) = \dim_K S.$$

But, by [26.1], we have

$$\begin{aligned}\dim_K S &= \dim_K(C^{\mathrm{op}} \otimes_K L)\dim_K \lambda(L)\\ &= (\dim_K C^{\mathrm{op}})(\dim_K L)(\dim_K \lambda(L))\\ &= (\dim_K C^{\mathrm{op}})(\dim_K \lambda(L))^2\end{aligned}$$

It follows that $\dim_K A = (\dim_K \lambda(L))^2$ and so, according to Theorem 26.4, $\lambda(L)$ is a maximal commutative subring of A.

The Corollary to Theorem 26.3 shows that R is isomorphic to the tensor product $(C^{\mathrm{op}} \otimes_K K) \otimes_K A$ and hence to $C^{\mathrm{op}} \otimes_K A$. Since $\mathrm{Sim}_K(R)$ is the identity element of $B(K)$, it follows easily (using Theorem 25.4) that $\mathrm{Sim}_K(A)$ and $\mathrm{Sim}_K(C^{\mathrm{op}}) = C^{-1}$ are inverses in $B(K)$; hence $\mathrm{Sim}_K(A) = C$.

(2) Conversely suppose there exists an algebra A such that $\mathrm{Sim}_K(A) = C$ and L is isomorphic to a maximal commutative subring of A, which we shall identify with L.

In the course of proving Theorem 25.3 we set up an isomorphism α from $A \otimes_K A^{\text{op}}$ onto $R = \text{Hom}_K(A, A)$ such that $\alpha(a \otimes b) = \lambda_a \rho_b$ for all elements a of A and b of A^{op}. Let B be the subalgebra $L \otimes_K K$ of $A \otimes_K A^{\text{op}}$; every element of B can be expressed uniquely in the form $l \otimes e_K$ where l is an element of L. Thus $\alpha(B)$ consists of all left translations of A by elements of L.

According to Theorem 26.4 L is its own centraliser in A; so it follows from Theorem 24.3 that the centraliser of B in $A \otimes_K A^{\text{op}}$ is $L \otimes_K A^{\text{op}}$. On the other hand the centraliser of $\alpha(B)$ in R is easily seen to be the ring of L-endomorphisms of A (considered as a vector space over L). Hence α sets up a K-algebra isomorphism from $L \otimes_K A^{\text{op}}$ onto $\text{Hom}_L(A, A)$, and we easily verify that this is in fact an L-algebra isomorphism.

Thus L is a splitting field for A^{op} and hence for the class C^{-1}. Hence L is a splitting field for C, as required.

COROLLARY 1. *Let D be a finite-dimensional central division algebra over a field K. Then every maximal subfield of D is a splitting field for* $\text{Sim}_K D$.

COROLLARY 2. *Let D be a central division algebra of dimension r^2 over its centre K. If L is any splitting field of D then* $\dim_K L$ *is a multiple of r.*

Note. We recall from Theorem 26.1 that $\dim_K D$ must be a square.

Proof. Let $C = \text{Sim}_K(D)$; let A be the algebra described in the theorem. Then there is a natural number s such that A is isomorphic to the matrix algebra $M_s(D)$. Hence $\dim_K A = r^2 s^2$.

Since L is isomorphic to a maximal commutative subring of A we have $(\dim_K L)^2 = \dim_K A$ and so $\dim_K L = rs$, as required.

Although every field which includes a maximal subfield of a division algebra D is plainly also a splitting field of D the converse does not hold: a splitting field of D need not include a maximal subfield of D. Nor is it the case that all maximal subfields of D are isomorphic.

We conclude this section by showing that every similarity class of central simple algebras over a field K has a splitting field which is a normal separable extension of finite degree over K. (For the technical terms and results concerning normal and separable field extensions we refer the reader to *I.F.T.*, §§13 and 15. It will be convenient to make a change in the terminology of §14 of *I.F.T.*

Namely, if K_1 and K_2 are fields with a common subfield F and τ is a monomorphism from K_1 to K_2 such that $\tau(x) = x$ for every element x of F, we shall call τ an *F-invariant monomorphism*, not an F-monomorphism as in *I.F.T.*)

Since every algebraic extension of a field of characteristic zero is separable (*I.F.T.*, Theorem 13.3) let us consider an extension E of finite degree over a subfield K of non-zero characteristic p. As in *I.F.T.*, we shall denote by E^p the subfield of E consisting of the pth powers of elements of E; we define similarly E^p, E^{p^2}, E^{p^3}, ... For each natural number n set $E_n = K(E^{p^n})$; then $(E_n)_{n\in\mathrm{N}}$ is a descending chain of fields, which must clearly level off since the degree of $E_0 = E$ over K is finite. Suppose the chain levels off at E_s; then we have $K(E_s^p) = K[(K(E^{p^s}))^p] = K(E^{p^{s+1}}) = E_{s+1} = E_s$ and hence (by *I.F.T.*, Theorem 13.5) E_s is a separable extension of K.

Now let σ and τ be distinct K-invariant monomorphisms from E to a field L including K; we shall show that σ and τ induce distinct K-invariant monomorphisms from E_s to L. Suppose, to the contrary, that $\sigma(a) = \tau(a)$ for every element a of E_s. If x is any element of E then x^{p^s} belongs to E_s and hence $\sigma(x^{p^s}) = \tau(x^{p^s}) = b$ say. Thus $\sigma(x)$ and $\tau(x)$ are both roots of the polynomial $X^{p^s} - b$ with coefficients in L. But $X^{p^s} - b = X^{p^s} - (\sigma(x))^{p^s} = (X - \sigma(x))^{p^s}$; so $\tau(x) = \sigma(x)$, i.e. $\tau = \sigma$, contradicting the hypothesis that σ and τ are distinct.

With these preliminaries we are ready to take a first step towards establishing the result stated above.

THEOREM 26.6. *Let D be a finite-dimensional central division algebra over a field K, distinct from K. Then D has a subfield which is separable over K and distinct from K.*

Proof. If K has characteristic zero then the maximal subfields of D are distinct from K and are all separable extensions of K (by *I.F.T.*, Theorem 13.3).

So suppose that K has non-zero characteristic p.

Let x be an element of D which does not belong to K; then the subfield $E = K(x)$ of D generated by x over K is of finite degree over K. In the notation of the discussion preceding this theorem, we have $E_n = K(E^{p^n}) = K(x^{p^n})$ for all natural numbers n. Suppose the descending chain $(E_n)_{n\in\mathrm{N}}$ levels off at E_s. It may happen that $E_s \neq K$; in this case E_s is a separable extension of K distinct from K.

Suppose on the other hand that $E_s = K$ and hence x^{p^s} belongs to K. Then there is a natural number $r \geqslant 1$ such that x^{p^r} belongs to K

but $x_1 = x^{p^{r-1}}$ does not. Let τ be the mapping from D to itself defined by setting $\tau(d) = x_1^{-1}dx_1$ for all elements d of D; then τ is a K-invariant automorphism of D and $\tau^p = I_D$, the identity automorphism of D. The mapping $\tau - I_D$ is a non-zero element of $\text{Hom}_K(D, D)$, where we regard D simply as a vector space over K, and $(\tau - I_D)^p = \tau^p - I_D$ is the zero homomorphism. Let q be the greatest integer such that $(\tau - I_D)^q$ is non-zero; clearly $q \geqslant 1$.

Since $(\tau - I_D)^q$ is not the zero endomorphism of D, there exists an element y of D such that $a = (\tau - I_D)^q(y)$ is non-zero. Then, according to the maximal property of q, we have $\tau(a) - a = (\tau - I_D)(a) = (\tau - I_D)^{q+1}(y) = 0$; so $\tau(a) = a$. Let now $b = (\tau - I_D)^{q-1}(y)$; then $\tau(b) - b = (\tau - I_D)(b) = (\tau - I_D)^q(y) = a$ and hence $\tau(b) = a + b$. If now we set $c = ba^{-1}$ we have $\tau(c) = \tau(b)\tau(a)^{-1} = (a + b)a^{-1} = e + ba^{-1} = e + c$.

Let $F = K(c)$ be the subfield of D generated over K by c. It follows from the preceding paragraph that $\tau(F)$ is included in F; since τ is a vector space automorphism we have $\dim_K\tau(F) = \dim_K F$ and hence $\tau(F) = F$. Thus τ induces a K-invariant automorphism τ' of the field F, distinct from the identity automorphism I_F. If we consider the descending chain of subfields $(F_n)_{n\in\mathrm{N}}$ of F where $F_n = K(c^{p^n})$ for each natural number n, we see that this chain levels off, at F_t say, and that F_t is a separable extension of F. The distinct automorphisms I_F and τ' of F induce distinct K-invariant automorphisms from F_t to F. Theorem 15.4 of *I.F.T.* implies that the number of K-invariant monomorphisms from the separable extension F_t of K to an extension of K cannot exceed the degree of F_t over K. It follows that the degree of F_t over K is at least 2.

This completes the proof.

THEOREM 26.7. *Let D be a finite-dimensional central division algebra over a field K. Then D has a maximal subfield which is separable over K.*

Proof. Let L be a subfield of D maximal among the subfields of D which are separable extensions of K.

It is easy to verify that L^*, the centraliser of L in D, is a division ring with centre L. If L^* is distinct from L, it follows from Theorem 26.6 that there is a subfield L' of L^* which is separable over L and distinct from L. Since L is separable over K, and L' is separable over L it follows from *I.F.T.*, Theorem 13.6 that L' is separable over K. This contradicts the maximal property of L since L' properly includes L. So $L = L^*$ and by Theorem 26.4 and its Corollary we deduce that L is a maximal subfield of D.

COROLLARY. *Let C be a similarity class of finite-dimensional central simple algebras over a field K. Then C has a splitting field which is a normal separable extension of finite degree over K.*

Proof. Let D be a central division algebra over K such that $\mathrm{Sim}_K(D) = C$. According to the theorem, D has a maximal subfield L which is separable over K, and by Corollary 1 to Theorem 26.5 this field L is a splitting field for D. Let N be a normal closure of L over K (see *I.F.T.*, §15); then N is a normal separable extension of finite degree over K. Since N is an extension of L it is a splitting field of D and hence of the similarity class C.

§27. Separable Algebras

Let M and N be subfields of a field C. We denote by $[MN]$ the subset of C consisting of all elements z which can be expressed in the form $z = \sum_{i \in I} x_i y_i$ where (x_i), (y_i) are families of elements of M, N respectively with the same finite index set I; clearly $[MN]$ is a subring of C. The smallest subfield of C which includes $[MN]$ is called the *compositum* of M and N and is denoted by (MN). This compositum may be obtained by adjoining N to M or by adjoining M to N, in the sense of *I.F.T.* §8; it is also easily seen to be isomorphic to the field of fractions of $[MN]$.

In general $[MN]$ and (MN) are distinct, but there is a useful special case in which they coincide.

THEOREM 27.1. *Let M and N be subfields of a field C. If there exists a subfield K of $M \cap N$ such that M is of finite degree over K then the compositum (MN) coincides with $[MN]$.*

Proof. It is clearly sufficient to show that under the condition stated $[MN]$ is a field.

Let $\{m_1, \ldots, m_r\}$ be a basis for M over K. If $z = \sum_{j=1}^{s} x_j y_j$ is an element of $[MN]$, where $x_1, \ldots, x_s$ are elements of M and $y_1, \ldots, y_s$ are elements of N, there exist elements a_{ij} of K ($i = 1, \ldots, r$; $j = 1, \ldots, s$) such that $x_j = \sum_{i=1}^{r} m_i a_{ij}$ ($j = 1, \ldots, s$); then we have

$$z = \sum_{i=1}^{r} \sum_{j=1}^{s} m_i a_{ij} y_j = \sum_{i=1}^{r} m_i y_i' \text{ where } y_i' = \sum_{j=1}^{s} a_{ij} y_j \ (i = 1, \ldots, r).$$

Since K is a subfield of N, the elements $y'_1, \ldots, y'_r$ belong to N. Thus $\{m_1, \ldots, m_r\}$ is a generating system for $[MN]$ considered as a vector space over N; so $[MN]$ is a finite-dimensional vector space over N.

If t is any non-zero element of $[MN]$, left translation λ_t by t is a vector space endomorphism of $[MN]$ which is injective (since $[MN]$ has no divisors of zero) and hence surjective (by the finite dimensionality). It follows that there exists an element t' of $[MN]$ such that $\lambda_t(t') = tt' = e$, i.e. t is invertible, and so $[MN]$ is a field.

Thus $[MN] = (MN)$, as required.

Now let K be a field, E and F extension fields of K which we may suppose to include K as a subfield. By a *composite extension* of E and F over K we mean a triple (C, α, β) consisting of an extension field C of K (again assumed to include K) and K-invariant monomorphisms α, β from E, F respectively to C such that C is the compositum of $\alpha(E)$ and $\beta(F)$. Two composite extensions (C, α, β) and (C', α', β') of E and F over K are said to be *equivalent* if there exists a K-invariant isomorphism λ from C onto C' such that $\lambda\alpha = \alpha'$ and $\lambda\beta = \beta'$.

If (C, α, β) is a composite extension of E and F over K we may define a mapping φ from $E \times F$ to C by setting $\varphi(x, y) = \alpha(x)\beta(y)$ for all elements x of E, y of F. This is easily seen to be bilinear; hence there is a vector space homomorphism γ from $E \otimes_K F$ to C such that $\gamma(x \otimes y) = \alpha(x)\beta(y)$ for all elements x of E, y of F. In fact γ is also a ring (and hence an algebra) homomorphism, which we shall denote by $\alpha . \beta$; the image of γ is clearly the ring $[\alpha(E)\beta(F)]$, which is an integral domain, and hence its kernel is a prime ideal P of $E \otimes_K F$. We call P the prime ideal of $E \otimes_K F$ *associated* with the composite extension (C, α, β) and denote it by $P(C, \alpha, \beta)$. We now show that the association of prime ideals and composite extensions is a harmonious one.

THEOREM 27.2. *Let E and F be extension fields of a field K. The prime ideals of $E \otimes_K F$ associated with equivalent composite extensions of E and F over K are equal. Further, every prime ideal of $E \otimes_K F$ is associated with a composite extension of E and F over K.*

Proof. (1) Let (C, α, β) and (C', α', β') be equivalent composite extensions of E and F over K; then there is a K-invariant isomorphism λ from C onto C' such that $\lambda\alpha = \alpha'$ and $\lambda\beta = \beta'$. It follows easily that if γ and γ' are the homomorphisms from $E \otimes_K F$ to C and C'

corresponding to (C, α, β) and (C', α', β') respectively then we have $\lambda\gamma = \gamma'$. Since λ is an isomorphism, it follows that the kernels of γ and γ' coincide, i.e. $P(C, \alpha, \beta) = P(C', \alpha', \beta')$ as required.

(2) Let P be any prime ideal of $E \otimes_K F$. Let η be the canonical epimorphism from $E \otimes_K F$ to the residue class ring $R = (E \otimes_K F)/P$; let C be the field of fractions of R and κ the canonical monomorphism from R to C. If α is the mapping from E to C defined by setting $\alpha(x) = \kappa\eta(x \otimes e_F)$ for all elements x of E we check at once that α is a K-invariant homomorphism, and since $\alpha(e_E)$ is non-zero α is not the zero homomorphism; so α is a K-invariant monomorphism (*I.F.T.*, Theorem 3.1). Similarly we have a K-invariant monomorphism β from F to C defined by setting $\beta(y) = \kappa\eta(e_E \otimes y)$ for all elements y of F. It is now a routine matter to check that (C, α, β) is a composite extension of E and F over K with associated prime ideal P.

Still let E and F be field extensions of a common subfield K and suppose now that F is an algebraic extension of K (see *I.F.T.*, §9). Let C be an algebraic closure of E which includes E as a subfield, with ι the canonical inclusion monomorphism from E to C; then C includes an algebraic closure of K (see *I.F.T.*, §12). We claim that in this situation every composite extension of E and F over K is equivalent to a composite extension of the form (G, ι, φ) where G is a subfield of C and φ is a K-invariant monomorphism from F to C.

To see this, let (L, α, β) be any composite extension of E and F over K. Since α is a K-invariant monomorphism from E to L there exists a K-invariant isomorphism α' from $\alpha(E)$ onto E such that $\alpha'\alpha = I_E$; then α' is a K-invariant monomorphism from $\alpha(E)$ to C and clearly $(C, \iota\alpha')$ is an algebraic closure of $\alpha(E)$. By hypothesis F is an algebraic extension of K and hence it follows easily that L is an algebraic extension of $\alpha(E)$. Thus, according to Theorem 12.3 of *I.F.T.*, there exists a monomorphism λ from L to C such that the restriction of λ to $\alpha(E)$ is $\iota\alpha'$. It is now clear that $(\lambda(L), \iota, \lambda\beta)$ is a composite extension of E and F over K which is equivalent to (L, α, β).

Specialising still further, let us suppose that F is not just algebraic over K but actually of finite degree over K. If (G, ι, φ) is a composite extension of the type just described then, according to Theorem 27.1, $G = [E\varphi(F)]$; hence the mapping $\gamma = \iota . \varphi$ from $E \otimes_K F$ to G is actually an epimorphism, i.e. Im γ is a field. It follows that the prime ideal $P(G, \iota, \varphi)$ associated with (G, ι, φ) is actually a maximal ideal. Consequently, as Theorem 27.2 shows, every prime ideal of $E \otimes_K F$ is a maximal ideal.

In order to exploit these remarks we need the following result on arbitrary commutative rings with identity.

THEOREM 27.3. *Let R be a commutative ring with identity and let $\{M_1, \ldots, M_n\}$ be a finite set of distinct maximal ideals of R. Then the residue class ring $R/\bigcap_{i=1}^{n} M_i$ is isomorphic to the direct sum of the family of fields (R/M_i).*

Proof. For $i = 1, \ldots, n$ let η_i be the canonical epimorphism from R onto R/M_i. We now define a mapping φ from R to the direct sum $\bigoplus_{i=1}^{n} R/M_i$ by setting

$$\varphi(a) = (\eta_1(a), \ldots, \eta_n(a))$$

for each element a of R. It is immediate that φ is a ring homomorphism with kernel $\bigcap_{i=1}^{n} M_i$; the theorem will be established if we can show that φ is an epimorphism.

To do this we remark first that for each pair of distinct integers i, j in $[1, n]$ the maximal ideals M_i and M_j are distinct and hence neither is included in the other; so there exist elements m_{ij} $(i \neq j;\ i,j = 1, \ldots, n)$ such that m_{ij} belongs to M_j but not to M_i. Thus if we set $b_i = \prod_{k \in [1,n], k \neq i} m_{ik}$ we see that $\eta_j(b_i) = 0$ for $j \neq i$ but $\eta_i(b_i)$ is non-zero and hence is invertible in the field R/M_i.

Now let $A = (A_1, \ldots, A_n)$ be any element of $\bigoplus_{i=1}^{n} R/M_r$. Since each of the mappings η_i is surjective, there exist elements a_i of R such that $\eta_i(a_i) = A_i(\eta_i(b_i))^{-1}$ $(i = 1, \ldots, n)$. Then if $a = \sum_{i=1}^{n} a_i b_i$ an easy calculation shows that $\varphi(a) = A$.

We apply this result at once to the situation we have been discussing.

THEOREM 27.4. *Let E and F be field extensions of a common subfield K. If F is of finite degree over K then the tensor product $F \otimes_K F$ has only finitely many prime ideals $M_1, \ldots, M_n$ all of which are maximal, and if J is the radical of $E \otimes_K F$ then $(E \otimes_K F)/J$ is isomorphic to the* *direct sum* $\bigotimes_{i=1}^{n} (E \otimes_K F)/M_i$.

Proof. We remarked a moment ago that under the conditions of the theorem all prime ideals of $R = E \otimes_K F$ are maximal.

Let $\{P_1, \ldots, P_r\}$ be a finite set of maximal ideals of R. According to Theorem 27.2 and the discussion preceding Theorem 27.3, each of the ideals $P_i (i = 1, . . , r)$ gives rise to a composite extension (G_i, ι, φ_i) where G_i is a subfield of a fixed algebraic closure C of E (with G_i isomorphic to R/P_i), ι is the canonical monomorphism from E to C and φ_i is a K-invariant monomorphism from F to C. It follows that Theorem 27.3 that $R/\bigcap_{i=1}^{r} P_i$ is isomorphic to the direct sum of the family (R/P_i) and hence to the direct sum of the family (G_i). Hence by a simple counting argument we have

$$\begin{aligned} \dim_K F = \dim_E R \geqslant \dim_E R/\bigcap_{i=1}^{r} P_i &= \dim_E \bigoplus_{i=1}^{r} G_i \\ &= \sum_{i=1}^{r} \dim_E G_i \geqslant r \qquad [27.1] \end{aligned}$$

This shows that there are are only finitely many prime ideals of R.

If $\{M_1, \ldots, M_n\}$ is the complete set of distinct prime (maximal) ideals of R, then $\bigcap_{i=1}^{n} M_i = J$ (by definition—see §23) and the final part of the theorem follows at once.

We may associate with each of the ideals $M_i (i = 1, \ldots, n)$ a composite extension (G_i, ι, φ_i) of the type described above. In terms of these extensions we have a useful consequence of our theorem.

COROLLARY. *In the situation of the theorem the tensor product* $E \otimes_K F$ *is semisimple if and only if* $\dim_K F = \sum_{i=1}^{n} \dim_E G_i$.

Proof. Referring to [27.1] we see that $\dim_K F = \sum_{i=1}^{n} \dim_E G_i$ if and only if $\dim_E R = \dim_E R/J$, i.e. if and only if J is the zero ideal.

If R is semisimple then, by Theorem 23.7, we have $J = \{0\}$ and hence $\dim_K F = \sum_{i=1}^{n} \dim_E G_i$.

Conversely, suppose this relation holds. Then $J = \{0\}$; since R is Artinian (every ideal of R is a subspace of the finite-dimensional vector space R over E) it follows from Theorem 23.8 that R is semisimple.

We now use the preceding discussion to give us a criterion for the field F to be a separable extension of K.

THEOREM 27.5. *Let F be an extension of finite degree over a subfield K. Then F is a separable extension of K if and only if* $E \otimes_K F$ *is semisimple for every field extension E of K.*

Proof. (1) Suppose F is a separable extension of K, of degree d say.

If E is any extension field of K which (we may suppose) includes K, let C be any algebraic closure of E which includes E. Then C includes an algebraic closure of K and according to *I.F.T.* Theorem 15.4 there exist d distinct K-invariant monomorphisms from F to C.

As before we associate with each of the maximal ideals M_i $(i = 1, \ldots, n)$ of $E \otimes_K F$ a composite extension (G_i, ι, φ_i). Then it is easy to see that since F is a separable extension of K each of the fields G_i is a separable extension of E. If G_i has degree d_i over E it follows again from *I.F.T.* Theorem 15.4 that there are d_i E-invariant monomorphisms from G_i to C, say $\lambda_{i1}, \ldots, \lambda_{id_i}$. Then the $\sum_{i=1}^{n} d_i$ mappings $\lambda_{ij}\varphi_i$ $(i = 1, \ldots, n; j = 1, \ldots, d_i)$ are distinct K-invariant monomorphisms from F to C; we claim that there are no others.

For suppose μ is a K-invariant monomorphism from F to C; if we set $G = (E\mu(F))$ we see that (G, ι, μ) is a composite extension of E and F over K and hence is equivalent to one of the composite extensions (G_i, ι, φ_i) $(i = 1, \ldots, n)$. If (G, ι, μ) is equivalent to (G_k, ι, φ_k) there exists a K-invariant isomorphism λ from G_k onto G such that $\lambda\iota = \iota$ and $\lambda\varphi_k = \mu$; thus λ is an E-invariant monomorphism from G_k to C. So μ is one of the monomorphisms $\lambda_{kj}\varphi_k$ as asserted.

It follows that $\sum_{i=1}^{n} d_i = d$ and hence $E \otimes_K F$ is semisimple, by the Corollary to Theorem 27.4.

(2) Conversely, suppose F is not a separable extension of K. Then K has non-zero characteristic, p say.

Since F is not separable over K there is an element x of F which is inseparable over F, i.e. such that the minimum polynomial $m = m_{x,K}$ of x over K is inseparable. Then if E is a splitting field for m over K there is a polynomial f with coefficients in E and a positive integer k such that $m = f^{p^k}$. Theorem 8.2 of *I.F.T.* shows essentially (though it is not expressed in such sophisticated terms) that the simple extension $K(x)$ of K is K-isomorphic to the residue class field $P(K)/(m)$ where $P(K)$ is the ring of polynomials with coefficients in K and (m) is the principal ideal of $P(K)$ generated by the polynomial m. We deduce easily that $E \otimes_K K(x)$ is K-isomorphic to the residue class ring $P(E)/(m)$ where $P(E)$ is the ring of polynomials with coefficients in E.

The residue class of the polynomial f in $P(E)$ modulo (m) is non-zero; but the p^kth power of this residue class is zero. It follows that $E \otimes_K K(x)$, and hence $E \otimes_K F$ has at least one non-zero nilpotent element z. Since $E \otimes_K F$ is a commutative ring the principal left ideal of $E \otimes_K F$ generated by z is a non-zero nilpotent ideal. We deduce from Theorem 23.5 that the radical of $E \otimes_K F$ is not the zero ideal and hence (by Theorem 23.7) $E \otimes_K F$ is not semisimple.

This completes the proof.

We now allow ourselves to be guided by the result of this theorem to generalise the notion of separability from field extensions to algebras. Namely, let A be an algebra of finite dimension over a field K. Then A is said to be *separable* over K if for every field extension E of K the tensor product $E \otimes_K A$ is a semisimple algebra over E. If A is a separable algebra over K then A itself (being isomorphic to $K \otimes_K A$) is semisimple. Our next theorem gives us two criteria for an algebra to be separable.

THEOREM 27.6. *Let A be an algebra of finite dimension over a field K. Then the following conditions are equivalent*:

(a) *A is a separable algebra over K;*

(b) *A is a semisimple algebra over K such that the centre of each simple component of A is a separable field extension of K;*

(c) *there is an extension field F of K such that $F \otimes_K A$ is isomorphic to a direct sum of matrix algebras over F.*

Proof. (1) Suppose A is a separable algebra over K.

We have already remarked that A is semisimple. Let $S_1, \ldots, S_n$ be the simple components of A, with centres $Z_1, \ldots, Z_n$. Suppose these centres are not all separable extensions of K; say Z_1 is not separable over K. Then, as we have just seen in the proof of Theorem 27.5, there exists an extension field E of K such that $E \otimes_K Z_1$ has at least one non-zero nilpotent element t_1. Since $E \otimes_K Z_1$ is the centre of $E \otimes_K S_1$, the left ideal of $E \otimes_K S_1$ generated by t_1 is nilpotent; hence $E \otimes_K S_1$, and so $E \otimes_K A$, is not semisimple. This contradicts the hypothesis that A is separable.

Hence $Z_1, \ldots, Z_n$ are separable extensions of K, as required.

(2) Let S be a simple algebra over K whose centre Z is a separable extension of K, of degree r say.

If N is a normal closure of Z over K then each of the r K-invariant monomorphisms $\varphi_1, \ldots, \varphi_r$ from Z to an algebraic closure C of N including N actually maps Z into N (see *I.F.T.* Theorem 15.3). Thus the corresponding composite extensions (G_i, ι, φ_i) all have $G_i = N$ and hence are pairwise inequivalent. It follows that $N \otimes_K Z$ is isomorphic to the direct sum of the family $(G_i)_{1 \leqslant i \leqslant r}$.

Now $Z \otimes_Z S$ is isomorphic to S (by Theorem 12.2 extended in an obvious way to algebras) and hence $N \otimes_K S$ is isomorphic to $(N \otimes_K Z) \otimes_Z S$ by an obvious extension of Theorem 24.2; consequently $N \otimes_K S$ is isomorphic to the direct sum of the family $(G_i \otimes_Z S)_{1 \leqslant i \leqslant r}$. But for $i = 1, \ldots, r$ we have $G_i \otimes_Z S = N \otimes_Z S$, which is a central simple algebra over N and so has a splitting field F, i.e. there is a field F such that $F \otimes_N (N \otimes_Z S)$ is isomorphic to a matrix algebra over F. But $F \otimes_N (N \otimes_Z S)$ is isomorphic to $(F \otimes_N N) \otimes_Z S$ and hence to $F \otimes_Z S$. So $F \otimes_K S$, which is isomorphic to $(F \otimes_N N) \otimes_K S$ and hence to $F \otimes_N (N \otimes_K S)$ is finally isomorphic to a direct sum of matrix algebras over F.

Suppose now that A is a semisimple algebra over K with simple components $S_1, \ldots, S_n$ whose centres are all separable extensions of K. It follows from what we have just proved that for $i = 1, \ldots, n$ there exists an extension field F_i of K such that $F_i \otimes_K S_i$ is a direct sum of matrix algebras over F_i. The fields F_i are all algebraic over K; so if L is an algebraic closure of K there exist K-invariant monomorphisms $\tau_1, \ldots, \tau_n$ from $F_1, \ldots, F_n$ respectively to L. If F is any subfield of L which includes all the subfields $\tau_1(F_1), \ldots, \tau_n(F_n)$ it follows that $F \otimes_K A$ is a direct sum of matrix algebras over F.

Thus (b) implies (c).

(3) Finally suppose there is an extension field F of K such that $F \otimes_K A$ is isomorphic to a direct sum of matrix algebras over F.

Let E be any extension field of K. We may suppose without loss of generality that there is an extension G of K which includes both E and F. Then, by an argument which we used in (2) above, $G \otimes_K A$ is isomorphic to $G \otimes_F (F \otimes_K A)$ and hence to a direct sum of matrix algebras over G; thus $G \otimes_K A$ is semisimple. But, by the same argument again, $G \otimes_K A$ is isomorphic to $G \otimes_E (E \otimes_K A)$. It follows that $E \otimes_K A$ is semisimple.

So A is separable.

Chapter 6

HOMOLOGY

§28. Complexes and Homology

Let C be a category with zero objects. A *chain complex* over C is an ordered pair $((X_p)_{p\in Z}, (\partial_p)_{p\in Z})$ consisting of a sequence of objects X_p of C and a sequence of morphisms ∂_p of C, where for each integer p the morphism ∂_p has domain X_p and codomain X_{p-1}, and $\partial_{p-1}\partial_p$ is the zero morphism from X_p to X_{p-2}. We shall often represent a chain complex $X = (X_p, \partial_p)$ by a diagram

$$\ldots\ldots \xrightarrow{\partial_3} X_2 \xrightarrow{\partial_2} X_1 \xrightarrow{\partial_1} X_0 \xrightarrow{\partial_0} X_{-1} \xrightarrow{\partial_{-1}} X_{-2} \xrightarrow{\partial_{-2}} \ldots\ldots \quad [28.1]$$

For each integer p we call X_p the p-dimensional *chain object* of X and ∂_p the p-dimensional *boundary morphism.* If the objects of the category C are sets then the elements of X_p are called the p-dimensional *chains* of X.

Let $X = (X_p, \partial_p)$ and $X' = (X'_p, \partial'_p)$ be two chain complexes over C. A *chain morphism* from X to X' is a sequence $\varphi = (\varphi_p)_{p\in Z}$ of morphisms of C, where for each integer p the morphism φ_p has domain X_p and codomain X'_p, and $\varphi_p\partial_{p+1} = \partial'_{p+1}\varphi_{p+1}$, i.e. the diagram

$$\begin{array}{ccccccccc}
\longrightarrow & X_{p+1} & \xrightarrow{\partial_{p+1}} & X_p & \xrightarrow{\partial_p} & X_{p-1} & \longrightarrow \\
 & \downarrow{\scriptstyle \varphi_{p+1}} & & \downarrow{\scriptstyle \varphi_p} & & \downarrow{\scriptstyle \varphi_{p-1}} & \\
\longrightarrow & X'_{p+1} & \xrightarrow{\partial'_{p+1}} & X'_p & \xrightarrow{\partial'_p} & X'_{p-1} & \longrightarrow
\end{array}$$

is commutative. It is easily verified that if $\psi = (\psi_p)$ is a chain morphism from X' to another chain complex X'' then the sequence $(\psi_p\varphi_p)$ is a chain morphism from X to X''; we naturally denote this chain morphism by $\psi\varphi$ and call it the *composition* of φ and ψ. Routine checking shows that if θ is a chain morphism from X'' to yet another chain complex over C then $\theta(\psi\varphi) = (\theta\psi)\varphi$, and that if $I_X = (I_{X_p})$ then $\varphi I_X = \varphi$ and $I_X\varphi' = \varphi'$ for all chain morphisms with X as domain and codomain respectively. Thus we have constructed a new category $K(C)$, the *category of chain complexes* over C; the objects of $K(C)$ are the chain complexes over C and the morphisms of $K(C)$ are the chain morphisms between such complexes. We now show how the properties of C are reflected in those of $K(C)$.

THEOREM 28.1. *Let C be a category with zero objects, $K = K(C)$ the category of chain complexes over C. If C is additive, then so is K; if C has products or coproducts, then so has K; if C is exact, then so is K.*

Proof. We notice first of all that if Z is a zero object of C then the complex whose chain objects are all equal to Z and whose boundary morphisms are all zero morphisms is clearly a zero object for K.

Now let $X = (X_p, \partial_p)$ and $Y = (Y_p, \varepsilon_p)$ be chain complexes over C.

(1) Suppose C is additive.

If $\alpha = (\alpha_p)$ and $\beta = (\beta_p)$ are chain morphisms from X to Y it is easy to show that the sequence $(\alpha_p + \beta_p)$ is also a chain morphism from X to Y, which we denote by $\alpha + \beta$. It is a trivial matter to verify that under the addition operation so defined the set $K(X, Y)$ forms an abelian group and that condition *AC3* is satisfied.

Thus K is additive.

(2) Next suppose that C has products.

For ease in writing we shall show that there exists a product in K for every pair of complexes X, Y over C; the method is perfectly general and can be applied to yield products of arbitrary families of complexes in K.

For each integer p let $(P_p; \varphi_p, \psi_p)$ be a product for (X_p, Y_p) in C (where φ_p and ψ_p are the projection morphisms from P_p to X_p and Y_p respectively). Then $\partial_p\varphi_p$ and $\varepsilon_p\psi_p$ are morphisms from P_p to X_{p-1} and Y_{p-1} respectively; it follows from the couniversal property of products that there exists a morphism Δ_p from P_p to P_{p-1} such that

$$\varphi_{p-1}\Delta_p = \partial_p\varphi_p \quad \text{and} \quad \psi_{p-1}\Delta_p = \varepsilon_p\psi_p. \qquad [28.2]$$

We claim that for every integer p we have $\Delta_{p-1}\Delta_p = \zeta$, the zero morphism from P_p to P_{p-2}. To see this, let ζ_X and ζ_Y be the zero morphisms from P_p to X_{p-2} and Y_{p-2} respectively. Using [28.2] we compute easily that

$$\varphi_{p-2}(\Delta_{p-1}\Delta_p) = \zeta_X \quad \text{and} \quad \psi_{p-2}(\Delta_{p-1}\Delta_p) = \zeta_Y.$$

But of course we also have

$$\varphi_{p-2}(\zeta) = \zeta_X \quad \text{and} \quad \psi_{p-2}(\zeta) = \zeta_Y.$$

Hence, using the couniversal property of $(P_{p-2}; \varphi_{p-2}, \psi_{p-2})$ we deduce that $\Delta_{p-1}\Delta_p = \zeta$. Thus $P = (P_p, \Delta_p)$ is a chain complex over $\mathbf{C}$, and [28.2] shows that $\varphi = (\varphi_p)$ and $\psi = (\psi_p)$ are chain morphisms from P to X and Y respectively.

We now show that $(P; \varphi, \psi)$ is a product for (X, Y) in $\mathbf{K}$. So let $\alpha = (\alpha_p)$ and $\beta = (\beta_p)$ be chain morphisms from a chain complex $T = (T_p, \tau_p)$ to X and Y respectively. For each integer p there exists a unique morphism γ_p from T_p to P_p such that $\varphi_p\gamma_p = \alpha_p$ and $\psi_p\gamma_p = \beta_p$; we contend that $\gamma = (\gamma_p)$ is a chain morphism from T to P. So we compute

$$\varphi_{p-1}\gamma_{p-1}\tau_p = \alpha_{p-1}\tau_p = \partial_p\alpha_p = \partial_p\varphi_p\gamma_p = \varphi_{p-1}\Delta_p\gamma_p$$

and similarly

$$\psi_{p-1}\gamma_{p-1}\tau_p = \psi_{p-1}\Delta_p\gamma_p.$$

Hence, by the couniversal property of $(P_{p-1}; \varphi_{p-1}, \psi_{p-1})$, we deduce that $\gamma_{p-1}\tau_p = \Delta_p\gamma_p$ for all integers p, i.e. that γ is a chain morphism from T to P such that $\varphi\gamma = \alpha$ and $\psi\gamma = \beta$; the uniqueness of the morphisms γ_p implies that γ is unique.

Thus every pair of objects in $\mathbf{K}$ has a product. A similar argument applies to coproducts.

(3) Finally, suppose $\mathbf{C}$ is exact.

Let $\alpha = (\alpha_p)$ be a chain morphism from X to Y. Since $\mathbf{C}$ is exact, each of the morphisms α_p has an analysis

$$K_p \xrightarrow{\kappa_p} X_p \xrightarrow{\alpha'_p} I_p \xrightarrow{\alpha''_p} Y_p \xrightarrow{\eta_p} L_p$$

as described in §16. It is an easy exercise to define boundary morphisms θ_p, φ_p, ψ_p such that $K = (K_p, \theta_p)$, $I = (I_p, \varphi_p)$ and $L = (L_p, \psi_p)$ are chain complexes and $\kappa = (\kappa_p)$, $\alpha' = (\alpha'_p)$, $\alpha'' = (\alpha''_p)$, $\eta = (\eta_p)$ are chain morphisms; we then verify that

$$K \xrightarrow{\kappa} X \xrightarrow{\alpha'} I \xrightarrow{\alpha''} Y \xrightarrow{\eta} L$$

is an analysis of α.

Hence $\mathbf{K}$ is exact.

We now consider a chain complex $X = (X_p, \partial_p)$ over a locally small abelian category $\mathbf{A}$. For each integer p let κ_p be a kernel of the morphism ∂_p, with domain C_p say. Since $\partial_p\partial_{p+1} = \zeta$ it follows from the universal property of kernels that there is a unique morphism θ_{p+1} from X_{p+1} to C_p such that $\partial_{p+1} = \kappa_p\theta_{p+1}$. Let η_p be a cokernel of θ_{p+1}; if H_p is the codomain of η_p we say that H_p is a p-dimensional *homology object* of the complex X. The situation is illustrated by the diagram

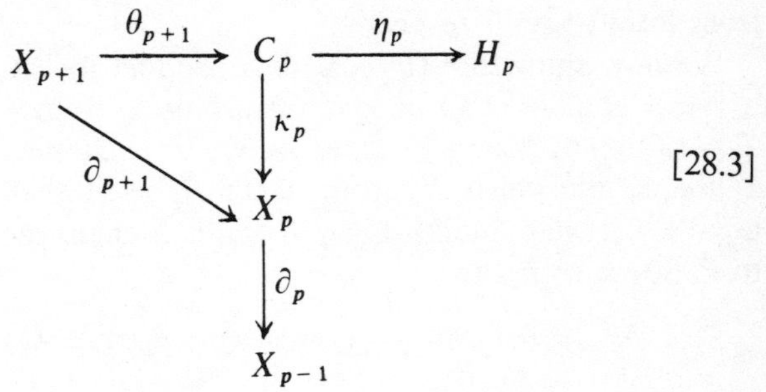

[28.3]

It is easy to see that all p-dimensional homology objects of X are isomorphic; we shall chose one of them, call it *the* p-dimensional homology object of X, and denote it by $H_p(X)$.

As an exercise the reader may care to show that the homology objects $H_p(X)$ are zero for every integer p if and only if the sequence [28.1] is exact; so we may think of the homology objects as measuring how far [28.1] is from being exact.

In the case where A is the category of left modules over a ring R we take $C_p = \text{Ker}\ \partial_p$ and $B_p = \text{Im}\ \partial_{p+1}$ (in the usual module sense) and call these the modules of *p-cycles* and *p-boundaries* of X respectively; then we may take $H_p(X)$ to be the factor module C_p/B_p.

We return now to the case of a general (locally small) abelian category A. Having just seen how to attach to each object X of $\mathsf{K}(\mathsf{A})$ an object $H_p(X)$ of A we now show how each morphism of $\mathsf{K}(\mathsf{A})$, i.e. each chain morphism $\alpha = (\alpha_p)$ from X to another chain complex $X' = (X'_p, \partial'_p)$, induces a morphism $H_p(\alpha)$ from $H_p(X)$ to $H_p(X')$ in such a way as to produce a covariant functor from $\mathsf{K}(\mathsf{A})$ to A. We use the obvious notation for objects and morphisms associated with X'; namely, let κ'_p (with domain C'_p) be a kernel for ∂'_p, let θ'_{p+1} be the unique morphism such that $\partial'_{p+1} = \kappa'_p\theta'_{p+1}$ and let η'_p (with codomain $H_p(X')$) be a cokernel of θ'_{p+1}.

We notice first that

$$\partial'_p\alpha_p\kappa_p = \alpha_{p-1}\partial_p\kappa_p = \zeta,$$

since κ_p is a kernel of ∂_p. Hence, since κ'_p is a kernel of ∂'_p, there exists a unique morphism τ_p from C_p to C'_p such that $\alpha_p\kappa_p = \kappa'_p\tau_p$.

Next we compute

$$\kappa'_p \tau_p \theta_{p+1} = \alpha_p \kappa_p \theta_{p+1} = \alpha_p \partial_{p+1} = \partial'_{p+1} \alpha_{p+1} = \kappa'_p \theta'_{p+1} \alpha_{p+1}$$

and deduce that $\tau_p \theta_{p+1} = \theta'_{p+1} \alpha_{p+1}$ since κ'_p is monic (being a kernel). Thus we have

$$\eta'_p \tau_p \theta_{p+1} = \eta'_p \theta'_{p+1} \alpha_{p+1} = \zeta,$$

since η'_p is a cokernel of θ'_{p+1}. Finally, since η_p is a cokernel of θ_{p+1}, there exists a unique morphism α_{*p} from $H_p(X)$ to $H_p(X')$ such that $\eta'_p \tau_p = \alpha*_p \eta_p$. The situation is illustrated in the commutative diagram

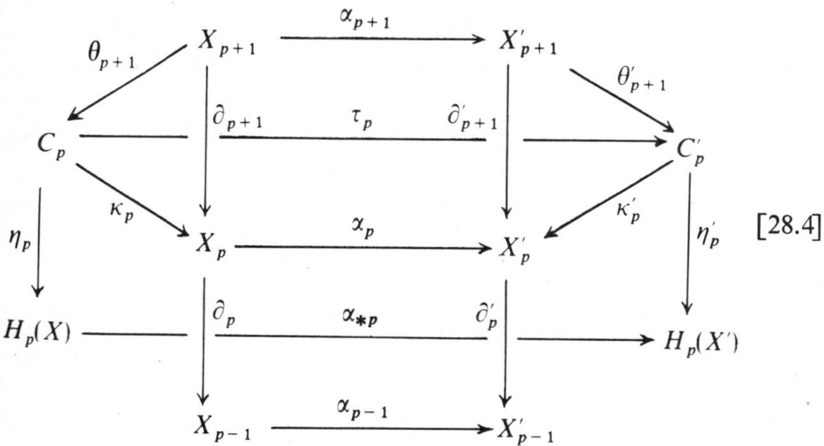

[28.4]

Let us now write $H_p(\alpha)$ instead of α_{*p}. Then we can check that if β is a chain morphism from X' to another chain complex then $H_p(\beta\alpha) = H_p(\beta)H_p(\alpha)$; and if I_X is the identity chain morphism of X then $H_p(I_X)$ is the identity morphism of $H_p(X)$. Thus, if we set $T(X) = H_p(X)$ for every chain complex X and $T(\alpha) = H_p(\alpha)$ for every chain morphism α, we see that we have defined a covariant functor T from $\mathsf{K}(\mathsf{A})$ to A, and we readily verify that this functor is additive. We call this the p-dimensional *homology functor* and denote it by H_p.

In the case where A is the category of left modules over a ring R we can describe explicitly how the homology (homo)morphism $H_p(\alpha)$ induced by a chain map α acts on elements of the homology module $H_p(X)$. Namely, let A be any element of $H_p(X)$, so that A is a coset of C_p modulo B_p; choose a p-cycle z from this coset; then

$\alpha_p(z)$ is a p-cycle for the complex X'. We set $[H_p(\alpha)](A)$ = the coset of this p-cycle modulo the p-boundary module of X'; this is independent of the choice of the cycle z.

Returning again to the general case, let α and β be chain morphisms from X to X'. A *chain homotopy* between α and β is a sequence $\sigma = (\sigma_p)_{p \in Z}$ of morphisms of $\boldsymbol{A}$ such that for each integer p the morphism σ_p has domain X_p and codomain X'_{p+1} and

$$\alpha_p - \beta_p = \partial'_{p+1}\sigma_p + \sigma_{p-1}\partial_p.$$

In this situation we write $\sigma : \alpha \simeq \beta$. If there exists a chain homotopy between chain morphisms α and β we say that α and β are *chain homotopic*. Two chain complexes X and X' are said to be of the same *homotopy type* if there exist chain morphisms α and α' from X to X' and from X' to X respectively such that $\alpha'\alpha$ and $\alpha\alpha'$ are chain homotopic to the identity chain morphisms of X and X' respectively. The importance of chain homotopy arises from the following theorem.

THEOREM 28.2. *Let X and X' be chain complexes over a locally small abelian category. If α and β are chain homotopic chain morphisms from X to X' then for every integer p we have $H_p(\alpha) = H_p(\beta)$.*

Proof. Let $\sigma = (\sigma_p)$ be a chain homotopy between α and β. Thus for every integer p we have $\alpha_p - \beta_p = \partial'_{p+1}\sigma_p + \sigma_{p-1}\partial_p$. It follows that

$$(\alpha_p - \beta_p)\kappa_p = \partial'_{p+1}\sigma_p\kappa_p + \sigma_{p-1}\partial_p\kappa_p = \partial'_{p+1}\sigma_p\kappa_p = \kappa'_p\theta'_{p+1}\sigma_p\kappa_p.$$

Now there are morphisms τ_p and ρ_p from C_p to C'_p such that $\alpha_p\kappa_p = \kappa'_p\tau_p$ and $\beta_p\kappa_p = \kappa'_p\rho_p$ (cf. the diagram [28.4]). Hence we have

$$\kappa'_p(\tau_p - \rho_p) = (\alpha_p - \beta_p)\kappa_p = \kappa'_p\theta'_{p+1}\sigma_p\kappa_p.$$

Since κ'_p is monic we deduce that $\tau_p - \rho_p = \theta'_{p+1}\sigma_p\kappa_p$ and so obtain

$$\eta'_p(\tau_p - \rho_p) = \eta'_p\theta'_{p+1}\sigma_p\kappa_p = \zeta$$

(because η'_p is a cokernel of θ'_{p+1}). But $\eta'_p\tau_p = H_p(\alpha)\eta_p$ (see [28.4]) and similarly $\eta'_p\rho_p = H_p(\beta)\eta_p$. Consequently we have

$$(H_p(\alpha) - H_p(\beta))\eta_p = \zeta.$$

Since η_p is epic it follows at once that $H_p(\alpha) = H_p(\beta)$.

This completes the proof.

COROLLARY. *Complexes of the same homotopy type have isomorphic homology objects in every dimension.*

Another useful way of looking at the morphism $H_p(\alpha)$ induced by a chain morphism α is obtained by passing to the category of relations $R(A)$ of A. If we identify the various A-morphisms in [28.4] with their graphs (which are R-morphisms) we have in particular the two equations

$$\alpha_p \circ \kappa_p = \kappa'_p \circ \tau_p \text{ and } \eta'_p \circ \tau_p = H_p(\alpha) \circ \eta_p$$

where we denote the composition by the circle notation to emphasize that we are now working in $R(A)$. Since κ'_p is monic and η_p is epic in A it follows from Theorem 18.2 that $(\kappa'_p)^* \circ \kappa_p$ and $\eta_p \circ \eta_p^*$ are identity morphisms in R. Thus we obtain

$$H_p(\alpha) = \eta'_p \circ (\kappa'_p)^* \circ \alpha_p \circ \kappa_p \circ \eta_p^*.$$

This discussion leads us to introduce for every integer p the p-dimensional *homology relation*; this is the morphism χ_p in $R(X_p, H_p(X))$ defined by setting $\chi_p = \eta_p \circ \kappa_p^*$. Then of course we have

$$H_p(\alpha) = \chi'_p \circ \alpha_p \circ \chi_p^*$$

where χ'_p is the p-dimensional homology relation for X'.

The next theorem establishes some basic properties of the homology relations.

THEOREM 28.3. *Let $X = (X_p, \partial_p)$ be a chain complex over a locally small abelian category A; let χ_p be the p-dimensional homology relation of X. Then we have*

$$\chi_p \circ \omega = \omega, \qquad \chi_p \circ \Omega = \Omega,$$
$$\chi_p^* \circ \omega = \partial_{p+1} \circ \Omega, \qquad \chi_p^* \circ \Omega = \partial_p^* \circ \omega.$$

Note. The codomains of the various morphisms denoted by ω, Ω are taken so that the various compositions are well defined. The A-morphisms $\partial_p, \partial_{p+1}$ are identified with their graphs in $R(A)$.

Proof. (1) By definition of χ_p we have $\chi_p \circ \omega = \eta_p \circ \kappa_p^* \circ \omega$.

Since κ_p is monic, its kernels are zero morphisms (by Theorem 16.4); so it follows from Theorem 18.3 that $\kappa_p^* \circ \omega = Sb(\zeta)$. We deduce easily from this that $\chi_p \circ \omega = Sb(\zeta) = \omega$.

(2) It follows from the definition that $\chi_p^* \circ \omega = \kappa_p \circ \eta_p^* \circ \omega$.

Since η_p is a cokernel of θ_{p+1}, each image θ of θ_{p+1} is a kernel of

η_p; so, using Theorem 18.3 again we see that $\eta_p^* \circ \omega = Sb(\theta)$, i.e. that (ζ, θ) is a representative pair for $\eta_p^* \circ \omega$. Hence so is (ζ, θ_{p+1}). By constructing a pullback in the usual way, we see that $(\zeta, \kappa_p \theta_{p+1}) = (\zeta, \partial_{p+1})$ is a representative pair for $\chi_p^* \circ \omega$. But it is easy to check that this is also a representative pair for $\partial_{p+1} \circ \Omega$.

We leave it to the reader as an exercise in manipulating pullbacks to verify the other two relations.

Let now

$$Z \to X' \xrightarrow{\alpha} X \xrightarrow{\beta} X'' \to Z \qquad [28.5]$$

be a short exact sequence of chain complexes over a locally small abelian category A where Z in each case denotes a chain complex all of whose chain objects are zero objects. For each integer p the sequence [28.5] gives rise to a sequence

$$H_p(X') \xrightarrow{H_p(\alpha)} H_p(X) \xrightarrow{H_p(\beta)} H_p(X''). \qquad [28.6]$$

We now propose to fit together the sequences [28.6] for all integers p so as to form a single infinite exact sequence. To this end we must define for each integer p a 'connecting morphism' ∂_{*p} from $H_p(X'')$ to $H_{p-1}(X')$.

We begin by considering the relation

$$\varepsilon_p = \chi'_{p-1} \circ \alpha_{p-1}^* \circ \partial_p \circ \beta_p^* \circ (\chi''_p)^*$$

and show that this is the graph of a morphism from $H_p(X'')$ to $H_{p-1}(X')$. For this purpose we remark that since [28.5] is exact the morphism β_{p+1} is epic and so the identity morphism of X''_{p+1} is an image for β_{p+1}; it follows from Theorem 18.3 that we have

$$\beta_{p+1} \circ \Omega = Sb(I_{X''}) = \Omega. \qquad [28.7]$$

Further, since [28.5] is exact, the morphism α_{p-1} is monic and so, according to Theorem 18.2, we have

$$\alpha_{p-1}^* \circ \alpha_{p-1} = J_{X_{p-1}}. \qquad [28.8]$$

Now we compute

$$\begin{aligned} \varepsilon_p \circ \omega(H_p(X'')) &= \chi'_{p-1} \circ \alpha_{p-1}^* \circ \partial_p \circ \beta_p^* \circ (\chi''_p)^* \circ \omega(H_p(X'')) \\ &= \chi'_{p-1} \circ \alpha_{p-1}^* \circ \partial_p \circ \beta_p^* \circ \partial''_{p+1} \circ \Omega(X''_{p+1}) \end{aligned}$$

using Theorem 28.3

$$= \chi'_{p-1} \circ \alpha^*_{p-1} \circ \partial_p \circ \beta^*_p \circ \partial''_{p+1} \circ \beta_{p+1} \circ \Omega(X_{p+1})$$

by [28.7]

$$= \chi'_{p-1} \circ \alpha^*_{p-1} \circ \partial_p \circ \beta^*_p \circ \beta_p \circ \partial_{p+1} \circ \Omega(X_{p+1})$$

since β is a chain morphism.

According to Theorem 18.3, $\partial_{p+1} \circ \Omega(X_{p+1}) = Sb(\iota)$, where ι is an image of ∂_{p+1} and $\partial^*_p \circ \omega(X_{p-1}) = Sb(\kappa_p)$. Since $\partial_p \partial_{p+1} = \zeta$ we deduce easily that $\partial_p \iota = \zeta$ and hence there is a morphism μ such that $\iota = \kappa_p \mu$; clearly μ is monic. We now apply Theorem 18.5 and deduce that there exists a monic morphism μ' such that

$$\varepsilon_p \circ \omega(H_p(X'')) = (\chi'_{p-1} \circ \alpha^*_{p-1} \circ \partial_p \circ \beta^*_p \circ \beta_p \circ \partial^*_p \circ \omega(X_{p-1}))\, \mu'$$

$$= (\chi'_{p-1} \circ \alpha^*_{p-1} \circ (\partial_p \ \beta^*_p) \circ (\partial_p \circ \beta^*_p)^* \circ \omega(X_{p-1}))\mu'$$

$$= (\chi'_{p-1} \circ \alpha^*_{p-1} \circ \partial_p \circ \beta^*_p \circ \omega(X_p))\mu'$$

using Theorem 18.4

$$= (\chi'_{p-1} \circ \alpha^*_{p-1} \circ \partial_p \circ \alpha_p \circ \Omega(X'_p))\mu'$$

by the Corollary of Theorem 18.3

$$= (\chi'_{p-1} \circ \alpha^*_{p-1} \circ \alpha_{p-1} \circ \partial'_p \circ \Omega(X'_p))\mu'$$

since α is a chain morphism

$$= [\chi'_{p-1} \circ (\chi'_{p-1})^* \circ \omega(H_{p-1}(X'))]\,\mu'$$

using [28.8] and Theorem 28.3

$$= (\chi'_{p-1} \circ \omega(X'_{p-1}))\mu'$$

by Theorem 18.4

$$= \omega(H_{p-1}(X'))\mu'$$

by Theorem 28.3

$$= \omega(H_{p-1}(X')).$$

A similar computation shows that $\varepsilon^*_p \circ \Omega(H_{p-1}(X')) = \Omega(H_p(X''))$. It follows from Theorem 18.1 that ε_p is indeed the graph of a morphism from $H_p(X'')$ to $H_{p-1}(X')$. We denote this morphism by ∂_{*p} and call it the *connecting morphism* from $H_p(X'')$ to $H_{p-1}(X')$.

In the category of left modules over a ring R the traditional definition of the connecting homomorphism ∂_{*p} from $H_p(X'')$ to $H_{p-1}(X')$ is as follows. Let A''_p be any element of $H_p(X'')$, so that A''_p is a coset of the module C''_p of p-cycles of X'' modulo B''_p, the module of p-boundaries; choose any p-cycle z''_p from this coset. Since β_p is an epimorphism there is an element x_p of X_p such that $\beta_p(x_p) = z''_p$. Then $\beta_{p-1}(\partial_p(x_p)) = \partial''_p(\beta_p(x_p)) = 0$; so $\partial_p(x_p) \in \operatorname{Ker} \beta_{p-1} = \operatorname{Im} \alpha_{p-1}$ and hence there is an element x'_{p-1} of X'_{p-1} such that $\alpha_{p-1}(x'_{p-1}) = \partial_p(x_p)$. It is easily seen that x'_{p-1} is in fact a $(p-1)$-cycle of X' and we define $\partial_{*p}(A''_p)$ to be the coset of x'_{p-1} modulo B'_{p-1}. It has to be verified that this coset does not depend on the two arbitrary choices in its definition—that of z''_p from A''_p and that of x_p such that $\beta_p(x_p) = z''_p$. The reader will find it easy to see how the steps of the traditional definition are represented by the factors of ε_p in our procedure.

We are now in a position to construct our infinite exact sequence from the pieces [28.6].

THEOREM 28.4. *Let $Z \to X' \xrightarrow{\alpha} X \xrightarrow{\beta} X'' \to Z$ be a short exact sequence of chain complexes over a locally small abelian category. Then, with the notation introduced above, the sequence*

$$\ldots\ldots \to H_p(X') \xrightarrow{H_p(\alpha)} H_p(X) \xrightarrow{H_p(\beta)} H_p(X'') \xrightarrow{\partial_{*p}} H_{p-1}(X') \to \ldots\ldots$$

is exact.

Proof. (1) To prove exactness at $H_p(X)$, we shall show that $H_p(\alpha) \circ \Omega(H_p(X')) = H_p(\beta)^* \circ \omega(H_p(X''))$ (cf. the Corollary to Theorem 18.3).

First we notice that since α_{p-1} is monic, it follows easily from Theorem 18.3 that $\alpha^*_{p-1} \circ \omega(X_{p-1}) = \omega(X'_{p-1})$. Now we examine

$$\begin{aligned} H_p(\alpha) \circ \Omega(H_p(X')) &= \chi_p \circ \alpha_p \circ (\chi'_p)^* \circ \Omega(H_p(X')) \\ &= \chi_p \circ \alpha_p \circ (\partial'_p)^* \circ \omega(X'_{p-1}) \\ &\qquad\qquad \text{by Theorem 28.3} \\ &= \chi_p \circ \alpha_p \circ (\partial'_p)^* \circ \alpha^*_{p-1} \circ \omega(X_{p-1}) \\ &= \chi_p \circ \alpha_p \circ \alpha^*_p \circ \partial^*_p \circ \omega(X_{p-1}) \\ &\qquad\qquad \text{since } \alpha \text{ is a chain morphism} \\ &= \chi_p \circ \alpha_p \circ \alpha^*_p \circ \chi^*_p \circ \Omega(H_p(X)) \\ &\qquad\qquad \text{by Theorem 28.3} \\ &= \chi_p \circ \alpha_p \circ \Omega(X'_p) \\ &\qquad\qquad \text{by Theorem 18.4.} \end{aligned}$$

Next, since β_{p+1} is epic, we have $\beta_{p+1} \circ \Omega(X_{p+1}) = \Omega(X'_{p+1})$; using this fact and Theorem 18.4, we obtain by an analogous computation the result that

$$H_p(\beta)^* \circ \omega(H_p(X'')) = \chi_p \circ \beta_p^* \circ \omega(X_p'').$$

But since the sequence

$$X_p' \to X_p \to X_p''$$

is exact, the Corollary to Theorem 18.3 shows that $\alpha_p \circ \Omega(X_p') = \beta_p^* \circ \omega(X_p'')$; hence we have the desired exactness at $H_p(X)$.

(2) Next, to establish exactness at $H_p(X'')$ we prove that $H_p(\beta) \circ \Omega(H_p(X)) = (\partial_{*p})^* \circ \omega(H_p(X'))$. So we compute

$$\begin{aligned} H_p(\beta) \circ \Omega(H_p(X)) &= \chi_p'' \circ \beta_p \circ \chi_p^* \circ \Omega(H_p(X)) \\ &= \chi_p'' \circ \beta_p \circ \partial_p^* \circ \omega(X_{p-1}) \\ &= \chi_p'' \circ \beta_p \circ \partial_p^* \circ \partial_p \circ \beta_p^* \circ \omega(X_p'') \\ &\qquad \text{by Theorem 18.4} \\ &= \chi_p'' \circ \beta_p \circ \partial_p^* \circ \partial_p \circ \alpha_p \circ \Omega(X_p') \\ &= \chi_p'' \circ \beta_p \circ \partial_p^* \circ \alpha_{p-1} \circ \partial_p' \circ \Omega(X_p') \\ &= \chi_p'' \circ \beta_p \circ \partial_p^* \circ \alpha_{p-1} \circ (\chi_{p-1}')^* \circ \omega(H_{p-1}(X')) \\ &= (\partial_{*p})^* \circ \omega(H_{p-1}(X')) \end{aligned}$$

as required.

(3) A similar calculation establishes exactness at $H_{p-1}(X')$.

The exact sequence of Theorem 28.4 is called the *homology exact sequence* associated with the short exact sequence [28.5].

Once again let $\boldsymbol{C}$ be an arbitrary category with zero objects. A *cochain complex* over $\boldsymbol{C}$ is an ordered pair $X^{\boldsymbol{\cdot}} = ((X^p)_{p\in\mathbf{Z}}, (\delta^p)_{p\in\mathbf{Z}})$ consisting of a sequence of objects X^p of $\boldsymbol{C}$ and a sequence of morphisms δ^p of $\boldsymbol{C}$, where for each integer p the morphism δ^p has domain X^p and codomain X^{p+1}, and $\delta^{p+1}\delta^p$ is the zero morphism from X^p to X^{p+2}. For each integer p we call X^p the p-dimensional *cochain object* of $X^{\boldsymbol{\cdot}}$ and δ^p the p-dimensional *coboundary morphism.* If the objects of $\boldsymbol{C}$ are sets, then the elements of X^p are called the p-dimensional *cochains* of $X^{\boldsymbol{\cdot}}$. It is clear how one would define cochain morphisms and set up the *category of cochain complexes* over $\boldsymbol{C}$.

We now consider a cochain complex $X^{\boldsymbol{\cdot}} = (X^p, \delta^p)$ over a locally small abelian category $\boldsymbol{A}$. For each integer p let κ^p be a kernel of δ^p, with domain C^p. Since $\delta^p\delta^{p-1} = \zeta$ there exists a unique morphism θ^{p-1} from X^{p-1} to C^p such that $\delta^{p-1} = \kappa^p\theta^{p-1}$. Let η^p be a cokernel of θ^{p-1}; if H^p is the codomain of η^p, we call H^p a p-dimensional *cohomology object* for $X^{\cdot}$. It is now easy to see how one would define for each integer p the p-dimensional *cohomology functor*. For each short exact sequence

$$Z^{\cdot} \to X'^{\cdot} \xrightarrow{\alpha^{\cdot}} X^{\cdot} \xrightarrow{\beta^{\cdot}} X''^{\cdot} \to Z^{\cdot}$$

of cochain complexes and cochain morphisms we can define connecting morphisms δ^{*p} by analogy with the connecting morphism ∂_{*p} for exact sequences of chain complexes, and we obtain a *cohomology exact sequence*

$$\ldots\ldots \to H^p(X'^{\cdot}) \xrightarrow{H^p(\alpha^{\cdot})} H^p(X^{\cdot}) \xrightarrow{H^p(\beta^{\cdot})} H^p(X''^{\cdot}) \xrightarrow{\delta^{*p}} H^{p+1}(X'^{\cdot}) \to \ldots.$$

We conclude this section with some concrete examples of cochain complexes. Let $X = (X_p, \partial_p)$ be a chain complex of left modules over a ring R and let M be a given left R-module. For each integer p set $X^p = \mathrm{Hom}_R(X_p, M)$ and $\delta^p = \mathrm{Hom}(\partial_{p+1}, I_M)$; thus for every R-homomorphism a^p from X_p to M, $\delta^p a^p$ is the R-homomorphism from X_{p+1} to M defined by setting $(\delta^p a^p)(x_{p+1}) = a^p(\partial_{p+1}(x_{p+1}))$ for every $(p+1)$-chain x_{p+1} of X. It is easily verified that $X^{\cdot} = (X^p, \delta^p)$ is indeed a cochain complex of abelian groups, which we may denote by $\mathrm{Hom}_R(X, M)$. The groups $X^p, C^p = \mathrm{Ker}\,\delta^p, B^p = \mathrm{Im}\,\delta^{p-1}$ are called respectively the groups of *p-cochains, p-cocycles, p-coboundaries* of the chain complex X *with coefficients in* the module M; we write $H^p(X^{\cdot}) = H^p(X, M)$ and call it the p-dimensional *cohomology group* of X *with coefficients in* M.

Let X and Y be chain complexes of left R-modules, M a left R-module, and let $X^{\cdot}$ and $Y^{\cdot}$ be the corresponding cochain complexes $\mathrm{Hom}_R(X, M)$ and $\mathrm{Hom}_R(Y, M)$ respectively, constructed as in the preceding paragraph. Let $\alpha = (\alpha_p)$ be a chain morphism from X to Y. Then it is easy to verify that $\alpha^{\cdot} = (\alpha^p) = (\mathrm{Hom}(\alpha_p, I_M))$ is a cochain morphism from $Y^{\cdot}$ to $X^{\cdot}$ and so induces homomorphisms, which we denote by $H^p(\alpha)$, from $H^p(Y, M)$ to $H^p(X, M)$ for every integer p.

Consider a short exact sequence

$$Z \to X' \xrightarrow{\alpha} X \xrightarrow{\beta} X'' \to Z \qquad [28.9]$$

of chain complexes of left R-modules. If M is a given left R-module and we form the corresponding complexes of cochains with coefficients in M, we obtain another sequence

$$Z \to \operatorname{Hom}(X'', M) \xrightarrow{\beta^{\cdot}} \operatorname{Hom}(X, M) \xrightarrow{\alpha^{\cdot}} \operatorname{Hom}(X', M) \to Z \quad [28.10]$$

which in general, however, is not exact. (We recall that if

$$0 \to X'_p \xrightarrow{\alpha_p} X_p \xrightarrow{\beta_p} X''_p \to 0 \quad [28.11]$$

is an exact sequence of left R-modules then all we can deduce from Theorem 8.4 is that

$$0 \to \operatorname{Hom}(X''_p, M) \xrightarrow{\beta^p} \operatorname{Hom}(X_p, M) \xrightarrow{\alpha^p} \operatorname{Hom}(X'_p, M)$$

is exact.) But, according to Theorem 9.5, if all the short exact sequences [28.11] are split, then all the sequences

$$0 \to \operatorname{Hom}(X''_p, M) \xrightarrow{\beta^p} \operatorname{Hom}(X_p, M) \xrightarrow{\alpha^p} \operatorname{Hom}(X'_p, M) \to 0 \quad [28.12]$$

are exact, and it follows easily that [28.10] is an exact sequence of cochain complexes. The following theorem is now an immediate consequence.

THEOREM 28.5. *Let* [28.9] *be a short exact sequence of chain complexes of left R-modules; if for every integer p the sequence* [28.11] *is split, then for each integer p there is a connecting homomorphism δ^{*p} from $H^p(X', M)$ to $H^{p+1}(X'', M)$ such that the sequence of cohomology groups*

$$\to H^p(X'', M) \xrightarrow{H^p(\beta)} H^p(X, M) \xrightarrow{H^p(\alpha)} H^p(X', M) \xrightarrow{\delta^{*p}} H^{p+1}(X'', M) \to$$

is exact.

We give an explicit description of the connecting homomorphism δ^{*p}. Let A'^p be any element of $H^p(X', M)$; choose any p-cocycle z'^p from A'^p. Since $z'^p \in \operatorname{Hom}(X'_p, M)$ and [28.12] is exact it follows that there is a p-cochain c^p in $X^p = \operatorname{Hom}(X_p, M)$ such that $z'^p = \alpha^p(c^p)$. Since $0 = \delta'^p(z'^p) = \delta'^p\alpha^p(c^p) = \alpha^{p+1}\delta^p(c^p)$ we see that $\delta^p(c^p) \in \operatorname{Ker} \alpha^{p+1} = \operatorname{Im} \beta^{p+1}$, i.e. there is an element c''^{p+1} in $\operatorname{Hom}(X''_{p+1}, M)$ such that $\delta^p(c^p) = \beta^{p+1}(c''^{p+1})$. Then c''^{p+1} is a $(p+1)$-cocycle and $\delta^{*p}(A'^p)$ is the coset of c''^{p+1}.

Finally, let X be a chain complex of left R-modules and let

$$0 \to M' \xrightarrow{\lambda} M \xrightarrow{\mu} M'' \to 0$$

be a short exact sequence of left R-modules. Then we may form the sequence

$$0 \to \mathrm{Hom}(X, M') \xrightarrow{\lambda_*} \mathrm{Hom}(X, M) \xrightarrow{\mu_*} \mathrm{Hom}(X, M'') \to 0 \quad [28.13]$$

where $\lambda_* = (\mathrm{Hom}(I_{X_p}, \lambda))$ and $\mu_* = (\mathrm{Hom}(I_{X_p}, \mu))$. This sequence is not in general exact, but it follows from Theorem 11.6 that if all the chain modules X_p of the complex X are projective then the sequence [28.13] is an exact sequence of cochain complexes. It therefore gives rise to an exact sequence of cohomology groups

$$\to H^p(X, M') \xrightarrow{H^p(\lambda_*)} H^p(X, M) \xrightarrow{H^p(\mu_*)} H^p(X, M'') \xrightarrow{\delta_{*p}} H^{p-1}(X, M') \to .$$

The reader should work out explicitly the action of the connecting homomorphism δ_{*p}.

§29. Resolutions and Coresolutions

Let C be a category with zero objects. A chain complex $X = (X_p, \partial_p)$ over C is called a *positive chain complex* if the objects X_p are zero objects for all negative integers p, and X is called a *negative chain complex* if the objects X_p are zero objects for all positive integers p. Positive and negative cochain complexes are defined in the obvious way. It is easy to see that a negative chain complex $X = (X_p, \partial_p)$ can be regarded as a positive cochain complex if we write $X^p = X_{-p}$ and $\delta^p = \partial_{-p}$ for every integer p.

Let A be any object of the category C. Then we can form a chain complex $C(A) = (C_p(A), \partial_p(A))$ by setting $C_0(A) = A$ and $C_p(A) = 0$, a zero object, for all non-zero integers p; the boundary morphisms must then be zero morphisms. A *chain complex over A* is a pair (X, ε) consisting of a positive chain complex $X = (X_p, \partial_p)$ and a chain morphism $\varepsilon = (\varepsilon_p)$ from X to the chain complex $C(A)$ associated with A. It is clear that for all non-zero integers p the morphisms ε_p must be zero morphisms, and that the composition $\varepsilon_0\partial_1$ must also be the zero morphism: the morphism ε_0 is called the *augmentation* of the chain complex (X, ε) over A. We shall often say loosely that 'X is a chain complex over A'.

We say that the chain complex (X, ε) over A is *acyclic* if the sequence

$$\ldots . \to X_n \xrightarrow{\partial_n} X_{n-1} \to \ldots . \to X_1 \xrightarrow{\partial_1} X_0 \xrightarrow{\varepsilon_0} A \to 0$$

is exact. The chain complex (X, ε) over A is said to be *projective* if all the objects X_p are projective. If (X, ε) is both acyclic and projective we call it a *projective resolution of A*.

A category C is said to *have enough projectives* if for every object A of C there exists a projective object P and an epic morphism η from P to A.

Example 1. The category of left R-modules has enough projectives. According to Theorem 11.3 every left R-module is an epimorphic image of a free left R-module, which is of course projective by Theorem 11.4.

THEOREM 29.1. *Let C be an exact category. If C has enough projectives then every object of C has a projective resolution.*

Proof. Let A be any object of C. Since C has enough projectives there exist a projective object X_0 of C and an epic morphism ε_0 from X_0 to A. Since C is exact, ε_0 has a kernel κ_0, with domain Y_0 say; then the sequence

$$0 \to Y_0 \xrightarrow{\kappa_0} X_0 \xrightarrow{\varepsilon_0} A \to 0 \qquad [29.1]$$

is exact.

We now proceed by induction. So suppose that for some integer n we have succeeded in constructing exact sequences

$$0 \to Y_k \xrightarrow{\kappa_k} X_k \xrightarrow{\varepsilon_k} Y_{k-1} \to 0 \qquad [29.2]$$

(for $k = 0, 1, \ldots, n$) in which the objects X_k are all projective. (If we set $Y_{-1} = A$ then [29.1] is a basis for our induction.) Since there are enough projectives, there exists a projective object X_{n+1} of C and an epic morphism ε_{n+1} from X_{n+1} to Y_n. Let κ_{n+1}, with domain Y_{n+1}, be a kernel of ε_{n+1}. Then the sequence

$$0 \to Y_{n+1} \xrightarrow{\kappa_{n+1}} X_{n+1} \xrightarrow{\varepsilon_{n+1}} Y_n \to 0$$

is exact. So we have exact sequences [29.2] for every non-negative integer k.

It is now a trivial matter to verify that if we set $\partial_p = \kappa_p \varepsilon_{p+1}$ for all positive integers p then the sequence

$$\ldots \to X_n \xrightarrow{\partial_n} X_{n-1} \to \ldots \to X_1 \xrightarrow{\partial_1} X_0 \xrightarrow{\varepsilon_0} A \to 0$$

is exact, and hence yields a projective resolution of A.

Theorem 29.1 thus establishes the existence of projective resolutions for all the objects of an exact category with enough projectives. The essential uniqueness of such resolutions will follow eventually from our next result.

First we introduce some new terminology. Let A and A' be objects of a category with zero objects, α_0 a morphism from A to A'. If we set $\alpha_p = \zeta \in \mathbf{C}(C_p(A), C_p(A'))$ for all non-zero integers p, it is clear that $\alpha = (\alpha_p)$ is a chain morphism from the complex $C(A)$ associated with A to the corresponding complex $C(A')$. Let (X, ε) and (X', ε') be chain complexes over A and A' respectively. A chain morphism φ from X to X' such that the diagram

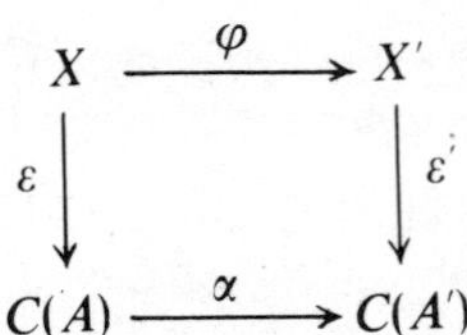

is commutative is said to be a *chain morphism over* α_0; we also say that φ *lifts* α_0.

THEOREM 29.2. *Let A and A' be objects of a category $\mathbf{C}$ with zero objects, α_0 a morphism from A to A'. If (X, ε) is a projective chain complex over A and (X', ε') is an acyclic chain complex over A' then there exists a chain morphism from X to X' which lifts α_0. If $\mathbf{C}$ is additive any two such chain morphisms are chain homotopic.*

Proof. (1) Since (X', ε') is acyclic, the sequence

$$X'_0 \xrightarrow{\varepsilon'_0} A' \to 0$$

is exact and hence ε'_0 is epic. Since (X, ε) is projective, the object X_0 is projective, and since $\alpha_0\varepsilon_0$ is a morphism from X_0 to A' it follows that there exists a morphism φ_0 from X_0 to X'_0 such that $\varepsilon'_0\varphi_0 = \alpha_0\varepsilon_0$, i.e. such that the diagram

$$\begin{array}{ccccc} X_0 & \xrightarrow{\varepsilon_0} & A & \longrightarrow & 0 \\ \downarrow{\scriptstyle \varphi_0} & & \downarrow{\scriptstyle \alpha_0} & & \\ X'_0 & \xrightarrow{\varepsilon'_0} & A' & \longrightarrow & 0 \end{array}$$

is commutative.

Now we proceed by induction. We suppose that for some integer n we have constructed morphisms φ_k from X_k to X'_k $(k = 0, 1, \ldots, n)$

such that the diagram

$$\begin{array}{ccccccccc} X_n & \xrightarrow{\partial_n} & X_{n-1} & \longrightarrow \cdots\cdots \longrightarrow & X_0 & \xrightarrow{\varepsilon_0} & A & \longrightarrow & 0 \\ \downarrow{\scriptstyle \varphi_n} & & \downarrow{\scriptstyle \varphi_{n-1}} & & \downarrow{\scriptstyle \varphi_0} & & \downarrow{\scriptstyle \alpha_0} & & \\ X'_n & \xrightarrow{\partial'_n} & X'_{n-1} & \longrightarrow \cdots\cdots \longrightarrow & X'_0 & \xrightarrow{\varepsilon'_0} & A' & \longrightarrow & 0 \end{array}$$

is commutative. Since (X', ε') is acyclic, ∂'_{n+1} has a factorisation

$$X'_{n+1} \xrightarrow{\lambda'_{n+1}} B'_n \xrightarrow{\kappa'_n} X'_n \qquad [29.3]$$

where $\kappa'_n\lambda'_{n+1} = \partial'_{n+1}$, λ'_{n+1} is epic and κ'_n is a kernel of ∂'_n. Since $\partial'_n\varphi_n\partial_{n+1} = \varphi_{n-1}\partial_n\partial_{n+1} = \zeta$ it follows that there exists a unique morphism β_{n+1} from X_{n+1} to B'_n such that $\varphi_n\partial_{n+1} = \kappa'_n\beta_{n+1}$. Now λ'_{n+1} is epic and X_{n+1} is projective; so there exists a morphism φ_{n+1} from X_{n+1} to X'_{n+1} such that $\lambda'_{n+1}\varphi_{n+1} = \beta_{n+1}$ and hence $\partial'_{n+1}\varphi_{n+1} = (\kappa'_n\lambda'_{n+1})\,\varphi_{n+1} = \kappa'_n(\lambda'_{n+1}\varphi_{n+1}) = \kappa'_n\beta_{n+1} = \varphi_n\partial_{n+1}$. (When $n = 0$ this argument requires a slight modification; namely we must replace ∂'_0 by ε_0.)

It follows by induction that we obtain a sequence $(\varphi_p)_{p\in\mathbf{N}}$ of morphisms of $\boldsymbol{C}$ such that for every natural number p we have $\varphi_p \in \boldsymbol{C}(X_p, X'_p)$ and $\partial'_{p+1}\varphi_{p+1} = \varphi_p\partial_{p+1}$, while $\varepsilon'_0\varphi_0 = \alpha_0\varepsilon_0$. If for every negative integer k we set $\varphi_k =$ the zero morphism from X_k to X'_k we see that $\varphi = (\varphi_p)_{p\in\mathbf{Z}}$ is a chain morphism from X to X' which lifts α_0.

(2) Now let $\boldsymbol{C}$ be an additive category, and suppose we have two chain morphisms φ and ψ from X to X' which lift α_0. We shall construct a chain homotopy Δ between φ and ψ.

Since X and X' are positive complexes it is clear that for all negative integers p we must define Δ_p to be the zero morphism from X_p to X'_{p+1}.

Because X' is acyclic, the sequence

$$X'_1 \xrightarrow{\partial'_1} X'_0 \xrightarrow{\varepsilon'_0} A'$$

is exact; that is to say, ∂'_1 has a factorisation

$$X'_1 \xrightarrow{\lambda'_1} B'_0 \xrightarrow{\kappa'_0} X'_0$$

where λ'_1 is epic and κ'_0 is a kernel of ε'_0. Now $\varepsilon'_0(\psi_0 - \varphi_0) = \alpha_0\varepsilon_0 - \alpha_0\varepsilon_0 = \zeta$; so there exists a unique morphism β_0 from X_0 to

B'_0 such that $\psi_0 - \varphi_0 = \kappa'_0\beta_0$. Since λ'_1 is epic and X_0 is projective, there exists a morphism Δ_0 from X_0 to X'_1 such that $\lambda'_1\Delta_0 = \beta_0$, whence $\Delta_{-1}\partial_0 + \partial'_1\Delta_0\,(= \partial'_1\Delta_0) = \psi_0 - \varphi_0$.

We now proceed by induction. Let n be a positive integer and suppose that for every integer $p < n$ we have constructed a morphism Δ_p from X_p to X'_{p+1} such that $\partial'_{p+1}\Delta_p + \Delta_{p-1}\partial_p = \psi_p - \varphi_p$. Set $\theta_n = \psi_n - \varphi_n - \Delta_{n-1}\partial_n$; then we have

$$\begin{aligned} \partial'_n\theta_n &= \partial'_n\psi_n - \partial'_n\varphi_n - \partial'_n\Delta_{n-1}\partial_n \\ &= \psi_{n-1}\partial_n - \varphi_{n-1}\partial_n - \partial'_n\Delta_{n-1}\partial_n \\ &= (\psi_{n-1} - \varphi_{n-1} - \partial'_n\Delta_{n-1})\,\partial_n \\ &= (\Delta_{n-2}\partial_{n-1})\,\partial_n = \zeta. \end{aligned}$$

Now using again the factorisation [29.3] for ∂'_{n+1}, we see that there exists a unique morphism γ_n from X_n to B'_n such that $\theta_n = \kappa'_n\gamma_n$. Since λ'_{n+1} is epic and X_n is projective there exists a morphism Δ_n from X_n to X'_{n+1} such that $\lambda'_{n+1}\Delta_n = \gamma_n$ and hence $\partial'_{n+1}\Delta_n = \kappa'_n\gamma_n = \theta_n$. Thus $\psi_n - \varphi_n = \partial'_{n+1}\Delta_n + \Delta_{n-1}\partial_n$.

It follows by induction that we obtain a chain homotopy Δ between φ and ψ.

COROLLARY 1. *Let A and A' be objects of a category* **C** *with zero objects, α_0 a morphism from A to A'. If (X, ε) and (X', ε') are projective resolutions of A and A' respectively then there exists a chain morphism from X to X' which lifts α_0. If* **C** *is additive any two such chain morphisms are chain homotopic.*

Proof. This is an immediate consequence of the theorem since X (being a projective resolution) is certainly projective and X' (for the same reason) is acyclic.

COROLLARY 2. *Let (X, ε) and (X', ε') be projective resolutions of an object of an additive category. Then X and X' have the same homotopy type.*

Proof. If we apply the theorem to the case where $A = A'$ and α_0 is the identity morphism I_A of A, we see that there exists a chain morphism φ from X to X' which lifts I_A. Similarly there is a chain morphism φ' from X' to X which lifts I_A. Then $\varphi'\varphi$ is a chain morphism from X to X which lifts I_A; clearly the identity chain morphism of X also lifts I_A. So, according to the theorem, $\varphi'\varphi$ is chain homotopic to the identity chain morphism of X. Similarly $\varphi\varphi'$ is chain homotopic to the identity chain morphism of X'. Thus X and X' have the same homotopy type.

Consider now a short exact sequence of objects of a category **C** with zero objects:

$$0 \to A' \xrightarrow{\alpha_0} A \xrightarrow{\beta_0} A'' \to 0. \qquad [29.4]$$

By an *exact sequence of chain complexes over* [29.4] we mean an exact sequence of chain complexes

$$Z \to X' \xrightarrow{\varphi} X \xrightarrow{\psi} X'' \to Z \qquad [29.5]$$

in which X', X, X'' are chain complexes over A', A, A'' respectively and φ and ψ are chain morphisms which lift α_0 and β_0 respectively. If X', X, X'' are projective resolutions of A', A, A'' respectively, we say that the exact sequence [29.5] is a *simultaneous projective resolution* of [29.4]. To establish the existence of such simultaneous projective resolutions we prove first the following result.

THEOREM 29.3. *Let* [29.4] *be an exact sequence of objects of an additive category* **C** *with finite products and coproducts. If* (X', ε') *is an acyclic chain complex over* A' *and* (X'', ε'') *is a projective chain complex over* A'' *then there exist a chain complex* (X, ε) *over* A *and chain morphisms* φ *and* ψ *from* X' *to* X *and* X *to* X'' *respectively such that* [29.5] *is an exact sequence of chain complexes over* [29.4].

Proof. For every integer n let X_n be an object of **C**, φ_n and ι_n morphisms from X'_n and X''_n respectively to X_n such that $(X_n; \varphi_n, \iota_n)$ is a coproduct of X'_n and X''_n. As we saw in §16, there exist morphisms π_n and ψ_n from X_n to X'_n and X''_n respectively such that $(X_n; \pi_n, \psi_n)$ is a product of X'_n and X''_n. We recall that in this situation $\pi_n\varphi_n = I_{X'_n}$, $\psi_n\iota_n = I_{X''_n}$, $\pi_n\iota_n = \zeta$, $\psi_n\varphi_n = \zeta$ and $\varphi_n\pi_n + \iota_n\psi_n = I_{X_n}$. Then for every integer n the sequence

$$0 \to X'_n \xrightarrow{\varphi_n} X_n \xrightarrow{\psi_n} X''_n \to 0$$

is exact (see Theorem 16.7).

We have now to define a morphism ε_0 from X_0 to A and for each integer n a morphism ∂_n from X_n to X_{n-1} such that (1) $\varepsilon_0\varphi_0 = \alpha_0\varepsilon'_0$ and $\varepsilon''_0\psi_0 = \beta_0\varepsilon_0$; (2) $\partial_n\varphi_n = \varphi_{n-1}\partial'_n$ and $\partial''_n\psi_n = \psi_{n-1}\partial_n$ for every integer n; (3) $\varepsilon_0\partial_1 = \zeta$ and (4) $\partial_{n-1}\partial_n = \zeta$ for every integer n.

In order to get a clue on how to set about constructing the morphism ε_0, let us suppose that we have actually succeeded in constructing a morphism ε_0 from X_0 to A which satisfies condition (1). If we then set $\tau = \varepsilon_0 \iota_0$ we have $\beta_0 \tau = \beta_0 \varepsilon_0 \iota_0 = \varepsilon_0'' \psi_0 \iota_0 = \varepsilon_0''$. Conversely, if there exists a morphism τ from X_0'' to A such that $\beta_0 \tau = \varepsilon_0''$ it is easy to verify that the morphism

$$\varepsilon_0 = \alpha_0 \varepsilon_0' \pi_0 + \tau \psi_0$$

satisfies condition (1). To see that there exists a morphism τ with the required property we need only examine the diagram

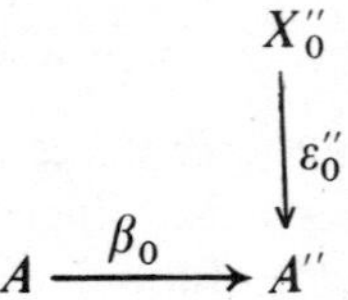

Since β_0 is epic and X_0'' is projective there exists a morphism τ from X_0'' to A such that $\beta_0 \tau = \varepsilon_0''$, as required.

Next we try to construct the morphism ∂_1; so let us suppose we have managed to construct a morphism ∂_1 from X_1 to X_0 satisfying conditions (2) and (3), and see whether we can get any useful hints about how the construction might be made. We have

$$\begin{aligned} \partial_1 = \partial_1(\varphi_1 \pi_1 + \iota_1 \psi_1) &= \partial_1 \varphi_1 \pi_1 + (\varphi_0 \pi_0 + \iota_0 \psi_0)\, \partial_1 \iota_1 \psi_1 \\ &= \varphi_0(\partial_1' \pi_1 + \pi_0 \partial_1 \iota_1 \psi_1) + \iota_0 \psi_0 \partial_1 \iota_1 \psi_1 \\ &= \varphi_0(\partial_1' \pi_1 + \theta_1 \psi_1) + \iota_0 \partial_1'' \psi_1 i_1 \psi_1 \\ &= \varphi_0(\partial_1' \pi_1 + \theta_1 \psi_1) + \iota_0 \partial_1'' \psi_1 . \end{aligned} \qquad [29.6]$$

where we write $\theta_1 = \pi_0 \partial_1 \iota_1$. Since $\varepsilon_0 \partial_1 = \zeta$ and $\varepsilon_0 = \alpha_0 \varepsilon_0' \pi_0 + \tau \psi_0$, we have

$$\begin{aligned} \zeta &= \varepsilon_0 \varphi_0(\partial_1' \pi_1 + \theta_1 \psi_1) + (\alpha_0 \varepsilon_0' \pi_0 + \tau \psi_0)(\iota_0 \partial_1'' \psi_1) \\ &= \alpha_0 \varepsilon_0'(\partial_1' \pi_1 + \theta_1 \psi_1) + \alpha_0 \varepsilon_0' \pi_0 \iota_0 \partial_1'' \psi_1 + \tau \psi_0 \iota_0 \partial_1'' \psi_1 \\ &= \alpha_0 \varepsilon_0' \partial_1' \pi_1 + \alpha_0 \varepsilon_0' \theta_1 \psi_1 + \alpha_0 \varepsilon_0' \pi_0 \iota_0 \partial_1'' \psi_1 + \tau \psi_0 \iota_0 \partial_1'' \psi_1 \\ &= \alpha_0 \varepsilon_0' \theta_1 \psi_1 + \tau \partial_1'' \psi_1, \end{aligned}$$

whence $\alpha_0 \varepsilon_0' \theta_1 + \tau \partial_1'' = \zeta$ since ψ_1 is epic. Conversely, now, it is easy to verify that if θ_1 is a morphism from X_1'' to X_0' such that $\alpha_0 \varepsilon_0' \theta_1 = - \tau \partial_1''$ then the morphism ∂_1 defined by [29.6] satisfies our requirements.

To see that there exists such a morphism θ_1 we remark first that since [29.4] is exact, α_0 has a factorisation

$$A' \xrightarrow{\lambda} I \xrightarrow{\kappa} A$$

such that $\alpha_0 = \kappa\lambda$, λ is epic and κ is a kernel of β_0. Now $\beta_0(-\tau\partial_1'') = -(\beta_0\tau)\partial_1'' = -\varepsilon_0''\partial_1'' = \zeta$. Hence there exists a unique morphism γ_1 from X_1'' to I such that $-\tau\partial_1'' = \kappa\gamma_1$. Since (X', ε') is acyclic, the morphism ε_0' is epic; hence $\lambda\varepsilon_0'$ is epic and so (since X_1'' is projective) there exists a morphism θ_1 from X_1'' to X_1' such that $\lambda\varepsilon_0'\theta_1 = \gamma_1$, whence $\alpha_0\varepsilon_0'\theta_1 = \kappa\gamma_1 = -\tau\partial_1''$, as required.

Suppose next that we have succeeded in constructing a morphism ∂_2 from X_2 to X_1 such that

$$\partial_2\varphi_2 = \varphi_1\partial_2', \quad \partial_2''\psi_2 = \psi_1\partial_2, \quad \partial_1\partial_2 = \zeta. \qquad [29.7]$$

Then, if we set $\theta_2 = \pi_1\partial_2\iota_2$ it turns out that

$$\partial_2 = \varphi_1(\partial_2'\varphi_2 + \theta_2\psi_2) + \iota_1\partial_2''\psi_2 \qquad [29.8]$$

and

$$\partial_1'\theta_2 + \theta_1\partial_2'' = \zeta. \qquad [29.9]$$

Conversely, if there exists a morphism θ_2 from X_2'' to X_1' satisfying [29.9], then the morphism ∂_2 defined by [29.8] satisfies the conditions [29.7].

To show that there does indeed exist such a morphism θ_2, we first use the fact that (X', ε') is acyclic to deduce that ∂_1' has a factorisation

$$X_1' \xrightarrow{\lambda_1} I_0 \xrightarrow{\kappa_0} X_0'$$

such that $\partial_1' = \kappa_0\lambda_1$, λ_1 is epic and κ_0 is a kernel of ε_0'. Now we have $\alpha_0\varepsilon_0'(-\theta_1\partial_2'') = -(\alpha_0\varepsilon_0'\theta_1)\partial_2'' = \tau\partial_1''\partial_2'' = \zeta = \alpha_0\zeta$, whence $\varepsilon_0'(-\theta_1\partial_2'') = \zeta$ (since α_0 is monic). Thus there exists a morphism γ_2 from X_2'' to I_0 such that $-\theta_1\partial_2'' = \kappa_0\gamma_2$. Since λ_1 is epic and X_2'' is projective there exists a morphism θ_2 from X_2'' to X_1' such that $\lambda_1\theta_2 = \gamma_2$, whence $\partial_1'\theta_2 = \kappa_0\gamma_2 = -\theta_1\partial_2''$, as required.

We now proceed by induction. So let n be any positive integer, $n > 2$, and suppose that we have constructed morphisms θ_k from X_k'' to X_{k-1}' $(1 < k < n)$ such that $\partial_{k-1}'\theta_k = -\theta_{k-1}\partial_k''$. If we define the morphisms ∂_k from X_k to X_{k-1} $(k < n)$ by setting $\partial_k = \zeta$ if $k < 0$, ∂_1 as above and

$$\partial_k = \varphi_{k-1}(\partial_k'\pi_k + \theta_k\psi_k) + \iota_{k-1}\partial_k''\psi_k \qquad [29.10]$$

if $1 < k < n$, then we have for all integers $k < n$

$$\partial_{k-1}\partial_k = \zeta, \partial_k\varphi_k = \varphi_{k-1}\partial'_k \text{ and } \partial''_k\psi_k = \psi_{k-1}\partial_k. \qquad [29.11]$$

Since (X', ε') is acyclic the morphism ∂'_{n-1} has a factorisation

$$X'_{n-1} \xrightarrow{\lambda_{n-1}} I_{n-2} \xrightarrow{\kappa_{n-2}} X'_{n-2}$$

such that $\partial'_{n-1} = \kappa_{n-2}\lambda_{n-1}$, λ_{n-1} is epic and κ_{n-2} is a kernel of ∂'_{n-2}. Now we have $\partial'_{n-2}(-\theta_{n-1}\partial''_n) = -(\partial'_{n-2}\theta_{n-1})\,\partial''_n = \zeta$. Hence there exists a unique morphism γ_n from X''_n to I_{n-2} such that $-\theta_{n-1}\partial''_n = \kappa_{n-2}\gamma_n$. Since X''_n is projective and λ_{n-1} is epic there exists a morphism θ_n from X''_n to X'_{n-1} such that $\lambda_{n-1}\theta_n = \gamma_n$ and hence $\partial'_{n-1}\theta_n = \kappa_{n-2}\gamma_n = -\theta_{n-1}\partial''_n$. The morphism ∂_n defined by replacing k by n in [29.10] is easily seen to satisfy the relations [29.11] with k replaced by n.

Thus the induction is complete and the theorem is established.

We now use this result to establish the existence of a simultaneous projective resolution for each short exact sequence of objects in a locally small abelian category with enough projectives.

THEOREM 29.4. *Let* **A** *be a locally small abelian category with enough projectives. Then every short exact sequence of objects of* **A** *has a simultaneous projective resolution.*

Proof. Let

$$0 \to A' \overset{\alpha_0}{\to} A \overset{\beta_0}{\to} A'' \to 0 \qquad [29.12]$$

be a short exact sequence of objects of **A**. According to Theorem 29.1 there exist projective resolutions (X', ε') and (X'', ε'') of A' and A'' respectively. Since (X', ε') is acyclic and (X'', ε'') is projective Theorem 29.3 assures us that there exists a chain complex (X, ε) over A and chain morphisms φ and ψ such that

$$Z \to X' \overset{\varphi}{\to} X \overset{\psi}{\to} X'' \to Z$$

is an exact sequence of chain complexes over [29.12]. Further, (in the notation of Theorem 29.3) for every integer n, $(X_n;\ \varphi_n,\ \iota_n)$ is a coproduct for (X'_n, X''_n). It follows from Theorem 15.2 that the objects X_n are all projective.

Let $Y' = (Y'_p, \tau'_p)$ be the chain complex defined by setting

$$Y'_p = \begin{cases} X'_p & (p \geqslant 0) \\ A' & (p = -1) \\ 0 & (p \leqslant -2) \end{cases} \quad \text{and} \quad \tau'_p = \begin{cases} \partial'_p & (p \geqslant 1) \\ \varepsilon_0 & (p = 0) \\ \zeta & (p \leqslant -1) \end{cases}$$

and let Y and Y'' be similarly defined. Define morphisms $\overline{\varphi}_p$, $\overline{\psi}_p$ from Y'_p to Y_p, Y_p to Y''_p respectively by setting

$$\overline{\varphi}_p = \begin{cases} \varphi_p & (p \geqslant 0) \\ \alpha_0 & (p = -1) \\ \zeta & (p \leqslant -2) \end{cases} \qquad \overline{\psi}_p = \begin{cases} \psi_p & (p \geqslant 0) \\ \beta_0 & (p = -1) \\ \zeta & (p \leqslant -2) \end{cases}$$

Then it is easy to check that $\overline{\varphi} = (\overline{\varphi}_p)$ and $\overline{\psi} = (\overline{\psi}_p)$ are chain morphisms from Y' to Y and Y to Y'' respectively. Since $\boldsymbol{A}$ is a locally small abelian category it follows from Theorem 28.4 that for every integer n we have an exact sequence

$$H_n(Y') \xrightarrow{H_n(\overline{\varphi})} H_n(Y) \xrightarrow{H_n(\overline{\psi})} H_n(Y'')$$

of homology objects. Since (X', ε') and (X'', ε'') are acyclic complexes over A' and A'' respectively, $H_n(Y')$ and $H_n(Y'')$ are zero objects for every integer n. Hence $H_n(Y)$ is a zero object for every integer n and so (X, ε) is an acyclic complex over A.

Thus (X, ε) is a projective resolution as required.

We can now undertake a discussion dual to all we have done so far in this section; we shall only sketch the procedure, leaving it to the reader (if he feels inclined) to fill in the details.

So let $\boldsymbol{C}$ be a category with zero objects, A any object of $\boldsymbol{C}$. Then we may form a cochain complex $C^{\boldsymbol{\cdot}}(A) = (C^p(A), \delta^p(A))$ by setting $C^0(A) = A$ and $C^p(A) = 0$, a zero object, for all non-zero integers p; the coboundary morphisms must then be zero morphisms. A *cochain complex over* A is a pair $(X^{\boldsymbol{\cdot}}, \varepsilon^{\boldsymbol{\cdot}})$ consisting of a positive cochain complex $X^{\boldsymbol{\cdot}} = (X^p, \delta^p)$ and a cochain morphism $\varepsilon^{\boldsymbol{\cdot}} = (\varepsilon^p)$ from the cochain complex $C^{\boldsymbol{\cdot}}(A)$ to $X^{\boldsymbol{\cdot}}$. Clearly $\varepsilon^p = \zeta$ for all non-zero integers p and $\delta^0\varepsilon^0 = \zeta$ also; ε^0 is called the *augmentation* of the cochain complex $(X^{\boldsymbol{\cdot}}, \varepsilon^{\boldsymbol{\cdot}})$ over A. We say that $(X^{\boldsymbol{\cdot}}, \varepsilon^{\boldsymbol{\cdot}})$ is *acyclic* if the sequence

$$0 \to A \xrightarrow{\varepsilon^0} X^0 \xrightarrow{\delta^0} X^1 \to \ldots . \to X^{n-1} \xrightarrow{\delta^{n-1}} X^n \to \ldots .$$

is exact. The cochain complex $(X^{\boldsymbol{\cdot}}, \varepsilon^{\boldsymbol{\cdot}})$ over A is said to be *injective* if all the objects X^p are injective. If $(X^{\boldsymbol{\cdot}}, \varepsilon^{\boldsymbol{\cdot}})$ is both acyclic and injective it is called an *injective coresolution* of A.

A category $\boldsymbol{C}$ is said to *have enough injectives* if for every object A of $\boldsymbol{C}$ there exists an injective object Q and a monic morphism ι from A to Q.

Example 2. Theorem 10.7 shows that the category of left R-modules has enough injectives.

We have the following result, dual to Theorem 29·1.

THEOREM 29.1*. *Let* $\mathbf{C}$ *be an exact category. If* $\mathbf{C}$ *has enough injectives then every object of* $\mathbf{C}$ *has an injective coresolution.*

Again let $\mathbf{C}$ be a category with zero objects, A and A' two objects of $\mathbf{C}$ and α^0 a morphism from A to A'. For all non-zero integers p set $\alpha^p = \zeta \in \mathbf{C}(C^p(A), C^p(A'))$; then $\alpha^\cdot = (\alpha^p)$ is a cochain morphism from $C^\cdot(A)$ to $C^\cdot(A')$. Let $(X^\cdot, \varepsilon^\cdot)$, $(X'^\cdot, \varepsilon'^\cdot)$ be cochain complexes over A and A' respectively. A cochain morphism $\varphi^\cdot$ from $X^\cdot$ to $X'^\cdot$ such that the diagram

$$\begin{array}{ccc} X^\cdot & \xrightarrow{\varphi^\cdot} & X'^\cdot \\ {\scriptstyle \varepsilon^\cdot}\uparrow & & \uparrow{\scriptstyle \varepsilon'^\cdot} \\ C^\cdot(A) & \xrightarrow{\alpha^\cdot} & C^\cdot(A') \end{array}$$

is commutative is called a *cochain morphism over* α^0; we also say that $\varphi^\cdot$ *lifts* α^0.

We have the following dual of Theorem 29.2.

THEOREM 29.2*. *Let* A *and* A' *be objects of a category* $\mathbf{C}$ *with zero objects,* α^0 *a morphism from* A *to* A'. *If* $(X^\cdot, \varepsilon^\cdot)$ *is an acyclic cochain complex over* A *and* $(X'^\cdot, \varepsilon'^\cdot)$ *is an injective cochain complex over* A' *then there exists a cochain morphism from* $X^\cdot$ *to* $X'^\cdot$ *which lifts* α^0. *If* $\mathbf{C}$ *is additive, any two such cochain morphisms are cochain homotopic.*

COROLLARY 1. *Let* A *and* A' *be objects of a category* $\mathbf{C}$ *with zero objects,* α^0 *a morphism from* A *to* A'. *If* $(X^\cdot, \varepsilon^\cdot)$ *and* $(X'^\cdot, \varepsilon'^\cdot)$ *are injective coresolutions of* A *and* A' *respectively then there exists a cochain morphism from* $X^\cdot$ *to* $X'^\cdot$ *which lifts* α^0. *If* $\mathbf{C}$ *is additive any two such cochain morphisms are cochain homotopic.*

COROLLARY 2. *Let* $(X^\cdot, \varepsilon^\cdot)$ *and* $(X'^\cdot, \varepsilon'^\cdot)$ *be injective coresolutions of an object of an additive category. Then* $X^\cdot$ *and* $X'^\cdot$ *have the same homotopy type.*

We may also consider short exact sequences

$$0 \to A' \xrightarrow{\alpha^0} A \xrightarrow{\beta^0} A'' \to 0 \qquad [29.13]$$

of objects in a category with zero objects. An exact sequence of cochain complexes

$$Z \to X'^{\bullet} \xrightarrow{\varphi^{\bullet}} X^{\bullet} \xrightarrow{\psi^{\bullet}} X''^{\bullet} \to Z \qquad [29.14]$$

is said to be *over* [29.13] if $X'^{\bullet}$, $X^{\bullet}$, $X''^{\bullet}$ are cochain complexes over A', A, A'' respectively and $\varphi^{\bullet}$, $\psi^{\bullet}$ are cochain morphisms which lift α^0, β^0 respectively. If $X'^{\bullet}$, $X^{\bullet}$, $X''^{\bullet}$ are injective coresolutions of A', A, A'' respectively we say that [29.14] is a *simultaneous injective coresolution* of [29.13].

THEOREM 29.4*. *Let $\mathbf{A}$ be a locally small abelian category with enough injectives. Then every short exact sequence of objects of $\mathbf{A}$ has a simultaneous injective coresolution.*

§30. Derived Functors

Let $\mathbf{A}_1$ be an abelian category with enough projectives; let T be a covariant additive functor from $\mathbf{A}_1$ to a locally small abelian category $\mathbf{A}_2$. Let A be any object of $\mathbf{A}_1$; according to Theorem 29.1, A has a projective resolution (X, ε) say, where $X = (X_p, \partial_p)$ is a positive chain complex over A. For every integer p we have

$$T(\partial_{p-1})T(\partial_p) = T(\partial_{p-1}\partial_p) = T(\zeta) = \zeta.$$

Hence the sequence $(T(X_p), T(\partial_p))$ of objects and morphisms of $\mathbf{A}_2$ is a chain complex over $\mathbf{A}_2$ (and clearly a positive one) which we denote by $T(X)$. For each integer n we write

$$L_{n,X}T(A) = H_n(T(X)).$$

If (X', ε') is another projective resolution of A, then X and X' have the same homotopy type (Theorem 29.2, Corollary 2). It follows easily that $T(X)$ and $T(X')$ have the same homotopy type and so for every integer n their n-dimensional homology objects are isomorphic (Corollary of Theorem 28.2); that is to say $L_{n,X}T(A)$ and $L_{n,X'}T(A)$ are isomorphic.

Let B be another object of $\mathbf{A}_1$, α a morphism from A to B. If (X, ε) and (Y, η) are projective resolutions of A and B respectively then, according to Corollary 1 of Theorem 29.2, there exists a chain morphism $\varphi = (\varphi_n)$ from X to Y which lifts α. It is easy to check that the sequence $T(\varphi) = (T(\varphi_n))$ is a chain morphism from $T(X)$ to $T(Y)$. Then $T(\varphi)$ induces homology morphisms $H_n(T(\varphi))$

from $H_n(T(X))$ to $H_n(T(Y))$ for every integer n. Let us write

$$L_{n,\varphi}T(\alpha) = H_n(T(\varphi)).$$

If φ' is another chain morphism from X to Y which lifts α, it follows from Corollary 1 of Theorem 29.2 that φ and φ' are chain homotopic. Then $T(\varphi)$ and $T(\varphi')$ are chain homotopic and so, by Theorem 28.2, $H_n(T(\varphi)) = H_n(T(\varphi'))$ for every integer n.

The preceding discussion shows how we may define a sequence $(L_n T)_{n \in \mathbf{Z}}$ of covariant additive functors from $\mathbf{A}_1$ to $\mathbf{A}_2$. Namely, let n be any integer; for each object A of $\mathbf{A}_1$ set

$$L_n T(A) = L_{n,X} T(A)$$

where (X, ε) is any projective resolution of A, and for each morphism α from A to another object B of $\mathbf{A}_1$, set

$$L_n T(\alpha) = L_{n,\varphi}\, T(\alpha)$$

where φ is any chain morphism from the projective resolution used to define $L_n T(A)$ to that used to define $L_n T(B)$. The functors $L_n T$ are called the *left derived functors* of T. The following theorem is an immediate consequence of the definition.

THEOREM 30.1. *Let T be a covariant additive functor from an abelian category $\mathbf{A}_1$ with enough projectives to a locally small abelian category $\mathbf{A}_2$. For every negative integer n the left derived functor $L_n T$ is the zero functor. The functors $L_n T$ are zero for all non-zero integers n if and only if T is an exact functor.*

Let us now consider a short exact sequence

$$0 \to A' \xrightarrow{\alpha_0} A \xrightarrow{\beta_0} A'' \to 0 \qquad [30.1]$$

of objects of $\mathbf{A}_1$. According to Theorem 29.4 this sequence has a simultaneous projective resolution

$$Z \to X' \xrightarrow{\varphi} X \xrightarrow{\psi} X'' \to Z \qquad [30.2]$$

where X', X, X'' are projective resolutions of A', A, A'' respectively and φ, ψ are chain morphisms which lift α_0, β_0 respectively. The exact sequence [30.2] gives rise to a sequence

$$Z \to T(X') \xrightarrow{T(\varphi)} T(X) \xrightarrow{T(\psi)} T(X'') \to Z \qquad [30.3]$$

of complexes of $\mathbf{A}_2$; we claim that this sequence is also exact. To see this let n be any integer and consider the exact sequence

$$0 \to X'_n \xrightarrow{\varphi_n} X_n \xrightarrow{\psi_n} X''_n \to 0.$$

Since X''_n is projective, this sequence is split exact (by Theorem 11.5); hence, according to Theorem 17.1, the sequence

$$0 \to T(X'_n) \xrightarrow{T(\varphi_n)} T(X_n) \xrightarrow{T(\psi_n)} T(X''_n) \to 0$$

is exact. Hence [30.3] is an exact sequence of complexes of $\mathbf{A}_2$. The general discussion of §28 now allows us to define connecting morphisms τ_{*n} from $H_n(T(X''))$ to $H_{n-1}(T(X'))$ (for every integer n) in such a way that the homology sequence

$$\to H_n(T(X')) \xrightarrow{H_n(T(\varphi))} H_n(T(X)) \xrightarrow{H_n(T(\psi))} H_n(T(X'')) \xrightarrow{\tau_{*n}} H_{n-1}(T(X')) \to$$

is exact. We now sum up this discussion in the following theorem.

THEOREM 30.2. *Let $\mathbf{A}_1$ be an abelian category with enough projectives, T a covariant additive functor from $\mathbf{A}_1$ to a locally small abelian category $\mathbf{A}_2$. Then for each exact sequence* [30.1] *of objects of $\mathbf{A}_1$ there exist morphisms τ_{*n} from $L_nT(A'')$ to $L_{n-1}T(A')$ (for all integers n) such that the sequence*

$$\to L_nT(A') \xrightarrow{L_nT(\alpha_0)} L_nT(A) \xrightarrow{L_nT(\beta_0)} L_nT(A'') \xrightarrow{\tau_{*n}} L_{n-1}T(A') \to$$

is exact.

COROLLARY 1. *In the situation of the theorem, the left derived functor L_0T is right exact.*

Proof. Let [30.1] be an exact sequence of objects of $\mathbf{A}_1$. If X' is a projective resolution of A', we have $X'_{-1} = 0$; whence $T(X'_{-1}) = 0$ and so $L_{-1}T(A') = H_{-1}(T(X')) = 0$. Thus the sequence

$$L_0T(A') \to L_0T(A) \to L_0T(A'') \to 0$$

is exact. Hence L_0T is right exact as required.

COROLLARY 2. *If, in the situation of the theorem, $L_{n+1}T$ is the zero functor, then L_nT is left exact.*

Let A be an object of $\mathbf{A}_1$, (X, ε) a projective resolution of A. We recall that ε is a chain morphism from the complex X to the complex $C(A)$ associated with A. Then $T(\varepsilon)$ is a chain morphism from $T(X)$ to $T(C(A)) = C(T(A))$ and hence induces homology object morphisms $T(\varepsilon)_{*n} = \lambda_n(A)$ from $L_nT(A) = H_n(T(X))$ to $H_n[C(T(A))]$. Since $H_n[C(T(A))] = 0$ for all non-zero integers n, the only one of these morphisms which is of interest is $\lambda_0(A)$, which is a morphism from $L_0T(A)$ to $H_0[C(T(A))] = T(A)$. It is easy to check that if α is a morphism from A to another object B of $\mathbf{A}_1$ then the diagram

$$\begin{array}{ccc} L_0T(A) & \xrightarrow{L_0T(\alpha)} & L_0T(B) \\ \downarrow{\scriptstyle \lambda_0(A)} & & \downarrow{\scriptstyle \lambda_0(B)} \\ T(A) & \xrightarrow{T(\alpha)} & T(B) \end{array}$$

is commutative. Thus we have a natural transformation λ_0 from the functor L_0T to the functor T. It is natural to ask for criteria under which λ_0 is an equivalence; our next theorem answers this question.

THEOREM 30.3. *Let $\mathbf{A}_1$ be an abelian category with enough projectives, T a covariant additive functor from $\mathbf{A}_1$ to a locally small abelian category $\mathbf{A}_2$. The natural transformation λ_0 from L_0T to T is an equivalence if and only if T is right exact.*

Proof. (1) Corollary 1 of the preceding theorem shows that L_0T is right exact. It follows at once that if λ_0 is an equivalence then T is right exact.

(2) Conversely suppose T is right exact.

Let A be any object of $\mathbf{A}_1$, (X, ε) a projective resolution of A, so that (in the usual notation) the sequence

$$\ldots\ldots \to X_1 \xrightarrow{\partial_1} X_0 \xrightarrow{\varepsilon_0} A \to 0$$

is exact. It follows that the sequence

$$\ldots\ldots \to T(X_1) \xrightarrow{T(\partial_1)} T(X_0) \xrightarrow{T(\varepsilon_0)} T(A) \to 0$$

is exact. Thus $T(\varepsilon_0)$ is a cokernel of $T(\partial_1)$.

Since $T(\partial_0) = \zeta$, we see that in the diagram used to define the homology object $H_0(T(X))$

$$\begin{array}{ccccc} T(X_1) & \xrightarrow{\theta_1} & C_0(T(X)) & \xrightarrow{\eta_0} & H_0(T(X)) \\ & \searrow^{T(\partial_1)} & \downarrow{\scriptstyle \kappa_0} & & \\ & & T(X_0) & & \\ & & \downarrow{\scriptstyle T(\partial_0)} & & \\ & & T(X_{-1}) & & \end{array}$$

we may take κ_0 to be the identity morphism of $T(X_0)$ and $\theta_1 = T(\partial_1)$. Thus η_0 is a cokernel of $T(\partial_1)$.

But $T(\varepsilon_0)$ is also a cokernel of $T(\partial_1)$; so there exists an isomorphism φ_0 from $H_0(T(X)) = L_0T(A)$ to $T(A)$ such that $\varphi_0\eta_0 = T(\varepsilon_0)$. Since $\lambda_0(A)\eta_0 = T(\varepsilon_0)$ also (by its definition) and η_0 is epic, it follows that $\lambda_0(A) = \varphi_0$. So $\lambda_0(A)$ is an isomorphism, and hence λ_0 is a natural equivalence between L_0T and T.

Example 1. Let $\mathbf{A}_1$ be the category of left R-modules, $\mathbf{A}_2$ the category of abelian groups. These are both abelian categories, $\mathbf{A}_1$ has enough projectives and $\mathbf{A}_2$ is locally small. Let A be a right R-module and consider the functor $\text{Ten}_{A\cdot}$ from $\mathbf{A}_1$ to $\mathbf{A}_2$ defined by setting

$$\text{Ten}_{A\cdot}(B) = A \otimes_R B$$

for each left R-module B and

$$\text{Ten}_{A\cdot}(\beta) = I_A \otimes \beta$$

for each homomorphism β from B to another left R-module B'. As we saw in Examples 1, 8, 11 of §17, $\text{Ten}_{A\cdot}$ is a right exact covariant additive functor from $\mathbf{A}_1$ to $\mathbf{A}_2$. We may apply all the preceding discussion to this functor and form its left derived functors.

If B is any left R-module we shall denote the abelian group $L_n\text{Ten}_{A\cdot}(B)$ by 2-$\text{Tor}_n(A, B)$. We shall very soon drop the prefix 2, which we have introduced temporarily to indicate that the group is obtained by taking a projective resolution $X' = (X'_p, \partial'_p)$ of the second module B and forming the n-dimensional homology group of the complex $A \otimes_R X' = (A \otimes X'_p, I_A \otimes \partial'_p)$.

The general discussion now yields results about the groups 2-$\text{Tor}_n(A, B)$; in the third statement we shall omit the prefix 2 for simplicity.

THEOREM 30.4. *Let A be a right R-module, B a left R-module. Then the group* 2-$\text{Tor}_0(A, B)$ *is isomorphic to the tensor product* $A \otimes_R B$.

Proof. Since $\text{Ten}_{A\cdot}$ is right exact, this result is an immediate consequence of Theorem 30.3.

THEOREM 30.5. *Let A be a right R-module. Then* 2-$\text{Tor}_n(A, B) = 0$ *for every left R-module B and every positive integer n if and only if A is flat.*

Proof. According to Example 11 of §17, $\text{Ten}_{A\cdot}$ is exact if and only if

A is flat. The result now follows from Theorem 30.1.

THEOREM 30.6. *Let A be a right R-module and let*

$$0 \to B' \to B \to B'' \to 0$$

be an exact sequence of left R-modules. Then for every integer n there exists a connecting homomorphism $\tau_n^{A\cdot}$ *from* $\mathrm{Tor}_n(A, B'')$ *to* $\mathrm{Tor}_{n-1}(A, B')$ *such that the sequence*

$$\ldots \to \mathrm{Tor}_n(A, B') \to \mathrm{Tor}_n(A, B) \to \mathrm{Tor}_n(A, B'') \xrightarrow{\tau_n^{A\cdot}} \mathrm{Tor}_{n-1}(A, B') \to \ldots$$

is exact.

Proof. This is simply the translation to the present case of Theorem 30.2.

Example 2. Let $\mathcal{A}_1$ be the category of right R-modules, $\mathcal{A}_2$ the category of abelian groups; these are both abelian categories, $\mathcal{A}_1$ has enough projectives and $\mathcal{A}_2$ is locally small. Let B be a left R-module and consider the functor $\mathrm{Ten}._B$ from $\mathcal{A}_1$ to $\mathcal{A}_2$ defined by setting

$$\mathrm{Ten}._B(A) = A \otimes_R B$$

for every right R-module A and

$$\mathrm{Ten}._B(\alpha) = \alpha \otimes I_B$$

for every homomorphism α from A to another right R-module A'. We saw in Examples 2, 8, 10 of §17 that $\mathrm{Ten}._B$ is a right exact covariant additive functor from $\mathcal{A}_1$ to $\mathcal{A}_2$. So, as in Example 1 above, we may form the left derived functors of $\mathrm{Ten}._B$. For each right R-module A and every integer n we shall denote the group $L_n\mathrm{Ten}._B(A)$ by $1\text{-}\mathrm{Tor}_n(A, B)$. Again the prefix 1 is introduced temporarily to indicate that the group is obtained by taking a projective resolution $X = (X_p, \partial_p)$ of the first module A and computing the n-dimensional homology group of the complex $X \otimes_R B = (X_p \otimes B, \partial_p \otimes I_B)$.

Applying the general discussion we obtain results analogous to the preceding three theorems. In stating the third of these we shall omit the prefix 1.

THEOREM 30.7. *Let A be a right R-module, B a left R-module. Then the group* $1\text{-}\mathrm{Tor}_0(A, B)$ *is isomorphic to the tensor product* $A \otimes_R B$.

THEOREM 30.8. *Let B be a left R-module. Then* $1\text{-}\mathrm{Tor}_n(A, B) = 0$ *for every right R-module A and every positive integer n if and only if B is flat.*

THEOREM 30.9. *Let B be a left R-module and let*

$$0 \to A' \to A \to A'' \to 0$$

be an exact sequence of right R-modules. Then for every integer n there exists a connecting homomorphism $\tau_n^{\cdot B}$ from $\mathrm{Tor}_n(A'', B)$ to $\mathrm{Tor}_{n-1}(A', B)$ such that the sequence

$$\ldots \to \mathrm{Tor}_n(A', B) \to \mathrm{Tor}_n(A, B) \to \mathrm{Tor}_n(A'', B) \xrightarrow{\tau_n^{\cdot B}} \mathrm{Tor}_{n-1}(A', B) \to \ldots$$

is exact.

It follows at once from Theorems 30.4 and 30.7 that for every right R-module A and every left R-module B the groups 1-$\mathrm{Tor}_0(A, B)$ and 2-$\mathrm{Tor}_0(A, B)$ are isomorphic. We now propose to show that in fact 1-$\mathrm{Tor}_n(A, B)$ and 2-$\mathrm{Tor}_n(A, B)$ are isomorphic for every integer n. Once we have established this result we shall allow ourselves to drop the prefixes 1 and 2 and denote the groups simply by $\mathrm{Tor}_n(A, B)$.

We first carry out an auxiliary construction. Let A and A' be a right R-module and a left R-module respectively; let (X, ε) and (X', ε') be projective resolutions for A and A' respectively. For each integer p let Y_p be the external direct sum of all the abelian groups $X_m \otimes_R X'_n$ such that $m + n = p$; let $\iota_{m,n}$ be the canonical injection monomorphism from $X_m \otimes X'_n$ to Y_{m+n}. If m and n are integers such that $m + n = p$, we have a homomorphism γ_{mn} from $X_m \otimes X'_n$ to Y_{p-1} defined by setting

$$\gamma_{mn} = \iota_{m-1,n}(\partial_m \otimes I'_n) + (-1)^m \iota_{m,n-1}(I_m \otimes \partial'_n)$$

where ∂_m, ∂'_n are the boundary homomorphisms of the complexes X and X' respectively, and I_m, I'_n are the identity mappings of X_m, X'_n respectively. Since Y_p is universal for homomorphisms from the family $(X_m \otimes X'_n)_{m+n=p}$ there exists a homomorphism Δ_p from Y_p to Y_{p-1} such that $\Delta_p \iota_{m,n} = \gamma_{mn}$ for all relevant pairs of integers m, n. A routine calculation shows that $\Delta_{p-1}\Delta_p = \zeta$ for every integer p; so $Y = (Y_p, \Delta_p)$ is a complex.

A rather more involved, but not impossibly difficult, argument (using the fact that any two projective resolutions of a module have the same homotopy type) shows that the homology groups of the complex Y depend only on A and A', not on the choice of the resolutions (X, ε) and (X', ε')—in the obvious sense that the homology groups arising from different resolutions are isomorphic. We shall thus denote the n-dimensional homology group of Y by $T_n(A, A')$.

THEOREM 30.10. *For every right R-module A, every left R-module B and every integer n the groups* 1-$\text{Tor}_n(A, B)$ *and* 2-$\text{Tor}_n(A, B)$ *are isomorphic.*

Proof. Our plan of campaign will be to show that for every right R-module A, every left R-module B and every integer n the groups 1-$\text{Tor}_n(A, B)$ and 2-$\text{Tor}_n(A, B)$ are both isomorphic to the group $T_n(A, B)$ which we have just defined.

Let A be any right R-module; let (X, ε) be a projective resolution of A. For each left R-module M let $(X'(M), \varepsilon'(M))$ be a projective resolution of M and let $Y(M)$ be the complex formed according to the procedure described above, using the resolutions (X, ε) and $(X'(M), \varepsilon'(M))$. Then 1-$\text{Tor}_n(A, M) = H_n(X \otimes_R M)$ and $T_n(A, M) = H_n(Y(M))$.

Our first move in proving that 1-$\text{Tor}_n(A, M)$ and $T_n(A, M)$ are isomorphic is to define a homomorphism $\varphi_{*n}(M)$ from $T_n(A, M)$ to 1-$\text{Tor}_n(A, M)$. We obtain this homomorphism from a chain map $\varphi(M)$ from $Y(M)$ to $X \otimes_R M$ which we construct as follows. For each integer p let $\varphi_p(M)$ be the homomorphism from $Y_p(M) = \bigoplus_{m+n=p} X_m \otimes X'_n(M)$ to $X_p \otimes M$ such that

$$\varphi_p(M)i_{p,0} = I_{X_p} \otimes \varepsilon'_0(M) \quad \text{and} \quad \varphi_p(M)i_{m,n} = \zeta \text{ for } (m, n) \neq (p, 0).$$

Routine computation shows that $\varphi_{p-1}(M)\Delta_p(M) = (\partial_p \otimes I_M)\varphi_p(M)$ for every integer p; so $\varphi(M) = (\varphi_p(M))$ is indeed a chain morphism and induces homology group homomorphisms $\varphi_{*n}(M)$.

Since $T_n(A, M) =$ 1-$\text{Tor}_n(A, M) = 0$ for every negative integer n, it follows that for every left R-module M and every negative integer n the homomorphism $\varphi_{*n}(M)$ is an isomorphism.

Next we contend that if P is a projective left R-module then the homomorphism $\varphi_{*n}(P)$ is an isomorphism for every integer n. For if P is projective we may choose a particularly simple projective resolution $(X'(P), \varepsilon'(P))$ for P: we take $X'_0(P) = P$, $X'_p(P) = 0$ for $p \neq 0$ and $\varepsilon_0(P) = I_P$. When we form the complex $Y(P)$ using this resolution we find that $Y_n = X_n \otimes P$ for every integer n and that the homomorphisms $\varphi_n(P)$ are all identity homomorphisms. Thus the induced homomorphisms $\varphi_{*n}(P)$ are all isomorphisms.

We now proceed by induction; let k be any integer for which we have established that for every left R-module A' the homomorphism $\varphi_{*k}(A')$ from $T_k(A, A')$ to 1-$\text{Tor}_k(A, A')$ is an isomorphism.

Let M be any left R-module. Then there exists a projective left

R-module P and a left R-module N such that the sequence

$$0 \to N \xrightarrow{\nu} P \xrightarrow{\pi} M \to 0$$

is exact. This sequence of modules gives rise to a diagram of complexes and chain morphisms

$$\begin{array}{ccccccccc}
Z & \longrightarrow & Y(N) & \longrightarrow & Y(P) & \longrightarrow & Y(M) & \longrightarrow & Z \\
 & & \Big\downarrow \varphi(N) & & \Big\downarrow \varphi(P) & & \Big\downarrow \varphi(M) & & \\
Z & \longrightarrow & X \otimes N & \longrightarrow & X \otimes P & \longrightarrow & X \otimes M & \longrightarrow & Z
\end{array} \qquad [30.4]$$

in which the rows are exact and the squares are commutative. (The chain morphisms in the upper row are obtained by lifting the homomorphisms ν and π to chain morphisms from $X'(N)$ to $X'(P)$ and $X'(P)$ to $X'(M)$ respectively and then extending in the obvious way to chain morphisms from $Y(N)$ to $Y(P)$ and $Y(P)$ to $Y(M)$.)

On passing to homology groups the diagram [30.4] gives rise to a commutative diagram

$$\begin{array}{ccccccc}
T_{k+1}(P) & \longrightarrow & T_{k+1}(M) & \longrightarrow & T_k(N) & \longrightarrow & T_k(P) \\
\Big\downarrow \varphi_{*(k+1)}(P) & & \Big\downarrow \varphi_{*(k+1)}(M) & & \Big\downarrow \varphi_{*k}(N) & & \Big\downarrow \varphi_{*k}(P) \\
T'_{k+1}(P) & \longrightarrow & T'_{k+1}(M) & \longrightarrow & T'_k(N) & \longrightarrow & T'_k(P)
\end{array} \qquad [30.5]$$

whose rows are the homology exact sequences associated with the rows of [30.4]. (To save space, we have suppressed the module A and have written T' as an abbreviation for 1-Tor.) In [30.5] $\varphi_{*k}(P)$ is a monomorphism and $\varphi_{*k}(N)$ and $\varphi_{*(k+1)}(P)$ are epimorphisms (actually all are isomorphisms); hence, by the Five Lemma (§7, Example 9), $\varphi_{*(k+1)}(M)$ is an epimorphism.

Let B be any left R-module. By what we have just proved, $\varphi_{*(k+1)}(B)$ is an epimorphism. Now there exists a projective left R-module P_1 and a left R-module M such that

$$0 \to M \to P_1 \to B \to 0$$

is exact. By an argument similar to that just used, we obtain a commutative diagram

$$\begin{array}{ccccccc}
T_{k+1}(M) & \longrightarrow & T_{k+1}(P_1) & \longrightarrow & T_{k+1}(B) & \longrightarrow & T_k(M) \\
\Big\downarrow \varphi_{*(k+1)}(M) & & \Big\downarrow \varphi_{*(k+1)}(P_1) & & \Big\downarrow \varphi_{*(k+1)}(B) & & \Big\downarrow \varphi_{*k}(M) \\
T'_{k+1}(M) & \longrightarrow & T'_{k+1}(P_1) & \longrightarrow & T'_{k+1}(B) & \longrightarrow & T'_k(M)
\end{array}$$

As we have just shown, $\varphi_{*(k+1)}(M)$ is an epimorphism; $\varphi_{*(k+1)}(P_1)$ and $\varphi_{*k}(M)$ are monomorphisms. Thus, by the second part of the Five Lemma, $\varphi_{*(k+1)}(B)$ is a monomorphism and hence an isomorphism.

This completes the induction and shows that for every natural number n, every right R-module A and every left R-module B the mapping $\varphi_{*n}(B)$ is an isomorphism from $T_n(A, B)$ onto 1-$\mathrm{Tor}_n(A, B)$. By a similar argument we can set up isomorphisms of the groups $T_n(A, B)$ onto 2-$\mathrm{Tor}_n(A, B)$. Hence the theorem is established.

Now let A_1 be an abelian category with enough injectives, and let T be a covariant additive functor from A_1 to a locally small abelian category A_2. Let A be any object of A_1; according to Theorem 29.1*, A has an injective coresolution $(X^{\cdot}, \varepsilon^{\cdot})$ say, where $X^{\cdot} = (X^p, \delta^p)$ is a positive cochain complex over A_1. For every integer p we have

$$T(\delta^{p+1})\, T(\delta^p) = T(\delta^{p+1}\delta^p) = T(\zeta) = \zeta.$$

Hence the sequence $(T(X^p), T(\delta^p))$ of objects and morphisms of A_2 is a positive cochain complex over A_2 which we shall denote by $T(X^{\cdot})$. By familiar arguments it follows that the cohomology objects of $T(X^{\cdot})$ depend only on A and are independent of the choice of the coresolution $(X^{\cdot}, \varepsilon^{\cdot})$, in the usual sense that the objects obtained by taking different coresolutions of A are all isomorphic.

Let B be another object of A_1, and let α be a morphism from A to B. Let $(X^{\cdot}, \varepsilon^{\cdot})$ and $(Y^{\cdot}, \eta^{\cdot})$ be injective coresolutions of A and B respectively. Then, according to Corollary 1 of Theorem 29.2*, there exists a cochain morphism $\varphi^{\cdot}$ from $X^{\cdot}$ to $Y^{\cdot}$ which lifts α and gives rise eventually to cohomology morphisms $H^n(T(\varphi^{\cdot}))$ from $H^n(T(X))$ to $H^n(T(Y))$. The usual arguments show that these morphisms are independent of the choice of the cochain morphism $\varphi^{\cdot}$ over α.

We now define a sequence $(R^nT)_{n\in\mathbf{Z}}$ of covariant additive functors from $\mathbf{A}_1$ to $\mathbf{A}_2$ as follows. Let n be any integer; then for each object A of $\mathbf{A}_1$ set

$$R^nT(A) = H^n(T(X^\bullet))$$

where $(X^\bullet, \varepsilon^\bullet)$ is any injective coresolution of A, and for every morphism α from A to another object B of $\mathbf{A}_1$ set

$$R^nT(\alpha) = H^n(T(\varphi^\bullet))$$

where $\varphi^\bullet$ is any cochain morphism from the injective coresolution used to define $R^nT(A)$ to that used to define $R^nT(B)$ such that $\varphi^\bullet$ lifts α. The functors R^nT are called the *right derived functors* of T. By arguments dual to those for the left derived functors we deduce the following results concerning the right derived functors.

THEOREM 30.11. *Let T be a covariant additive functor from an abelian category $\mathbf{A}_1$ with enough injectives to a locally small abelian category $\mathbf{A}_2$. For every negative integer n the right derived functor R^nT is the zero functor. The functors R^nT are zero for all non-zero integers n if and only if T is an exact functor.*

THEOREM 30.12. *In the situation just described, if*

$$0 \to A' \overset{\alpha^0}{\to} A \overset{\beta^0}{\to} A'' \to 0$$

is a short exact sequence of objects of $\mathbf{A}_1$ there exist morphisms τ^{*n} from $R^nT(A'')$ to $R^{n+1}T(A')$ (for all integers n) such that the sequence

$$\to R^nT(A') \xrightarrow{R^nT(\alpha^0)} R^nT(A) \xrightarrow{R^nT(\beta^0)} R^nT(A'') \xrightarrow{\alpha^{*n}} R^{n+1}T(A') \to$$

is exact.

COROLLARY 1. *In the situation of Theorem 30.11, R^0T is left exact.*

COROLLARY 2. *In the situation of Theorem 30.11, if $R^{n+1}T$ is the zero functor then the functor R^nT is right exact.*

Let A be an object of $\mathbf{A}_1$, $(X^\bullet, \varepsilon^\bullet)$ an injective coresolution of A and $C^\bullet(A)$ the cochain complex associated with A. Then $T(\varepsilon^\bullet)$ induces a morphism $\rho^0(A)$ from $R^0T(A)$ to $T(A)$; we are thus able to define a natural transformation ρ^0 from the functor R^0T to the functor T and to prove the following analogue of Theorem 30.3.

THEOREM 30.13. *Let A_1 be an abelian category with enough injectives, T a covariant additive functor from A_1 to a locally small abelian category A_2. The natural transformation ρ^0 from R^0T to T is an equivalence if and only if T is left exact.*

Example 3. Let A_1 be the category of left R-modules, A_2 the category of abelian groups; both these categories are abelian, A_1 has enough injectives and A_2 is locally small. Let A be a left R-module and consider the functor $\text{Hom}_{A\cdot}$ from A_1 to A_2 defined by setting

$$\text{Hom}_{A\cdot}(B) = \text{Hom}(A, B)$$

for each left R-module B and

$$\text{Hom}_{A\cdot}(\beta) = \text{Hom}(I_A, \beta)$$

for each homomorphism β from B to another left R-module B'. According to Examples 3, 9, 12 of §17, $\text{Hom}_{A\cdot}$ is a left exact covariant additive functor from A_1 to A_2. Thus the preceding discussion may be applied to the functor $\text{Hom}_{A\cdot}$, yielding its right derived functors.

If B is any left R-module, we shall denote the abelian group $R^n\text{Hom}_{A\cdot}(B)$ by 2-$\text{Ext}^n(A, B)$. The prefix 2 is introduced temporarily to show that the group is obtained by taking an injective coresolution $X^\cdot = (X^p, \delta^p)$ of the second module B and forming the n-dimensional cohomology group of the complex $\text{Hom}(A, X^\cdot) = (\text{Hom}(A, X^p), \text{Hom}(I_A, \delta^p))$. The general discussion, applied to this example, gives us the following results about the groups 2-$\text{Ext}^n(A, B)$; in the third statement we omit the prefix 2 in order to save space.

THEOREM 30.14. *Let A and B be left R-modules. Then the group* 2-$\text{Ext}^0(A, B)$ *is isomorphic to the homomorphism group* $\text{Hom}(A, B)$.

Proof. Since $\text{Hom}_{A\cdot}$ is left exact, this result follows at once from Theorem 30.13.

THEOREM 30.15. *Let A be a left R-module. Then* 2-$\text{Ext}^n(A, B) = 0$ *for every left R-module B and every positive integer n if and only if A is projective.*

Proof. By Example 12 of §17, $\text{Hom}_{A\cdot}$ is exact if and only if A is projective. So the result is a special case of Theorem 30.11.

THEOREM 30.16. *Let A be a left R-module and let*

$$0 \to B' \to B \to B'' \to 0$$

be an exact sequence of left R-modules. Then for every integer n there exists a connecting homomorphism ξ_A^n. from $\mathrm{Ext}^n(A, B'')$ *to* $\mathrm{Ext}^{n+1}(A, B')$ *such that the sequence*

$$\ldots \to \mathrm{Ext}^n(A, B') \to \mathrm{Ext}^n(A, B) \to \mathrm{Ext}^n(A, B'') \xrightarrow{\xi_A^n} \mathrm{Ext}^{n+1}(A, B') \to \ldots$$

is exact.

So far in this section we have been dealing exclusively with covariant functors. We have now to consider briefly the case of contravariant functors—only briefly, because contravariant functors from a category $\boldsymbol{C}$ to another category $\boldsymbol{C}'$ are of course (according to their definition) covariant functors from the dual category $\boldsymbol{C}^*$ to $\boldsymbol{C}'$. Thus in dealing with contravariant functors, all the preceding theory is available for use.

Let $\boldsymbol{A}_1$ be an abelian category with enough injectives and projectives; then the dual category $\boldsymbol{A}_1^*$ is also an abelian category with enough injectives and projectives. (We recall that the objects of $\boldsymbol{A}_1^*$ are the same as the objects of $\boldsymbol{A}_1$, but since $\boldsymbol{A}_1^*(A, B) = \boldsymbol{A}_1(B, A)$ the injective objects of $\boldsymbol{A}_1$ are projective in $\boldsymbol{A}_1^*$ and *vice versa.*) Let T be a contravariant functor from $\boldsymbol{A}_1$ to a locally small abelian category $\boldsymbol{A}_2$. Then we define the left and right derived functors of T to be the left and right derived functors of T (considered as covariant functor from $\boldsymbol{A}_1^*$ to $\boldsymbol{A}_2$) as defined above.

We describe briefly how these are formed; let A be any object of $\boldsymbol{A}_1$; take a projective resolution X of A in $\boldsymbol{A}_1^*$—which is the same thing as an injective coresolution of A in $\boldsymbol{A}_1$; then $T(X)$ is a chain complex over $\boldsymbol{A}_2$ and the left derived functors of T are defined by setting $L_nT(A) = H_n(T(X))$ and making an appropriate definition of $L_nT(\alpha)$ for each morphism α. Similarly, to form the right derived functors R^nT of T we take an injective coresolution of A in $\boldsymbol{A}_1^*$, i.e. a projective resolution of A in $\boldsymbol{A}_1$ and form the cohomology objects of the corresponding cochain complex over $\boldsymbol{A}_2$.

We leave to the reader the task of formulating theorems about the derived functors of contravariant functors analogous to those which we established earlier for the derived functors of covariant functors.

Example 4. Let $\boldsymbol{A}_1$ be the category of left R-modules, $\boldsymbol{A}_2$ the category of abelian groups. Both these categories are abelian, $\boldsymbol{A}_1$ has enough projectives and $\boldsymbol{A}_2$ is locally small. Let B be a left R-module and consider the functor $\mathrm{Hom}_{\cdot B}$ from $\mathbf{A}_1$ to $\mathbf{A}_2$ defined by setting

$$\mathrm{Hom}_{\cdot B}(A) = \mathrm{Hom}(A, B)$$

for each left R-module A and

$$\mathrm{Hom}_{\cdot B}(\alpha) = \mathrm{Hom}(\alpha, I_B)$$

for each homomorphism α from A to another left R-module A'. We have already seen (in Examples 4, 9, 13 of §17) that $\mathrm{Hom}_{\cdot B}$ is a left exact contravariant additive functor from $\mathbf{A}_1$ to $\mathbf{A}_2$. We denote the right derived functors of $\mathrm{Hom}_{\cdot B}(A)$ by $1\text{-}\mathrm{Ext}^n(A, B)$. Once again the (temporary) prefix 1 indicates the method of constructing the groups: namely, we take a projective resolution $X = (X_p, \partial_p)$ of the first module A (X is an injective coresolution of A in $\mathbf{A}_1^*$) and form the cohomology groups of the cochain complex $\mathrm{Hom}(X, B) = (\mathrm{Hom}(X_p, B), \mathrm{Hom}(\partial_{p+1}, I_B))$. We obtain results for the groups $1\text{-}\mathrm{Ext}^n(A, B)$ corresponding to those for $2\text{-}\mathrm{Ext}^n(A, B)$.

THEOREM 30.17. *Let A and B be left R-modules. Then* $1\text{-}\mathrm{Ext}^0(A, B)$ *is isomorphic to* $\mathrm{Hom}(A, B)$.

THEOREM 30.18. *Let B be a left R-module. Then* $1\text{-}\mathrm{Ext}^n(A, B) = 0$ *for every left R-module A and every positive integer n if and only if B is injective.*

In stating the third result we omit the prefix 1.

THEOREM 30.19. *Let B be a left R-module and let*

$$0 \to A' \to A \to A'' \to 0$$

be an exact sequence of left R-modules. Then for every integer n there exists a connecting homomorphism $\xi^n_{\cdot B}$ *from* $\mathrm{Ext}^n(A', B)$ *to* $\mathrm{Ext}^{n+1}(A'', B)$ *such that the sequence*

$$\ldots \to \mathrm{Ext}^n(A'', B) \to \mathrm{Ext}^n(A, B) \to \mathrm{Ext}^n(A', B) \xrightarrow{\xi^n_{\cdot B}} \mathrm{Ext}^{n+1}(A'', B) \to \ldots$$

is exact.

Just as we proved that $1\text{-}\mathrm{Tor}_n(A, B)$ and $2\text{-}\mathrm{Tor}_n(A, B)$ are isomorphic for every integer n and every pair of modules A, B, so we would like to show that $1\text{-}\mathrm{Ext}^n(A, B)$ and $2\text{-}\mathrm{Ext}^n(A, B)$ are isomorphic—again for every integer n and every pair of modules A, B. To do this we begin with an auxiliary construction. Let A and B be left R-modules; let (X, ε) be a projective resolution for A and $(X^{\cdot}, \varepsilon^{\cdot})$ an injective coresolution for B. For each integer p let Y^p be the external direct sum of the family of abelian groups $\mathrm{Hom}(X_m, X^n)$ such that $m + n = p$; let ι_m^n be the canonical injection monomorphism from $\mathrm{Hom}(X_m, X^n)$ to Y^p. If m and n are integers such that $m + n = p$ there is a homomorphism γ_m^n from $\mathrm{Hom}(X_m, X^n)$ to Y^{p+1} defined by setting

$$\gamma_m^n = \iota_{m+1}^n \mathrm{Hom}(\partial_{m+1}, I^n) + (-1)^m \iota_m^{n+1} \mathrm{Hom}(I_m, \delta^n)$$

where ∂_{m+1}, δ^n are the boundary and coboundary homomorphisms of the complexes X and $X^{\bullet}$ respectively and I^n, I_m are the identity mappings of X^n, X_m respectively. Since Y^p is universal for homomorphisms from the family $(\mathrm{Hom}(X_m, X^n))_{m+n=p}$ there exists a homomorphism Δ^p from Y^p to Y^{p+1} such that $\Delta^p\iota_m^n = \gamma_m^n$ for all relevant pairs of integers m, n. It turns out that $Y = (Y^p, \Delta^p)$ is a cochain complex whose cohomology groups depend only on A and B, i.e. are independent of the choice of resolutions. We denote the nth cohomology group of Y by $E^n(A, B)$.

Adapting the method of Theorem 30.10 we show that 1-$\mathrm{Ext}^n(A, B)$ and 2-$\mathrm{Ext}^n(A, B)$ are both isomorphic to $E^n(A, B)$; we omit the details, which the reader may like to supply for himself, and content ourselves with stating the final result.

THEOREM 30.20. *For all left R-modules A, B and every integer n the groups* 1-$\mathrm{Ext}^n(A, B)$ *and* 2-$\mathrm{Ext}^n(A, B)$ *are isomorphic.*

We now allow ourselves to drop the prefixes 1 and 2 and to denote the groups simply by $\mathrm{Ext}^n(A, B)$.

§31. Cohomology of Algebras

Let A be an algebra over a field K. By an *A-bimodule* we shall mean an abelian group M equipped with left and right scalar multiplications by elements of A in such a way that M is both a left A-module and a right A-module and such that, in addition, we have $(a_1x)a_2 = a_1(xa_2)$ for all elements a_1, a_2 of A and x of M.

To deal with A-bimodules, we find it convenient to introduce the *enveloping algebra* A^e of A, defined by setting $A^e = A \otimes_K A^{\mathrm{op}}$. We shall now show how to give each A-bimodule the structure of a left A^e-module. So let M be an A-bimodule. Then for each element x of M we may define a mapping φ_x from $A \times A^{\mathrm{op}}$ to M by setting $\varphi_x(a, a') = (ax)a'$ for all elements a of A and a' of A^{op} (we recall that the elements of A^{op} are the same as those of A). It is easy to check that this is a bilinear mapping. So there exists a K-homomorphism α_x from $A^e = A \otimes_K A^{op}$ to M such that $\alpha_x(a \otimes a') = (ax)a'$ for all elements a of A and a' of A^{op}. We may now define a left scalar multiplication of M by elements of A^e: for each element x of M and each element t of A^e we set $t.x = \alpha_x(t)$. With this scalar multiplication M becomes a left A^e-module; the only condition which is not immediately obvious is that $(t_1t_2)x = t_1(t_2x)$ for all elements t_1, t_2 of A^e and all elements x of M. To establish this it is clearly sufficient to consider the case where t_1 and t_2 have the simple forms $a_1 \otimes a_1'$ and $a_2 \otimes a_2'$ respectively. But since $(a_1 \otimes a_1')(a_2 \otimes a_2') = a_1a_2 \otimes (a_1' * a_2') = a_1a_2 \otimes a_2'a_1'$ (where $*$ denotes the multiplication in A^{op}) we have

$$(a_1 \otimes a_1')[(a_2 \otimes a_2')x] = (a_1 \otimes a_1')(a_2 x a_2') = a_1(a_2 x a_2')a_1'$$

and

$$(a_1a_2 \otimes a_2'a_1')x = (a_1a_2)x(a_2'a_1') = a_1(a_2 x a_2')a_1'.$$

Thus M is a left A^e-module as required.

Let now M be an A-bimodule. For each natural number n we define the n-th *cohomology group* of A *with coefficients in* M to be the group

$$H^n(A, M) = \mathrm{Ext}^n_{A^e}(A, M).$$

According to Example 4 of §30, we may compute these groups by taking any projective resolution (X, ε) of A and forming the cohomology groups of the cochain complex $\mathrm{Hom}_{A^e}(X, M)$.

According to Theorem 30.17, $H^0(A, M) = \mathrm{Ext}^0_{A^e}(A, M)$ is isomorphic to the group $\mathrm{Hom}_{A^e}(A, M)$. This remark allows us to give another description of $H^0(A, M)$ in terms of the so-called *invariant elements* of M; these are the elements x of M such that $ax = xa$ for all elements a of A. The invariant elements of M clearly form a subgroup M_0 of M.

THEOREM 31.1. *Let A be an algebra over a field K and let M be an A-bimodule. Then $H^0(A, M)$ is isomorphic to the subgroup M_0 of invariant elements of M.*

Proof. First let α be any A^e-homomorphism from A to M and set $\alpha(e) = x_0$. Then for each element a of A we have

$$ax_0 = (a \otimes e)x_0 = (a \otimes e)\alpha(e) = \alpha((a \otimes e)e) = \alpha(aee) = \alpha(a)$$

and also

$$x_0 a = (e \otimes a)x_0 = (e \otimes a)\alpha(e) = \alpha((e \otimes a)e) = \alpha(eea) = \alpha(a).$$

Thus we may define a mapping μ from $\mathrm{Hom}_{A^e}(A, M)$ to M_0 by setting $\mu(\alpha) = \alpha(e)$ for each A^e-homomorphism α from A to M; clearly μ is a monomorphism. In fact it is also an epimorphism; for if x_0 is any element of M_0 the mapping α_0 from A to M defined by setting $\alpha_0(a) = ax_0$ for all elements a of A is obviously an A^e-homomorphism such that $\mu(\alpha_0) = x_0$.

Hence $H^0(A, M)$, which is isomorphic to $\mathrm{Hom}_{A^e}(A, M)$, is isomorphic also to M_0.

In order to describe the 1-dimensional cohomology group $H^1(A, M)$ we introduce the notion of a *crossed homomorphism.* Namely, let A be any algebra over a field K, M any A-bimodule; then a crossed homomorphism from A to M is a K-homomorphism χ from A to M such that

$$\chi(ab) = a\chi(b) + \chi(a)b$$

for all elements a, b of A. The crossed homomorphisms from A to M form a subgroup of $\mathrm{Hom}_K(A, M)$. If x is an element of M then the mapping χ_x from A to M given by setting

$$\chi_x(a) = ax - xa$$

is easily seen to be a crossed homomorphism. The mappings of this type are called *principal crossed homomorphisms*; they form a subgroup of the group of all crossed homomorphisms. It is clear that if χ is any crossed homomorphism we have

$$\chi(e) = \chi(e^2) = e\chi(e) + \chi(e)e = \chi(e) + \chi(e)$$

and hence $\chi(e) = 0$.

Consider now the mapping φ from $A \times A^{\mathrm{op}}$ to A defined by setting $\varphi(a, a') = aa'$ for all elements a of A, a' of A^{op}; since φ is bilinear there is a K-homomorphism ε from $A^e = A \otimes_K A^{\mathrm{op}}$ to A such that $\varepsilon(a \otimes a') = aa'$ for all elements a of A, a' of A^{op}, and in fact it is an easy matter to show that ε is actually a left A^e-module homomorphism. Let J be the kernel of ε, so that J is an A^e-submodule, i.e. a left ideal of A^e.

From the short exact sequence

$$0 \to J \to A^e \xrightarrow{\varepsilon} A \to 0$$

Theorem 30.19 yields an exact sequence

$$\ldots \to \mathrm{Ext}^n(A^e, M) \to \mathrm{Ext}^n(J, M) \to \mathrm{Ext}^{n+1}(A, M) \to \mathrm{Ext}^{n+1}(A^e, M) \to \ldots \qquad [31.1]$$

Since A^e is a projective left A^e-module, it follows from Theorem 30.15 that for all positive integers n we have $\mathrm{Ext}^n(A^e, M) = 0$. In particular, the sequence

$$\mathrm{Ext}^0(A^e, M) \to \mathrm{Ext}^0(J, M) \to \mathrm{Ext}^1(A, M) \to 0$$

is exact. Using Theorem 30.17 again we deduce that

$$\mathrm{Hom}_{A^e}(A^e, M) \to \mathrm{Hom}_{A^e}(J, M) \to H^1(A, M) \to 0$$

is exact; finally, since $\mathrm{Hom}_{A^e}(A^e, M)$ is isomorphic to M itself (by Theorem 8.1), the sequence

$$M \xrightarrow{\beta} \mathrm{Hom}_{A^e}(J, M) \to H^1(A, M) \to 0$$

is exact. From this it follows that $H^1(A, M)$ is isomorphic to the factor group $\mathrm{Hom}_{A^e}(J, M)/\beta(M)$. We now transform this result into another shape.

THEOREM 31.2. *Let A be an algebra over a field* K *and let M be an A-bimodule. Then $H^1(A, M)$ is isomorphic to the factor group of the group of crossed homomorphisms from A to M modulo the subgroup of principal crossed homomorphisms.*

Proof. Let X be the group of crossed homomorphisms from A to M, X_0 the subgroup of principal crossed homomorphisms.

We define a mapping φ from $\mathrm{Hom}_{A^e}(J, M)$ to X as follows: for each A^e-homomorphism α from J to M let $\varphi(\alpha)$ be the mapping from A to M given by setting

$$[\varphi(\alpha)]\,(a) = \alpha(a \otimes e - e \otimes a)$$

for each element a of A. (Clearly $a \otimes e - e \otimes a$ belongs to J since $\varepsilon(a \otimes e) = a = \varepsilon(e \otimes a)$. It is plain that $\varphi(\alpha)$ is a K-homomorphism from A to M; to show that it is a crossed homomorphism, we let a and b any two elements of A and compute

$$\begin{aligned}([\varphi(\alpha)](a))b &= (e \otimes b)\,\alpha(a \otimes e - e \otimes a) = \alpha((e \otimes b)\,(a \otimes e - e \otimes a)) \\ &= \alpha(ea \otimes (b * e) - ee \otimes (b * a)) \\ &= \alpha(a \otimes b - e \otimes ab);\end{aligned}$$

similarly

$$a([\varphi(\alpha)](b)) = \alpha(ab \otimes e - a \otimes b).$$

It follows that

$$a([\varphi(\alpha)](b)) + ([\varphi(\alpha)]\,(a))b = \alpha(ab \otimes e - e \otimes ab) = [\varphi(\alpha)]\,(ab).$$

We claim that φ is actually an isomorphism from $\mathrm{Hom}_{A^e}(J, M)$ onto X. It is clearly a homomorphism. To show that φ is a monomorphism we remark first that the left ideal J of A^e is generated by the set of elements of the form $a \otimes e - e \otimes a$: for if $t = \sum a_i \otimes a_i'$ is an element of J we have $\varepsilon(t) = \sum a_i a_i' = 0$ and hence

$$t = \sum(a_i \otimes e)(e \otimes a_i' - a_i' \otimes e).$$

This result clearly implies that if $\varphi(\alpha) = \varphi(\beta)$ then $\alpha = \beta$.

Now let χ be any crossed homomorphism from A to M. Let α_χ be the mapping from J to M defined by setting

$$\alpha_\chi(t) = -\sum a_i \chi(a_i')$$

for every element $t = \sum(a_i \otimes a_i')$ of J. Then α_χ is an A^e-homomorphism; to see this it is sufficient to compute

$$(b \otimes b')\alpha_\chi(t) = -\sum b a_i \chi(a_i')b'$$

and

$$\begin{aligned} \alpha_\chi((b \otimes b')t) &= \alpha_\chi(\textstyle\sum b a_i \otimes a_i' b') \\ &= -\textstyle\sum b a_i \chi(a_i' b') \\ &= -\textstyle\sum b a_i a_i' \chi(b') - \sum b a_i \chi(a_i')b' \\ &= -\textstyle\sum b a_i \chi(a_i')b' \end{aligned}$$

(since $\sum a_i a_i' = 0$ because t is in J). Now it is easy to see that $\varphi(\alpha_\chi) = \chi$; for if a is any element of A we have

$$[\varphi(\alpha_\chi)]\,(a) = \alpha_\chi(a \otimes e - e \otimes a) = -a\chi(e) + e\chi(a) = \chi(a)$$

since $\chi(e) = 0$.

Thus φ is an isomorphism from $\mathrm{Hom}_{A^e}(J, M)$ onto X.

To complete the proof we show that $\varphi\beta(M) = X_0$. So let x be any element of M; then $\beta(x)$ is the A^e-homomorphism from J to M defined by setting $[\beta(x)](t) = tx$ for all elements t of J. Thus for every element a of A we have

$$\begin{aligned}[\varphi\beta(x)]\,(a) &= [\beta(x)]\,(a \otimes e - e \otimes a)\\ &= (a \otimes e)x - (e \otimes a)x\\ &= ax - xa = \chi_x(a)\end{aligned}$$

So it follows directly that $\varphi\beta(M) = X_0$.

This completes the proof.

We are interested in finding a description of the 2-dimensional cohomology group of an algebra A with coefficients in a bimodule M. To do this we shall give first a general method for computing the cohomology groups $H^n(A, M)$ for all natural numbers n by constructing a special A^e-projective resolution $(S(A), \varepsilon)$ of A.

We define the modules $S_k(A)$ of this resolution $S(A)$ inductively by setting $S_0(A) = A \otimes_K A$ and then, for all positive integers k, $S_k(A) = A \otimes_K S_{k-1}(A)$. By a familiar procedure we can define a scalar multiplication on the left of $S_k(A)$ by elements of A^e in such a way that

$$(a \otimes a')\,(a_0 \otimes a_1 \otimes \ldots \otimes a_k \otimes a_{k+1}) = aa_0 \otimes a_1 \otimes \ldots \otimes a_k \otimes a_{k+1}a'$$

for all elements $a, a_0, a_1, \ldots, a_{k+1}$ of A and a' of A^{op}; with this scalar multiplication $S_k(A)$ becomes a left A^e-module. We show also that there exist A^e-homomorphisms ε_0 from $S_0(A)$ to A and ∂_k from $S_k(A)$ to $S_{k-1}(A)$ $(k = 1, 2, \ldots)$ such that

$$\varepsilon_0(a_0 \otimes a_1) = a_0a_1$$

and

$$\partial_k(a_0 \otimes \ldots \otimes a_{k+1}) = \sum_{i=0}^{k} (-1)^i\, a_0 \otimes \ldots \otimes a_ia_{i+1} \otimes \ldots \otimes a_{k+1}.$$

Consider now the sequence

$$\ldots \to S_k(A) \xrightarrow{\partial_k} S_{k-1}(A) \to \ldots \to S_1(A) \xrightarrow{\partial_1} S_0(A) \xrightarrow{\varepsilon_0} A \to 0. \qquad [31.2]$$

If we can prove that this sequence is exact it will follow that $S(A) = (S_k(A), \partial_k)$ is a positive complex and also that $(S(A), \varepsilon)$ is an acyclic complex over A. It is clear that ε_0 is an epimorphism, and direct computation shows that $\varepsilon_0\partial_1 = \zeta$. We now define mappings σ from A to $S_0(A)$ and σ_k from $S_k(A)$ to $S_{k+1}(A)$ (for all natural numbers k) by setting

$$\sigma(a) = e \otimes a$$

and

$$\sigma_k(x_k) = e \otimes x_k$$

for all elements a of A and x_k of $S_k(A)$. These mappings σ, σ_k are clearly right A-module homomorphisms, and the left A-modules generated by Im σ, Im σ_k are $S_0(A)$, $S_{k+1}(A)$ respectively. We now discover by another simple computation that

$$(\partial_1\sigma_0 + \sigma\varepsilon_0)(x_0) = x_0 \qquad [31.3]$$

for every element x_0 of $S_0(A)$ and that for all positive integers k

$$(\partial_{k+1}\sigma_k + \sigma_{k-1}\partial_k)(x_k) = x_k \qquad [31.4]$$

for every element x_k of $S_k(A)$.

The relations [31.3] and [31.4] show that Ker ε_0 is included in Im ∂_1 and that for all positive integers k Ker ∂_k is included in Im ∂_{k+1}. We have already remarked that $\varepsilon_0\partial_1 = \zeta$; so we have exactness at $S_0(A)$. Next let x_1 be any element of $S_1(A)$; then we have

$$\begin{aligned}\partial_1\partial_2\sigma_1(x_1) &= \partial_1(x_1) - \partial_1\sigma_0\partial_1(x_1)\\ &= \partial_1(x_1) - \partial_1(x_1) + \sigma\varepsilon_0\partial_1(x_1) = 0\end{aligned}$$

It follows that $\partial_1\partial_2 = \zeta$ (since Im σ_1 generates $S_2(A)$ and $\partial_1\partial_2$ is an A^e-homomorphism). A simple inductive argument now shows that $\partial_k\partial_{k+1} = \zeta$, i.e. that Im ∂_{k+1} is included in Ker ∂_k for all positive integers k. Thus $(S(A), \varepsilon_0)$ is an acyclic complex over A.

We claim now that $(S(A), \varepsilon_0)$ is in fact a projective resolution of A; so we must show that all the left A^e-modules $S_k(A)$ are projective. To do this we define a sequence $(V_k)_{k\in\mathbf{N}}$ of vector spaces over K by setting $V_0 = K$ and $V_{k+1} = A \otimes_K V_k$ $(k \geqslant 0)$. Then $S_k(A)$ is isomorphic to the left A^e-module $A^e \otimes_K V_k$ $(k \geqslant 0)$ under an isomorphism φ such that

$$\varphi(a_0 \otimes a_1 \otimes \ldots \otimes a_k \otimes a_{k+1}) = (a_0 \otimes a_{k+1}) \otimes (a_1 \otimes \ldots \otimes a_k).$$

If $k = 0$ the factor $a_1 \otimes \ldots \otimes a_k$ is to be understood as the identity element e of K. We remark that φ is indeed an A^e-isomorphism, since the left scalar multiplication of $A^e \otimes_K V_k$ by A^e is given by the formula

$$\begin{aligned}(a \otimes a')\,[(a_0 \otimes a_{k+1}) \otimes (a_1 \otimes \ldots \otimes a_k)]&\\ = [(a \otimes a')(a_0 \otimes a_{k+1})] \otimes (a_1 \otimes \ldots \otimes a_k)&\\ = [aa_0 \otimes (a' * a_{k+1})] \otimes (a_1 \otimes \ldots \otimes a_k)&\end{aligned}$$

$$= (aa_0 \otimes a_{k+1}a') \otimes (a_1 \otimes \ldots \otimes a_k)$$
$$= (aa_0 \otimes a_1 \otimes \ldots \otimes a_k \otimes a_{k+1}a').$$

(As usual $*$ denotes the multiplication in A^{op}.)

To show that the left A^e-module $S_k(A)$ is projective we use the criterion of Theorem 11.6; so let

$$0 \to M' \xrightarrow{\alpha} M \xrightarrow{\beta} M'' \to 0 \qquad [31.5]$$

be any exact sequence of left A^e-modules. This may of course be regarded as an exact sequence of vector spaces over K. Since V_k is a vector space over K, and hence projective, the sequence of abelian groups

$$0 \to \text{Hom}_K(V_k, M') \xrightarrow{\alpha*} \text{Hom}_K(V_k, M) \xrightarrow{\beta*} \text{Hom}_K(V_k, M'') \to 0 \quad [31.6]$$

is exact, where α_* and β_* are the homomorphisms induced by α and β respectively. The groups in this last sequence can be turned into left A^e-modules under a left scalar multiplication such that

$$[(a \otimes a')\gamma]\,(a_1 \otimes \ldots \otimes a_k) = \gamma(aa_1 \otimes a_2 \otimes \ldots \otimes a_{k-1} \otimes a_k a')$$

for all elements $a, a_1, \ldots, a_k$ of A, a' of A^{op} and all homomorphisms γ from V_k. The homomorphisms α_*, β_* are then easily shown to be A^e-homomorphisms; so [31.6] is an exact sequence of left A^e-modules. Since A^e is a projective left A^e-module, it follows again from Theorem 11.6 that the sequence

$$0 \to \text{Hom}_{A^e}(A^e, \text{Hom}_K(V_k, M')) \xrightarrow{\alpha_{**}} \text{Hom}_{A^e}(A^e, \text{Hom}_K(V_k, M))$$
$$\xrightarrow{\beta_{**}} \text{Hom}_{A^e}(A^e, \text{Hom}_K(V_k, M'')) \to 0 \qquad [31.7]$$

is exact, where a_{**} and β_{**} are the homomorphisms induced by α_* and β_* respectively. By adapting the argument of Theorem 12.9 we can show that there is an isomorphism ψ from $X = \text{Hom}_{A^e}(A^e, \text{Hom}_K(V_k, M))$ onto $\text{Hom}_{A^e}(A^e \otimes_K V_k, M)$ such that for every element μ of X we have

$$[\psi(\mu)]\,(b \otimes x_k) = [\mu(b)]\,(x_k)$$

for all elements b of A^e and x_k of V_k; hence there is an isomorphism θ from X onto $\text{Hom}_{A^e}(S_k(A), M)$. Similarly there are isomorphisms θ' and θ'' (the notation is obvious) and we check that $\theta\alpha_{**} = \alpha.\theta'$ and $\theta''\beta_{**} = \beta.\theta$ where $\alpha.$ and $\beta.$ are the homomorphisms of the sequence

$$0 \to \mathrm{Hom}_{A^e}(S_k(A), M') \xrightarrow{\alpha_\cdot} \mathrm{Hom}_{A^e}(S_k(A), M) \xrightarrow{\beta_\cdot} \mathrm{Hom}_{A^e}(S_k(A), M'') \to 0$$

induced by α and β. This sequence is therefore exact, and so (by Theorem 11.6 again) $S_k(A)$ is a projective left A^e-module; thus $(S_k(A), \varepsilon_0)$ is a projective resolution of A, as required.

Now let M be any A-bimodule. By definition, the cohomology groups of A with coefficients in M are the groups $H^n(A, M) = \mathrm{Ext}^n_{A^e}(A, M)$ and hence may be obtained as the cohomology groups of the cochain complex $\mathrm{Hom}_{A^e}(S(A), M)$. To compute these, we remark that for each natural number k there is an isomorphism η_k from $\mathrm{Hom}_{A^e}(S_k(A), M)$ onto $\mathrm{Hom}_K(V_k, M)$ such that

$$[\eta_k(\tau)]\,(x_k) = \tau((e_A \otimes e_A) \otimes x_k)$$

for all A^e-homomorphisms τ from $S_k(A)$ to M and all elements x_k of V_k. We write $C^k(A, M)$ instead of $\mathrm{Hom}_K(V_k, M)$ and call it the group of k-dimensional *cochains of A with coefficients in M*; for each natural number k we define a K-homomorphism δ^k from $C^k(A, M)$ to $C^{k+1}(A, M)$ such that

$$(\delta^k\gamma)\,(a_1 \otimes \ldots \otimes a_{k+1}) = a_1\gamma(a_2 \otimes \ldots \otimes a_{k+1})$$
$$+ \sum_{i=1}^{k} (-1)^i\gamma(a_1 \otimes \ldots \otimes a_{i-1} \otimes a_ia_{i+1} \otimes a_{i+2} \otimes \ldots \otimes a_{k+1})$$
$$+ (-1)^{k+1}\gamma(a_1 \otimes \ldots \otimes a_k)a_{k+1}$$

for all k-cochains γ and all elements $a_1, \ldots, a_{k+1}$ of A. We easily verify that $\delta^{k+1}\delta^k = \zeta$ for all natural numbers k, so that $C^\cdot(A, M) = (C^k(A, M), \delta^k)$ is a cochain complex, and that the mappings η_k induce isomorphisms between the cohomology groups $H^k(A, M)$ of the cochain complex $\mathrm{Hom}_{A^e}(S(A), M)$ and those of the cochain complex $C^\cdot(A, M)$. Thus $H^k(A, M)$ as originally defined is isomorphic to the factor group $Z^k(A, M)/B^k(A, M)$ of k-cocycles modulo k-coboundaries for the complex $C^\cdot(A, M)$.

In §33 we shall use this discussion to give a 'concrete' interpretation of the 2-dimensional cohomology group $H^2(A, M)$.

§32. Algebras of Dimension Zero

As usual let A be an algebra over a field K and let A^e be its enveloping algebra. Let I be the set of natural numbers r such that for every A-bimodule M the cohomology group $H^r(A, M)$ is zero. If I is non-empty, then it has a least element; if this element is $n + 1$, we say that A has *homological dimension* n as algebra over K and write $\text{h.dim}_K A = n$ or simply $\text{h.dim}\, A = n$. If I is empty, we say that A has infinite homological dimension as algebra over K and we write $\text{h.dim}_K A = \infty$.

It follows from these definitions that if $\text{h.dim}\, A = -1$ then $H^0(A, M) = 0$ for every A-bimodule M and in particular $H^0(A, A_b) = 0$ where by A_b we mean A itself considered as an A-bimodule. By Theorem 31.1, $H^0(A, A_b)$ is isomorphic to the subgroup of invariant elements of A_b, which certainly contains the identity element e of A; thus $e = 0$ and so A consists of the zero element alone. Since the converse is obvious, we have established that $\text{h.dim}\, A = -1$ if and only if $A = \{0\}$ and so $H^r(A, M) = 0$ for every A-bimodule M and every natural number r.

Suppose now that $0 \leqslant \text{h.dim}\, A = n < \infty$. Then there exists an A-bimodule N such that $H^n(A, N)$ is non-zero, but for every A-bimodule M we have $H^{n+1}(A, M) = 0$. We claim that in fact we have $H^r(A, M) = 0$ for every A-bimodule M and every natural number r greater than n. To establish this we proceed by induction, for which a basis is provided by the fact that $H^{n+1}(A, M) = 0$ for every A-bimodule M. Suppose we have shown for some natural number $r > n + 1$ that $H^r(A, M) = 0$ for all A-bimodules M. Let N be any A-bimodule; if we consider N as a left A^e-module in the usual way and use Theorem 11.9, we see that there exists an exact sequence

$$0 \to N \to Q \to N'' \to 0$$

of left A^e-modules, with Q injective. Theorem 30.16 yields an exact sequence

$$\ldots \to \text{Ext}^r(A, Q) \to \text{Ext}^r(A, N'') \xrightarrow{\xi^r_A} \text{Ext}^{r+1}(A, N) \to \text{Ext}^{r+1}(A, Q) \to \ldots$$

According to Theorem 30.18 we have $\text{Ext}^r(A, Q) = \text{Ext}^{r+1}(A, Q) = 0$ and consequently ξ^r_A. is an isomorphism; by the inductive hypothesis we have $\text{Ext}^r(A, N'') = H^r(A, N'') = 0$ and hence $H^{r+1}(A, N) = \text{Ext}^{r+1}(A, N) = 0$ also. Thus the induction is completed.

Our main object in this section is to prove that an algebra A over a field K has homological dimension zero if and only if A is a separable algebra of finite dimension (as a vector space) over K. We begin by establishing several criteria for an algebra to have homological dimension zero.

THEOREM 32.1. *Let A be an algebra over a field K. Then the following* *conditions are equivalent*:

(a) $\text{h.dim}_K A = 0$;

(b) *A is a projective left A^e-module*;

(c) *there exists a left A^e-module homomorphism ε' from A to A^e such that $\varepsilon\varepsilon' = I_A$ (where ε is the homomorphism such that $\varepsilon(a \otimes a') = aa'$)*;

(d) *there exists an invariant element b of the A-bimodule A^e such that $\varepsilon(b) = e_A$.*

Proof. (1) Suppose $\text{h.dim}_K A = 0$.

Then for every positive integer n and every left A^e-module (A-bimodule) M we have $H^n(A, M) = 0$, i.e. $\text{Ext}^n(A, M) = 0$. It follows from Theorem 30.15 that A is a projective left A^e-module.

So (a) implies (b).

(2) Suppose conversely that A is a projective left A^e-module.

If we set $X_0 = A$ and $X_p = 0$ for all non-zero integers p, and define all the boundary homomorphisms ∂_p to be zero homomorphisms, then (X_p, ∂_p) is a projective resolution of A. Using this resolution to compute the cohomology groups, we discover at once that for each positive integer n and left A^e-module M we have $H^n(A, M) = \text{Ext}^n(A, M) = 0$.

Thus $\text{h.dim}_K A = 0$, and so (b) implies (a).

(3) Suppose again that A is a projective left A^e-module.

Since ε is an epimorphism from A^e onto A and I_A is a homomorphism from the projective module A to A, there exists an A^e-module homomorphism ε' from A to A^e such that $\varepsilon\varepsilon' = I_A$.

Thus (b) implies (c).

(4) Suppose there exists an A^e-homomorphism ε' from A to A^e such that $\varepsilon\varepsilon' = I_A$.

Then the exact sequence

$$0 \to \text{Ker}\,\varepsilon \to A^e \xrightarrow{\varepsilon} A \to 0$$

of left A^e-modules is split and hence, according to Theorem 9.4, A is isomorphic to a direct summand of A^e. Since A^e is of course a free left A^e-module, Theorem 11.5 shows that A is a projective left A^e-module.

So (c) implies (b).

(5) Suppose again that there is an A^e-homomorphism ε' from A to A^e such that $\varepsilon\varepsilon' = I_A$.

Let $b = \varepsilon'(e_A)$. Then $\varepsilon(b) = e_A$, and since ε' is an A^e-homomorphism it follows that for every element a of A we have $ab = a\varepsilon'(e_A) = \varepsilon'(ae_A) = \varepsilon'(e_A a) = \varepsilon'(e_A)a = ba$, as required.

Thus (c) implies (d).

(6) Finally suppose there is an invariant element b of A^e such that $\varepsilon(b) = e_A$.

Define a mapping ε' from A to A^e by setting $\varepsilon'(a) = ab$ for every element a of A. Then ε' is an A^e-homomorphism and $\varepsilon\varepsilon' = I_A$ as required.

So (d) implies (c) and the proof is completed.

We now make use of the criterion (d) to determine the homological dimension of a matrix algebra over a field.

THEOREM 32.2. *Let F be any field, $M_n(F)$ the algebra of $n \times n$ matrices with coefficients in F. Then* $\text{h.dim}_F M_n(F) = 0$.

Proof. As in §24, Example 5, let E_{ij} be the $n \times n$ matrix which has the identity element of F in the (i, j)-th position and zero in every other position $(i, j = 1, \ldots, n)$.

Consider the element $b = \sum_{i=1}^{n} E_{i1} \otimes E_{1i}$ of $(M_n(F))^e$. With the usual notation we have

$$\varepsilon(b) = \sum_{i=1}^{n} E_{i1}E_{1i} = \sum_{i=1}^{n} E_{ii},$$

which is the identity $n \times n$ matrix. Further, for every pair of indices $r, s = 1, \ldots, n$, we have

$$E_{rs}b = \sum_{i=1}^{n} E_{rs}E_{i1} \otimes E_{1i} = E_{r1} \otimes E_{1s}$$

and

$$bE_{rs} = \sum_{i=1}^{n} E_{i1} \otimes E_{1i}E_{rs} = E_{r1} \otimes E_{1s}$$

also. Since the matrices E_{rs} $(r, s = 1, \ldots, n)$ make up a basis for $M_n(F)$ over F it follows easily that b is an invariant element.

Hence, according to Theorem 32.1, we have $\text{h.dim}_F M_n(F) = 0$.

In dealing with separable algebras we found ourselves considering direct sums of matrix algebras over a field; so we should investigate whether the direct sum of algebras with homological dimension zero again has homological dimension zero. This is in fact the case, as our next theorem shows.

THEOREM 32.3. *Let A_1 and A_2 be algebras over a field K and let A be the external direct sum of A_1 and A_2. If* $\text{h.dim}_K A_1 = \text{h.dim}_K A_2 = 0$ *then* $\text{h.dim}_K A = 0$ *also.*

Proof. As usual let ε be the A^e-homomorphism from A^e to A such that $\varepsilon(a \otimes a') = aa'$ for all elements a, a' of A; let ε_1 and ε_2 be the corresponding A_1^e- and A_2^e-homomorphisms from A_1^e to A_1 and A_2^e to A_2 respectively. Since $\text{h.dim}_K A_1 = \text{h.dim}_K A_2 = 0$, it follows from criterion (c) of Theorem 32.1 that there are A_1^e- and A_2^e-homomorphisms ε_1' and ε_2' from A_1 to A_1^e and A_2 to A_2^e respectively such that $\varepsilon_1\varepsilon_1' = I_{A_1}$ and $\varepsilon_2\varepsilon_2' = I_{A_2}$. When we regard A_1, A_2, A_1^e and A_2^e as A-bimodules (A^e-modules) in the obvious way, the mappings $\varepsilon_1, \varepsilon_2, \varepsilon_1'$ and ε_2' are all A^e-homomorphisms.

It is not hard to check that there is an A^e-isomorphism φ from $A^e = A \otimes_K A^{op} = (A_1 \oplus A_2) \otimes_K (A_1 \oplus A_2)^{op}$ onto the direct sum

$$(A_1 \otimes_K A_1^{op}) \oplus (A_1 \otimes_K A_2^{op}) \oplus (A_2 \otimes_K A_1^{op}) \oplus (A_2 \otimes_K A_2^{op}).$$

If we now define the mapping ε' from A to A^e by setting

$$\varepsilon'(a_1, a_2) = \varphi^{-1}(\varepsilon_1'(a_1), 0, 0, \varepsilon_2'(a_2))$$

for all elements (a_1, a_2) of $A = A_1 \oplus A_2$, we can easily verify that ε' is an A^e-homomorphism and $\varepsilon\varepsilon' = I_A$.

Hence $\text{h.dim}_K A = 0$, as required.

COROLLARY. *Let $(A_i)_{1 \leqslant i \leqslant n}$ be a finite family of algebras over a field K and let A be the external direct sum of this family. If* $\text{h.dim}_K A_i = 0$ $(i = 1, \ldots, n)$ *then* $\text{h.dim}_K A = 0$ *also.*

The direct sum of matrix algebras which turns up when we study a separable algebra A over K is obtained when we form the tensor product of A with a suitable extension field F of K. To descend from this tensor product to the original algebra A we shall use the following result.

THEOREM 32.4. *Let A be an algebra over a field K. If F is an extension field of K such that* $\text{h.dim}_F(F \otimes_K A) = 0$ *then* $\text{h.dim}_K A = 0$ *also.*

Proof. Let M be any A-bimodule. Then $F \otimes_K M$ is an $(F \otimes_K A)$-bimodule under left and right scalar multiplications such that

$$(r \otimes a)(s \otimes x) = rs \otimes ax \text{ and } (s \otimes x)(r \otimes a) = sr \otimes xa$$

for all elements r, s of F, a of A, x of M.

Now let χ be any crossed homomorphism from A to M. We define a mapping φ from $F \times A$ to $F \otimes_K M$ by setting

$$\varphi(r, a) = r \otimes \chi(a)$$

for all elements r of F and a of A. It is easy to see that φ is bilinear; hence there exists a K-homomorphism $\bar{\chi}$ from $F \otimes_K A$ to $F \otimes_K M$ such that

$$\bar{\chi}(r \otimes a) = r \otimes \chi(a)$$

for all elements r of F and a of A. We claim that $\bar{\chi}$ is in fact a crossed homomorphism. There is no trouble in showing that $\bar{\chi}$ is an F-homomorphism; to complete the proof that it is a crossed homomorphism it is clearly sufficient to verify the relation $\bar{\chi}(tt') = t\bar{\chi}(t') + \bar{\chi}(t)t'$ for elements of $F \otimes_K A$ of the form $t = r \otimes a, t' = r' \otimes a'$. But for such elements we have

$$\begin{aligned} \bar{\chi}(tt') = \bar{\chi}(rr' \otimes aa') &= rr' \otimes \chi(aa') \\ &= rr' \otimes (a\chi(a') + \chi(a)a') \\ &= (rr' \otimes a\chi(a')) + (rr' \otimes \chi(a)a') \\ &= t\bar{\chi}(t') + \bar{\chi}(t)t' \end{aligned}$$

as required.

By hypothesis, we have $H^1(F \otimes_K A, F \otimes_K M) = 0$; according to Theorem 31.2 it follows that $\bar{\chi}$ is a principal crossed homomorphism. That is to say, there is an element b of $F \otimes_K M$ such that

$$\bar{\chi}(t) = tb - bt \qquad [32.1]$$

for every element t of $F \otimes_K A$.

Let $(a_i)_{i \in I}$ and $(x_j)_{j \in J}$ be bases for A and M respectively as vector spaces over K. For each index i in I and each index j in J we set

$$\chi(a_i) = \sum_{k \in J} c_{ik} x_k, \qquad [32.2]$$

$$a_i x_j = \sum_{k \in J} l_{ij,k} x_k \text{ and } x_j a_i = \sum_{k \in J} l'_{ij,k} x_k \qquad [32.3]$$

where the coefficients c_{ik}, $l_{ij,k}$, $l'_{ij,k}$ $(i \in I, j, k \in J)$ all belong to the field K. Every element of $F \otimes_K A$ can be expressed uniquely in the form $\sum_{i \in I} r_i \otimes a_i$ and every element of $F \otimes_K M$ can be expressed uniquely in the form $\sum_{j \in J} s_j \otimes x_j$ where $(r_i)_{i \in I}$, $(s_j)_{j \in J}$ are quasi-finite families of elements of F. In particular, suppose $b = \sum_{j \in J} p_j \otimes x_j$.

If i is any index in I, then, taking $t = e_F \otimes a_i$ in [32.1] and using [32.2] and [32.3], we obtain (after some elementary computation)

$$\sum_{k\in J} c_{ik} \otimes x_k = \sum_{k\in J} \Big(\sum_{j\in J} (l_{ij,k} - l'_{ij,k})p_j\Big) \otimes x_k$$

and hence for all indices i in I and k in J we have

$$\sum_{j\in J} (l_{ij,k} - l'_{ij,k})p_j = c_{ik}. \qquad [32.4]$$

Now we take a basis for F over K, one of whose elements is the identity element e_K of K; if d_j is the coefficient of e_K when p_j is expressed as a linear combination of elements of this basis (so that d_j is an element of K) we deduce from [32.4] that

$$\sum_{j\in J} (l_{ij,k} - l'_{ij,k})d_j = c_{ik}$$

for all indices i in I and k in J. Let us put $d = \sum_{j\in J} d_j x_j$. Then for every index i in I we have

$$\begin{aligned}\chi(a_i) = \sum_{k\in J} c_{ik}x_k &= \sum_{j\in J} d_j\Big(\sum_{k\in J} (l_{ij,k} - l'_{ij,k})x_k\Big)\\ &= \sum_{j\in J} d_j(a_i x_j - x_j a_i)\\ &= a_i d - d a_i.\end{aligned}$$

It follows at once that for every element a of A we have

$$\chi(a) = ad - da.$$

So χ is a principal crossed homomorphism.

Thus $H^1(A, M) = 0$ for every A-bimodule M and so $\text{h.dim}_K A = 0$ as required.

We can now establish the first part of the main result of this section.

THEOREM 32.5. *Let A be a separable algebra of finite dimension (as a vector space) over a field K. Then the homological dimension of A over K is zero.*

Proof. According to Theorem 27.6 there exists an extension field F of K such that $F \otimes_K A$ is isomorphic to a direct sum of matrix algebras over F. By Theorem 32.2 and the Corollary to Theorem 32.3 we have $\text{h.dim}_F(F \otimes_K A) = 0$. Hence, by Theorem 32.4, $\text{h.dim}_K A = 0$ also.

To prove the converse result we shall require the converse of Theorem 32.4.

THEOREM 32.6. *Let A be an algebra over a field K such that* $\text{h.dim}_K A = 0$. *If F is any extension of K then* $\text{h.dim}_F(F \otimes_K A) = 0$ *also.*

Proof. Since $\text{h.dim}_K A = 0$ it follows from the criterion (d) of Theorem 32.1 that there is an invariant element $b = \sum_{i \in I} a_i \otimes a_i'$ of A^e such that $\varepsilon(b) = \sum_{i \in I} a_i a_i' = e_A$.

Consider now the enveloping algebra $(F \otimes_K A)^e = (F \otimes_K A) \otimes_F (F \otimes_K A)^{\text{op}}$. If $c = \sum_{i \in I} (e_F \otimes a_i) \otimes (e_F \otimes a_i')$, it is clear that c is an invariant element of the $(F \otimes_K A)$-bimodule $(F \otimes_K A)^e$ and further

$$\begin{aligned}\sum_{i \in I} (e_F \otimes a_i)(e_F \otimes a_i') &= \sum_{i \in I} e_F \otimes a_i a_i' \\ &= e_F \otimes \sum_{i \in I} a_i a_i' \\ &= e_F \otimes e_A \\ &= e_{F \otimes A}.\end{aligned}$$

Hence, by criterion (d) of Theorem 32.1, we have $\text{h.dim}_F(F \otimes_K A) = 0$.

Finally we have the second part of the main result.

THEOREM 32.7. *Let A be an algebra over a field K. If the homological dimension of A over K is zero then A is a separable algebra and has finite dimension as a vector space over K.*

Proof. Let E be any extension field of K. We shall prove that $B = E \otimes_K A$ is a semisimple algebra over E by showing that every left B-module is projective (cf. Corollary 1 of Theorem 22.1).

So let V be any left B-module, and let

$$0 \to W' \xrightarrow{\alpha} W \xrightarrow{\beta} W'' \to 0$$

be a short exact sequence of left B-modules. This is also, of course, an exact sequence of vector spaces over K; since V, considered as a vector space over K, is free (and hence projective), the sequence

$$0 \to \text{Hom}_K(V, W') \xrightarrow{\alpha_*} \text{Hom}_K(V, W) \xrightarrow{\beta_*} \text{Hom}_K(V, W'') \to 0 \quad [32.5]$$

of abelian groups is also exact (where α_* and β_* are the homomorphisms induced by α and β respectively).

We may define left and right scalar multiplications of $\mathrm{Hom}_K(V, W)$ by elements in B; namely, for each element φ of $\mathrm{Hom}_K(V, W)$ and each element t of B we set

$$(\varphi t)(x) = \varphi(tx) \text{ and } (t\varphi)(x) = t\varphi(x)$$

for every element x of V. It is easy to check that under these multiplications $\mathrm{Hom}_K(V, W)$ becomes a B-bimodule and hence in the usual way a left B^e-module. Similarly $\mathrm{Hom}_K(V, W')$ and $\mathrm{Hom}_K(V, W'')$ are left B^e-modules. Furthermore we can easily verify that α_* and β_* are B^e-homomorphisms.

Now since $\mathrm{h.dim}_K A = 0$ it follows from Theorem 32.6 that $\mathrm{h.dim}_E B = 0$ also, and hence, by the criterion (b) of Theorem 32.1, B is a projective left B^e-module. Hence, according to Theorem 11.6, [32.5] gives rise to an exact sequence

$$0 \to \mathrm{Hom}_{B^e}(B, \mathrm{Hom}_K(V, W')) \xrightarrow{\alpha_{**}} \mathrm{Hom}_{B^e}(B, \mathrm{Hom}_K(V, W))$$
$$\xrightarrow{\beta_{**}} \mathrm{Hom}_{B^e}(B, \mathrm{Hom}_K(V, W'')) \to 0$$

where α_{**} and β_{**} are the homomorphisms induced by α_* and β_* respectively.

Adapting once again the argument of Theorem 12.9 we can show that there is an isomorphism ψ from $X = \mathrm{Hom}_{B^e}(B, \mathrm{Hom}_K(V, W))$ onto $\mathrm{Hom}_B(B \otimes_B V, W)$ such that for every element μ of X we have

$$[\psi(\mu)](b \otimes x) = [\mu(b)](x)$$

for all elements b of B and x of V. Since the B-module $B \otimes_B V$ is isomorphic to V itself (by Theorem 12.2) it follows that there is an isomorphism θ from X onto $\mathrm{Hom}_B(V, W)$. Similarly, in an obvious notation, there are isomorphisms θ' and θ'' for which we have $\theta\alpha_{**} = \alpha_.\theta'$ and $\theta''\beta_{**} = \beta_.\theta$ where $\alpha_.$ and $\beta_.$ are the homomorphisms induced by α and β in the sequence

$$0 \to \mathrm{Hom}_B(V, W') \xrightarrow{\alpha_.} \mathrm{Hom}_B(V, W) \xrightarrow{\beta_.} \mathrm{Hom}_B(V, W'') \to 0.$$

This sequence is therefore exact and hence, by Theorem 11.6, we deduce that V is projective.

It follows from Corollary 1 of Theorem 22.1 that $B = E \otimes_K A$ is semisimple. Hence A is a separable algebra over K.

To show that A has finite dimension as a vector space over F we make use of the criterion (d) of Theorem 32.1, according to which there exists an invariant element $b = \sum_{j \in J} b_j \otimes b'_j$ of the A-bimodule A^e such that $\sum_{j \in J} b_j b'_j = e_A$.

Let $(a_i)_{i \in I}$ be a basis for A, considered as a right vector space over K. Then for each index j in the (finite) set J there exists a quasi-finite family $(k_{ij})_{i \in I}$ of elements of K such that $b_j = \sum_{i \in I} a_i k_{ij}$. Then we have

$$\begin{aligned} b &= \sum_{j \in J} (\sum_{i \in I} a_i k_{ij}) \otimes b'_j \\ &= \sum_{i \in I} a_i \otimes (\sum_{j \in J} k_{ij} b'_j) \\ &= \sum_{i \in I} a_i \otimes \bar{a}_i \end{aligned}$$

say; clearly only finitely many of the elements $\bar{a}_i$ are non-zero. Let I_0 be the subset of I consisting of those indices i for which $\bar{a}_i$ is non-zero. Then we have

$$\sum_{i \in I_0} a_i \bar{a}_i = e_A, \tag{32.6}$$

and, since b is invariant, for every element a of A,

$$\sum_{i \in I_0} a a_i \otimes \bar{a}_i = \sum_{i \in I_0} a_i \otimes \bar{a}_i a. \tag{32.7}$$

If a is any element of A then for each index j in I_0 there exists a quasi-finite family $(l_{ij})_{i \in I}$ of elements of K such that $aa_j = \sum_{i \in I_0} a_i l_{ij}$. Hence we deduce from [32.7] that

$$\begin{aligned} \sum_{i \in I_0} a_i \otimes \bar{a}_i a &= \sum_{j \in I_0} a a_j \otimes \bar{a}_j \\ &= \sum_{j \in I_0} (\sum_{i \in I} a_i l_{ij}) \otimes \bar{a}_j \\ &= \sum_{i \in I} a_i \otimes (\sum_{j \in I_0} l_{ij} \bar{a}_j). \end{aligned}$$

From this it follows that for each index i in I_0 we have

$$\bar{a}_i a = \sum_{j \in I_0} l_{ij} \bar{a}_j.$$

Finally we deduce from [32.6] that for every element a of A we have

$$\begin{aligned} a = e_A a = \sum_{i \in I_0} a_i \bar{a}_i a &= \sum_{i \in I_0} a_i (\sum_{j \in I_0} l_{ij} \bar{a}_j) \\ &= \sum_{(i,j) \in I_0 \times I_0} l_{ij} a_i \bar{a}_j. \end{aligned}$$

Thus A is generated as a vector space over K by the finite family $(a_i\bar{a}_j)_{(i,j)\in I_0\times I_0}$. So A is a finite-dimensional vector space over K.

§33. Extensions of Algebras

Let A be an algebra over a field K. By an *extension* of the algebra A we shall mean an ordered pair (B, σ) consisting of an algebra B over K and an (algebra) epimorphism σ from B onto A. If (B, σ) is an extension of A then A is a B-bimodule under the left and right scalar multiplications defined by setting

$$b \, . \, a = \sigma(b)a \quad \text{and} \quad a \, . \, b = a\sigma(b)$$

for all elements a of A and b of B; the products on the right are of course formed under the algebra multiplication in A. The kernel N of the epimorphism σ, which we also call the *kernel of the extension* (B, σ) is a two-sided ideal of B and so is also a B-bimodule.

If (B, σ) is an extension of A and there exists an algebra homomorphism τ from A to B such that $\sigma\tau = I_A$ then we say that A is *segregated* in (B, σ) and that (B, σ) is a *cleft* or *inessential* extension.

An extension (B, σ) of A with kernel N is said to be *singular* if $N^2 = \{0\}$, that is to say if the product of every pair of elements of N is zero. In such a situation we may give N the structure of an A-bimodule. Namely, let x be any element of N, a any element of A. Since σ is an epimorphism there exists an element b of B such that $\sigma(b) = a$; set $ax = bx$ (the algebra product of b and x in B). This does not depend on the choice of b; for if $\sigma(b') = a$ also then we have $\sigma(b - b') = 0$, whence $b - b'$ is in N and so $(b - b')x = 0$, i.e. $bx = b'x$; the element bx is of course in N since N is a two-sided ideal of B. In the same way, we define xa to be xb where b is any element of B such that $\sigma(b) = a$.

Conversely, let N be an A-bimodule and (B, σ) an extension of A. Suppose there exists a monomorphism κ from N to B, considered as vector spaces over K, such that

$$0 \to N \xrightarrow{\kappa} B \xrightarrow{\sigma} A \to 0 \qquad [33.1]$$

is exact and for every element x of N and b of B we have

$$b\kappa(x) = \kappa(\sigma(b)x) \quad \text{and} \quad \kappa(x)b = \kappa(x\sigma(b)). \qquad [33.2]$$

If we take $b = \kappa(x')$ where x' is any element of N then

$$\kappa(x)\kappa(x') = \kappa[x\sigma(\kappa(x'))] = \kappa(0) = 0. \qquad [33.3]$$

Hence, if we identify N with its image under κ then [33.2] shows that N is a two-sided ideal of B, and according to [33.3] we have $N^2 = \{0\}$. Thus [33.1] gives us a singular extension of A with kernel N.

We say that two singular extensions (B, σ) and (B', σ') of A with kernel N are *congruent* if there exists an algebra homomorphism φ from B to B' such that the diagram

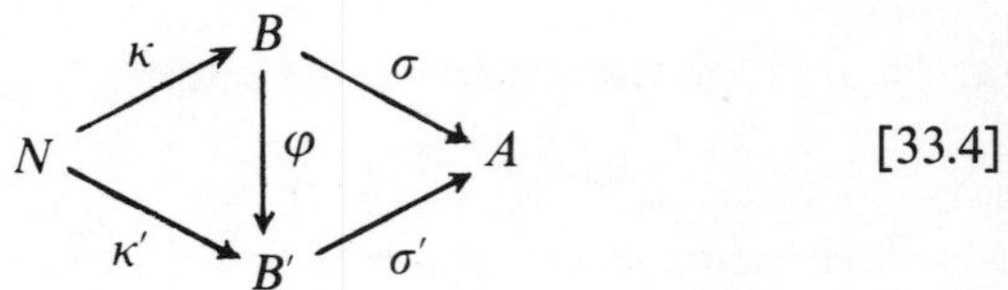

[33.4]

is commutative (where κ and κ' are the monomorphisms from the kernel N to B and B' respectively). It is an easy exercise to show that φ must be an isomorphism.

We are now ready to give the 'concrete' interpretation of the 2-dimensional cohomology groups of an algebra which we promised at the end of §31.

So let A be an algebra over K, N an A-bimodule; let (B, σ) be a singular extension of A with kernel N. Since every vector space over a field is free and hence projective, the sequence [33.1] splits when regarded as an exact sequence of vector spaces over K. Thus we may assume without loss of generality that B is the external direct sum of N and A and that the mappings κ and σ are given by

$$\kappa(x) = (x, 0) \quad \text{and} \quad \sigma(x, a) = a \qquad [33.5]$$

for all elements x of N and a of A. Having regard to [33.2] and [33.3] we see that for all elements x_1, x_2 of N and all elements a_1, a_2 of A we have

$$(x_1, 0)(x_2, 0) = (0, 0),$$
$$(x_1, 0)(0, a_2) = (x_1 a_2, 0),$$
$$(0, a_1)(x_2, 0) = (a_1 x_2, 0).$$

Now since σ is an algebra homomorphism we have

$$\sigma[(0, a_1)(0, a_2)] = \sigma(0, a_1)\sigma(0, a_2) = a_1 a_2.$$

So there is a mapping φ from $A \times A$ to N such that for all elements

a_1, a_2 of A we have

$$(0, a_1)(0, a_2) = (\varphi(a_1, a_2), a_1a_2).$$

This mapping is clearly bilinear and so there exists a K-homomorphism β from $A \otimes_K A$ to N, i.e. a 2-dimensional cochain of A with coefficients in N, such that

$$(0, a_1)(0, a_2) = (\beta(a_1 \otimes a_2), a_1a_2)$$

for all elements a_1, a_2 of A It follows that the multiplication in B is given by

$$(x_1, a_1)(x_2, a_2) = (x_1a_2 + a_1x_2 + \beta(a_1 \otimes a_2), a_1a_2) \quad [33.6]$$

for all elements x_1, x_2 of N and all elements a_1, a_2 of A.

The 2-cochain β is in fact a cocycle. For if a_1, a_2, a_3 are any three elements of A then, since the multiplication in B is associative, we have

$$\begin{aligned}(0, 0) &= ((0, a_1)(0, a_2))(0, a_3) - (0, a_1)((0, a_2)(0, a_3)) \\ &= (\beta(a_1 \otimes a_2), a_1a_2)(0, a_3) - (0, a_1)(\beta(a_2 \otimes a_3), a_2a_3) \\ &= (\beta(a_1 \otimes a_2)a_3 + \beta(a_1a_2 \otimes a_3), (a_1a_2)a_3) \\ &\qquad - (a_1\beta(a_2 \otimes a_3) + \beta(a_1 \otimes a_2a_3), a_1(a_2a_3)) \\ &= (\delta^2\beta(a_1 \otimes a_2 \otimes a_3), 0).\end{aligned}$$

Thus $\delta^2\beta(a_1 \otimes a_2 \otimes a_3) = 0$.

We have shown so far that every singular extension (B, σ) of A with kernel N determines a 2-cocycle β of A with coefficients in N. Suppose now that (B, σ) and (B', σ') are congruent singular extensions of A with kernel N. Referring to the diagram [33.4] we see that for every element x of N we have

$$\varphi(x, 0) = \varphi\kappa(x) = \kappa'(x) = (x, 0)$$

and for each element a of A we have

$$\sigma'\varphi(0, a) = \sigma(0, a) = a;$$

so there is a mapping γ from A to N such that

$$\varphi(0, a) = (\gamma(a), a)$$

for each element a of A. The mapping γ is clearly a K-homomorphism, i.e. a 1-dimensional cochain of A with coefficients in N.

Since φ is an algebra homomorphism it follows that for each pair of elements (x_1, a_1), (x_2, a_2) of B we have

$$\varphi(x_1, a_1)\varphi(x_2, a_2) = \varphi[(x_1, a_1)(x_2, a_2)].$$

Now

$$\begin{aligned}\varphi(x_1, a_1)\varphi(x_2, a_2) &= (x_1 + \gamma(a_1), a_1)(x_2 + \gamma(a_2), a_2) \\ &= (x_1 a_2 + \gamma(a_1)a_2 + a_1 x_2 + a_1\gamma(a_2) \\ &\qquad + \beta'(a_1 \otimes a_2), a_1 a_2) \qquad [33.7]\end{aligned}$$

where β' is the 2-cocycle determined by the extension (B', σ'); on the other hand,

$$\begin{aligned}\varphi[(x_1, a_1)(x_2, a_2)] &= \varphi(x_1 a_2 + a_1 x_2 + \beta(a_1 \otimes a_2), a_1 a_2) \\ &= (x_1 a_2 + a_1 x_2 + \beta(a_1 \otimes a_2) + \gamma(a_1 a_2), a_1 a_2). \qquad [33.8]\end{aligned}$$

It follows that

$$\begin{aligned}\beta'(a_1 \otimes a_2) &= \beta(a_1 \otimes a_2) + \gamma(a_1 a_2) - a_1\gamma(a_2) - \gamma(a_1)a_2 \\ &= \beta(a_1 \otimes a_2) - \delta^1\gamma(a_1 \otimes a_2).\end{aligned}$$

Thus the 2-cocycles corresponding to congruent extensions belong to the same coset of $Z^2(A, N)$ modulo $B^2(A, N)$ and hence determine the same element of the cohomology group $H^2(A, N)$.

Conversely, let β be any 2-cocycle of A with coefficients in N. Let B be the external direct sum of N and A considered as vector spaces over K. Define an operation of multiplication in B by means of the formula [33.6]; it is easy to verify that this multiplication is distributive over the addition in B, and a simple computation (using the fact that β is a cocycle) shows that the multiplication is associative. To show that there is an identity element for the multiplication we notice first that since β is a cocycle then for every element a of A we have

$$\begin{aligned}0 &= (\delta^2\beta)(e_A \otimes e_A \otimes a) \\ &= e_A\beta(e_A \otimes a) - \beta(e_A e_A \otimes a) + \beta(e_A \otimes e_A a) - \beta(e_A \otimes e_A)a,\end{aligned}$$

whence $\beta(e_A \otimes e_A)a = \beta(e_A \otimes a)$. Similarly $a\beta(e_A \otimes e_A) = \beta(a \otimes e_A)$. Then for every element (x, a) of B we have

$$\begin{aligned}(x, a)(-\beta(e_A \otimes e_A), e_A) &= (xe_A - a\beta(e_A \otimes e_A) + \beta(a \otimes e_A), ae_A) \\ &= (x, a)\end{aligned}$$

and similarly

$$(-\beta(e_A \otimes e_A), e_A)(x, a) = (x, a).$$

Thus $(-\beta(e_A \otimes e_A), e_A)$ is an identity for the multiplication in B. It follows that B is an algebra over K and we easily verify that if κ and σ are defined by [33.5] then [33.1] is exact and the relations [33.2] are satisfied. Hence (B, σ) is a singular extension of A with kernel N.

Let γ be any 1-cochain of A with coefficients in N; set $\beta' = \beta - \delta^1\gamma$ and let (B', σ') be the singular extension of A with kernel N which corresponds to β' in the same way as (B, σ) corresponds to β. Consider the mapping φ from B to B' defined by setting

$$\varphi(x, a) = (x + \gamma(a), a)$$

for all elements (x, a) of B. Then the diagram [33.4] is commutative and equations [33.7] and [33.8] show that φ is an algebra homomorphism. So (B, σ) and (B', σ') are congruent extensions.

The preceding discussion can be summed up in the following description of $H^2(A, N)$.

THEOREM 33.1. *Let A be an algebra over a field K and let N be an A-bimodule. Then there exists a one-to-one correspondence between $H^2(A, N)$ and the set of congruence classes of singular extensions of A with kernel N.*

Now let (B, σ) be a cleft singular extension of A with kernel N; let β be the 2-cocycle determined by this extension. Since the extension is cleft there exists an algebra homomorphism τ from A to B such that $\sigma\tau = I_A$. Thus we may define a mapping γ from A to N by setting

$$\tau(a) = (\gamma(a), a)$$

for every element a of A. Clearly γ is a K-homomorphism from A to N, i.e. a1-cochain of A with coefficients in N. Since τ is an algebra homomorphism it follows that for all pairs of elements a_1, a_2 of A we have

$$\begin{aligned}(\gamma(a_1a_2), a_1a_2) &= \tau(a_1a_2) \\ &= \tau(a_1)\tau(a_2) \\ &= (\gamma(a_1), a_1)(\gamma(a_2), a_2) \\ &= (a_1\gamma(a_2) + \gamma(a_1)a_2 + \beta(a_1 \otimes a_2), a_1a_2),\end{aligned}$$

whence we have

$$\beta(a_1 \otimes a_2) = -(a_1\gamma(a_2) - \gamma(a_1a_2) + \gamma(a_1)a_2) = -\delta^1\gamma(a_1 \otimes a_2).$$

Thus the 2-cocycle β determines the zero element of $H^2(A, N)$. Conversely, it is clear (by reversing the steps of the argument) that if a 2-cocycle β in $Z^2(A, N)$ is actually a 2-coboundary then the singular extension corresponding to β is cleft.

This discussion gives us the following complement of the preceding theorem.

THEOREM 33.2. *Let A be an algebra over a field K and N an A-bimodule. All cleft singular extensions of A with kernel N are congruent. In the correspondence of Theorem* 33.1 *the congruence class of cleft singular extensions corresponds to the zero element of $H^2(A, N)$.*

COROLLARY 1. *With the notation of the theorem, if $H^2(A, N) = \{0\}$ then every singular extension of A with kernel N is cleft.*

COROLLARY 2. *Let A be an algebra over a field K. If* $\text{h.dim}_K A < 2$ *then every singular extension of A is cleft.*

Proof. If $\text{h.dim}_K A < 2$ then for every A-bimodule N we have $H^2(A, N) = 0$ and so, according to Corollary 1, every singular extension of A with kernel N is cleft.

We now prove a result which generalises Corollary 2 and includes the so-called Wedderburn Principal Theorem on algebras.

THEOREM 33.3. *Let A be an algebra over a field K. If* $\text{h.dim}_K A < 2$ *then every extension of A with nilpotent kernel is cleft.*

Proof. Let (B, σ) be an extension of A with nilpotent kernel N. If $N = \{0\}$ then B and A are isomorphic and hence the extension (B, σ) is cleft. If $N^2 = \{0\}$ but $N \neq \{0\}$ then (B, σ) is a singular extension and hence, according to Corollary 2 of Theorem 33.2, is cleft. We now proceed by induction. Suppose we have established for some natural number $s \geqslant 2$ that every extension of A whose kernel N_1 satisfies $N_1^s = \{0\}$, $N_1^{s-1} \neq \{0\}$ is cleft, and suppose that the kernel N of (B, σ) satisfies $N^{s+1} = \{0\}$, $N^s \neq \{0\}$.

Then $N^2 \subset N$; certainly $N^2 \subseteq N$, but if we had $N^2 = N$ it would follow that $\{0\} = N^{s+1} = N \neq \{0\}$, which is a contradiction. Let η be the canonical epimorphism from B onto the residue class algebra B/N^2; since Ker $\eta = N^2$ is included in Ker $\sigma = N$ there exists an epimorphism σ' from B/N^2 onto A such that $\sigma'\eta = \sigma$. Clearly Ker $\sigma' = \eta(N)$, and since $(\eta(N))^2 = \eta(N^2) = \{0\}$ we see that $(B/N^2, \sigma')$ is a singular extension of A.

By Corollary 2 of Theorem 33.2 this extension is cleft, i.e. there is an algebra homomorphism τ' from A to B/N^2 such that $\sigma'\tau' = I_A$. Let $B_1 = \eta^{-1}(\tau'(A))$; then B_1 is a subalgebra of B. If we set $\sigma_1 = \sigma\iota$ where ι is the inclusion monomorphism from B_1 to B we see that (B_1, σ_1) is an extension of A with kernel N_2. Since $(N^2)^s = \{0\}$ we may apply the inductive hypothesis and deduce that this extension is cleft. Hence there is an algebra homomorphism τ_1 from A to B_1 such that $\sigma_1\tau_1 = I_A$; then $\sigma(\iota\tau_1) = (\sigma\iota)\tau_1 = \sigma_1\tau_1 = I_A$ and so the original extension (B, σ) is cleft.

This completes the induction and the theorem is established.

As a corollary we obtain the *Wedderburn Principal Theorem.*

COROLLARY. *Let A be an algebra of finite dimension (as a vector space) over a field K. Let J be the radical of A and η the canonical epimorphism from A onto the residue class algebra A/J. If A/J is a separable algebra over K then the extension (A, η) is cleft and A includes a subalgebra isomorphic to A/J.*

Proof. Since A is finite-dimensional over K it follows that A is an Artinian ring and hence, by Theorem 23.6, the radical J (which is the kernel of the extension (A, η)) is nilpotent. By Theorem 32.5 we have $\text{h.dim}_K A/J = 0 < 2$. So the result is now an immediate consequence of the theorem.

READING LIST

ADAMSON, I. T. *Introduction to field theory*. Edinburgh, 1964.

ALBERT, A. A. *Structure of algebras*. New York, 1939.

ARTIN, E., NESBITT, C. J. and THRALL, R. M., *Rings with minimum condition*. Ann Arbor, Mich., 1948.

BOURBAKI, N. *Algèbre* (Ch. 2, Algèbre linéaire). Paris, 1962.

BOURBAKI, N. *Algèbre* (Ch. 8, Modules et anneaux semi-simples). Paris, 1958.

CARTAN, H. and EILENBERG, S. *Homological algebra*. Princeton, N.J., 1956.

DIVINSKY, N. J. *Rings and radicals*. London, 1965.

FREYD, P. *Abelian categories*. New York, 1964.

HERSTEIN, I. N. *Non-commutative rings*. Carus Mathematical Monograph No. 15, 1968.

JACOBSON, N. *Structure of rings*. Providence, R.I., 1956.

JANS, J. P. *Rings and homology*. New York, 1964.

LAMBEK, J. *Lectures on rings and modules*. Waltham, Mass., 1966.

McCOY, N. H. *The theory of rings*. New York, 1964.

MACLANE, S. *Homology*. Berlin, 1963.

MITCHELL, B. *Theory of categories*. New York, 1965.

NORTHCOTT, D. G. *An introduction to homological algebra*. Cambridge, 1960.

NORTHCOTT, D. G. *Lessons on rings, modules and multiplicities*. Cambridge, 1968.

PUPPE, D. 'Korrespondenzen in abelsche Kategorien', *Math. Ann.* **148** (1–30), 1962.

INDEX